Methods in Molecular Biology

Series Editor
John M. Walker
School of Life Sciences
University of Hertfordshire
Hatfield, Hertfordshire, AL10 9AB, UK

For further volumes:
http://www.springer.com/series/7651

DNA Vaccines

Methods and Protocols

Third Edition

Edited by

Monica Rinaldi

Section of Medical Biotechnology, Institute of Translational Pharmacology (IFT), National Research Council (CNR), Rome, Italy

Daniela Fioretti

Section of Medical Biotechnology, Institute of Translational Pharmacology (IFT), National Research Council (CNR), Rome, Italy

Sandra Iurescia

Section of Medical Biotechnology, Institute of Translational Pharmacology (IFT), National Research Council (CNR), Rome, Italy

Humana Press

Editors
Monica Rinaldi
Section of Medical Biotechnology
Institute of Translational Pharmacology (IFT)
National Research Council (CNR)
Rome, Italy

Daniela Fioretti
Section of Medical Biotechnology
Institute of Translational Pharmacology (IFT)
National Research Council (CNR)
Rome, Italy

Sandra Iurescia
Section of Medical Biotechnology
Institute of Translational Pharmacology (IFT)
National Research Council (CNR)
Rome, Italy

ISSN 1064-3745 ISSN 1940-6029 (electronic)
ISBN 978-1-4939-0409-9 ISBN 978-1-4939-0410-5 (eBook)
DOI 10.1007/978-1-4939-0410-5
Springer New York Heidelberg Dordrecht London

Library of Congress Control Number: 2014934311

Printed on acid-free paper

Humana Press is a brand of Springer
Springer is part of Springer Science+Business Media (www.springer.com)

Preface

Vaccination has had a profound positive effect on the quality of public health. Vaccines have long been used to combat infectious disease; however, the last decade has witnessed a revolution in the approach to vaccine design and development. Several groundbreaking studies demonstrated that immunological responses could be generated against antigenic transgenes delivered via DNA vaccination. Since then new sophisticated technologies, advances in molecular biology techniques, and new bioinformatics analysis tools to study and manipulate the basic elements of an organism's genome have been used for the rational design and production of DNA vaccines.

Nowadays, DNA vaccination is the most important early application of nonviral gene therapy, and it seems clear that the route of DNA vaccine and the methods of vaccine preparation have strong effects on the immune response and the effectiveness of that response in preventing or treating disease.

DNA Vaccines: Methods and Protocols, Third Edition reviews innovative approaches and technologies used to design, deliver, and enhance the efficacy of DNA vaccines. In this book, expert international authors critically review the current cutting-edge research in DNA vaccine design and development. Topics also include methods of production and purification. The book also has chapters on recent DNA vaccine applications which should be of great value in moving vaccines from research to clinic. All of these chapters, as well as the others presented in the previous *DNA Vaccines* editions, have the important role of further documenting the potential of the DNA vaccination as a platform technology for treatment and prevention of human diseases suitable also for developing nations. Several peculiar features of DNA vaccines (i.e., preparation and purification, stability, cost-effectiveness and non-requirement of cold chain) emphasize this prospect.

The current status of three gene vaccines licensed for veterinary use (i.e., the West Nile virus DNA vaccine for horses, a fish DNA vaccine against the Infectious Haematopoietic Necrosis virus, and the Canine Malignant Melanoma vaccine (ONCEPT™)) will pave the way for future application in humans.

To date, no human DNA vaccine has been licensed; however, during recent years, more than 100 clinical trials have been undertaken worldwide on DNA vaccines covering the full range of prophylactic through to therapeutic vaccines against infections, cancers, and a range of other disorders (details at: http://www.dnavaccine.com/; http://clinicaltrials.gov; http://www.cancer.gov/clinicaltrials).

DNA-based vaccine technology has moved from pioneering animal studies to clinical testing quite rapidly. However, more work is still required on design and delivery to lift the immunogenicity of DNA vaccines to the levels required for human regulatory approval and commercial exploitation.

Consistent with the approach of the *Methods in Molecular Biology* series, *DNA Vaccines, Third Edition* contains detailed practical procedures on the latest DNA vaccine technology and is recommended to microbiologists and vaccinologists, immunologists, infectious diseases and public health physicians, and to the many scientists working on vaccine development (e.g., biochemists, and molecular biologists).

In conclusion, we hope that this book will be a productive opportunity to further push the recent improvements in DNA vaccine technology to their full clinical potential, moving from the benchtop to the patient.

Rome, Italy

Monica Rinaldi
Daniela Fioretti
Sandra Iurescia

Contents

Contributors

FATIN NAWWAB AL-DEEN • *Department of Chemical Engineering, Monash University, Clayton, VIC, Australia*
STEPHEN I. ALEXANDER • *Centre for Kidney Research, Children's Hospital at Westmead, The University of Sydney, Westmead, NSW, Australia*
SUNGHYUN BAEK • *Department of Chemical and Biomolecular Engineering, Korea Advanced Institute of Science and Technology (KAIST), Daejeon, Korea*
JOOST H. VAN DEN BERG • *Division of Immunology, The Netherlands Cancer Institute, Amsterdam, The Netherlands*
ADRIAAN D. BINS • *Division of Immunology, The Netherlands Cancer Institute, Amsterdam, The Netherlands*
DAVID B. BOYLE • *Australian Animal Health Laboratory, CSIRO Animal, Food, and Health Sciences, Geelong, Victoria, Australia*
KATE E. BRODERICK • *Inovio Pharmaceuticals, Blue Bell, PA, USA*
AARON E. CARNES • *Nature Technology Corporation, Lincoln, NE, USA*
ROSS L. COPPEL • *Department of Microbiology, Monash University, Clayton, VIC, Australia*
HAYK DAVTYAN • *Department of Molecular Immunology, Institute for Molecular Medicine, Huntington Beach, CA, USA*
MIGUEL FEVEREIRO • *INIAV, Instituto Nacional de Investigação Agrária e Veterinária, Lisboa, Portugal*
DANIELA FIORETTI • *Section of Medical Biotechnology, Institute of Translational Pharmacology (IFT), National Research Council (CNR), Rome, Italy*
ELISABETE BORGES FREITAS • *Institute of Biotechnology and Bioengineering, Centre for Chemical and Biological Engineering, Instituto Superior Técnico, Lisboa, Portugal*
JULIE GEHL • *Center for Experimental Drug and Gene Electrotransfer (C*EDGE), Department of Oncology, Copenhagen University Hospital Herlev, Herlev, Denmark*
ANAHIT GHOCHIKYAN • *Department of Molecular Immunology, Institute for Molecular Medicine, Huntington Beach, CA, USA*
ANITA GOTHELF • *Center for Experimental Drug and Gene Electrotransfer (C*EDGE), Department of Oncology, Copenhagen University Hospital Herlev, Herlev, Denmark*
JOHN B.A.G. HAANEN • *Division of Immunology, The Netherlands Cancer Institute, Amsterdam, The Netherlands*
ANA MARGARIDA HENRIQUES • *INIAV, Instituto Nacional de Investigação Agrária e Veterinária, Lisboa, Portugal*
ALMEDINA ISAKOVIC • *Division of Allergy and Immunology, Department of Molecular Biology, University of Salzburg, Salzburg, Austria*
SANDRA IURESCIA • *Section of Medical Biotechnology, Institute of Translational Pharmacology (IFT), National Research Council (CNR), Rome, Italy*
RONALD J. JACKSON • *Molecular Mucosal Vaccine Immunology Group, Department of Immunology, The John Curtin School of Medical Research, The Australian National University, Canberra, ACT, Australia*

AMIR S. KHAN • *Inovio Pharmaceuticals, Blue Bell, PA, USA*
YEU-CHUN KIM • *Department of Chemical and Biomolecular Engineering, Korea Advanced Institute of Science and Technology (KAIST), Daejeon, Korea*
KWANG LEE • *Department of Biological and Environmental Engineering, Cornell University, Ithaca, NY, USA*
JEREMY M. LUKE • *Nature Technology Corporation, Lincoln, NE, USA*
DAN LUO • *Department of Biological and Environmental Engineering, Cornell University, Ithaca, NY, USA*
CHARLES MA • *Department of Microbiology, Monash University, Clayton, VIC, Australia*
GABRIEL AMARO MONTEIRO • *Institute of Biotechnology and Bioengineering, Centre for Chemical and Biological Engineering, Instituto Superior Técnico, Lisboa, Portugal; Department of Bioengineering, Instituto Superior Técnico, University of Lisbon, Lisboa, Portugal*
KOEN OOSTERHUIS • *Division of Immunology, The Netherlands Cancer Institute, Amsterdam, The Netherlands*
IRINA PETRUSHINA • *Institute for Memory Impairments and Neurological Disorders, University of California, Irvine, CA, USA*
KEVIN R. PORTER • *Infectious Diseases Directorate, Naval Medical Research Center, Silver Spring, MD, USA*
DUARTE MIGUEL PRAZERES • *Institute of Biotechnology and Bioengineering, Centre for Chemical and Biological Engineering, Instituto Superior Técnico, Lisboa, Portugal; Department of Bioengineering, Instituto Superior Técnico, University of Lisbon, Lisboa, Portugal*
CHARANI RANASINGHE • *Molecular Mucosal Vaccine Immunology Group, Department of Immunology, The John Curtin School of Medical Research, The Australian National University, Canberra, ACT, Australia*
KANAKATTE RAVIPRAKASH • *Viral and Rickettsial Diseases Department, Infectious Disease Directorate, Naval Medical Research Center, Silver Spring, MD, USA*
MONICA RINALDI • *Section of Medical Biotechnology, Institute of Translational Pharmacology (IFT), National Research Council (CNR), Rome, Italy*
YOUNG HOON ROH • *David H. Koch Institute for Integrative Cancer Research, Massachusetts Institute of Technology, Cambridge, MA, USA*
NIRANJAN Y. SARDESAI • *Inovio Pharmaceuticals, Blue Bell, PA, USA*
SANDRA SCHEIBLHOFER • *Division of Allergy and Immunology, Department of Molecular Biology, University of Salzburg, Salzburg, Austria*
MARTIN SCHLEEF • *PlasmidFactory GmbH & Co. KG, Bielefeld, Germany*
MARCO SCHMEER • *PlasmidFactory GmbH & Co. KG, Bielefeld, Germany*
TON N.M. SCHUMACHER • *Division of Immunology, The Netherlands Cancer Institute, Amsterdam, The Netherlands*
CORDELIA SELOMULYA • *Department of Chemical Engineering, Monash University, Clayton, VIC, Australia*
HAE-YONG SEOK • *Department of Chemical and Biomolecular Engineering, Korea Advanced Institute of Science and Technology (KAIST), Daejeon, Korea*
HYEMEE SUH • *Department of Chemical and Biomolecular Engineering, Korea Advanced Institute of Science and Technology (KAIST), Daejeon, Korea*
NIMFA TENEZA-MORA • *Infectious Diseases Directorate, Naval Medical Research Center, Silver Spring, MD, USA*

JOSEF THALHAMER • *Division of Allergy and Immunology, Department of Molecular Biology, University of Salzburg, Salzburg, Austria*
YUAN MIN WANG • *Centre for Kidney Research, Children's Hospital at Westmead, The University of Sydney, Westmead, NSW, Australia*
RICHARD WEISS • *Division of Allergy and Immunology, Department of Molecular Biology, University of Salzburg, Salzburg, Austria*
JAMES A. WILLIAMS • *Nature Technology Corporation, Lincoln, NE, USA*
JESSICA JANE YE • *School of Medicine, Yale University, New Haven, CT, USA*

Part I

DNA Vaccine Design and Enhancement

Chapter 1

A Blueprint for DNA Vaccine Design

Sandra Iurescia, Daniela Fioretti, and Monica Rinaldi

Abstract

Although safety concerns have been overcome, lower immunogenicity profiles of DNA vaccines have hindered their progress in humans. DNA vaccines need to make up for this limitation by altering plasmid construction through vector design innovations intended for enhancement of transgene expression and immunogenicity. The next-generation vectors also address safety issues such as selection markers. This chapter discusses (a) plasmid backbone design, (b) enhancement of antigenic protein expression and immunogenicity, and (c) vector modification to increase innate immunity. Modifications of the basic design, when combined with improved delivery devices and/or prime/boost regimens, may enhance DNA vaccine performance and clinical outcomes.

Key words DNA vaccines, Plasmid design, Regulatory elements, Antigen, Immunogenicity, Transgene expression, Marker selection, DNA uptake, Innate immunity

1 Introduction

DNA vaccine efficacy is hampered by low level of transgene expression. Basically, a DNA vaccine consists in a self-replicating, circular double-stranded DNA molecule (the plasmid) joining bacterial regions necessary for selection and replication in *E. coli* host with eukaryotic sequences that regulate expression of the encoded antigen in the target tissue.

Ongoing efforts are focused to optimize the DNA vaccine platform to improve antigen (Ag) expression and immunogenicity as increased antigen expression correlates with improved immunogenicity and augmented levels of immune response in humans and large animal models [1].

This blueprint presents an overview of the issues facing the development of a platform technology for the production of improved DNA vaccines. Specifically, this chapter deals with strategies relevant to (a) engineering the plasmid backbone to obtain a

Sandra Iurescia and Daniela Fioretti contributed equally to this work.

Monica Rinaldi et al. (eds.), *DNA Vaccines: Methods and Protocols*, Methods in Molecular Biology, vol. 1143,
DOI 10.1007/978-1-4939-0410-5_1, © Springer Science+Business Media New York 2014

more efficient expression plasmid, (b) enhancement of antigenic protein expression and immunogenicity, and (c) complementary plasmid design modifications to increase innate immunity. All modifications outlined herein are made at the level of molecular cloning (*see* **Note 1**).

The aspects next described depict potential design strategies and technologies necessary to carry out the development of improved DNA vaccines. A selection of relevant patents employed to improve DNA vaccine immunogenicity through several strategies such as the use of tissue-specific transcriptional elements, nuclear localization signaling, and codon optimization are reported in Fioretti et al. [2].

2 Materials

1. Plasmid DNA backbone.
2. DNA coding for the transgene.
3. Oligonucleotide primers to amplify target transgene from the DNA template by polymerase chain reaction (PCR) (*see* **Note 2**).
4. Regulatory elements addressing antigen expression and immunogenicity augmentation.

3 Methods

3.1 Plasmid DNA Backbone

DNA vaccine design is relatively simple. Basically, the construction of a DNA vaccine requires a plasmid backbone such as pcDNA™3.1 or pVAX1© (Invitrogen, Life Technologies Corporation). An important component of the plasmid backbone is the promoter that drives expression of the transgene of interest.

For most expression plasmids, the human cytomegalovirus (CMV) promoter (reviewed in ref. 3) is the usual choice as it promotes high-level constitutive expression in a wide range of mammalian cells. Some early investigations reported that CMV immediate/early enhancer/promoter activity was consistently the highest among several constructs tested in mammalian somatic tissues [4, 5]. Otherwise, the use of host tissue-specific promoters prevents antigen expression in inappropriate tissues yet leading to sufficient stimulation of immune responses [6]. A Kozak sequence conforming to the consensus gccgccRcc<u>AUG</u>G (R = G or A, AUG start codon underlined, critical residues in caps) is included immediately prior to the ATG start codon ensuring efficient translation. Approaches to increase transcription and translation thereby improve DNA vaccine immunogenicity. This can be achieved by optimization of regulatory elements in the plasmid backbone [2]. The inclusion of an intron (i.e., the intronA of the CMV immediate–early gene) in the vector backbone downstream of the promoter can enhance the stability of mRNA and may improve gene expression. Furthermore, the presence of the polyadenylation (polyA) signal site, which contains accessory sequences upstream and downstream of the polyA tail, ensures proper termination of transcription, increased mRNA levels, and export of the mRNA from the nucleus resulting in improved transgene expression. Both enhancer elements and transcriptional transactivators may increase promoter efficiency. For example, incorporation of the human T-cell leukemia virus type I R 5′ untranslated region (UTR) (HTLV-I R-U5) downstream of the CMV promoter increased mRNA translation efficiency [7] and immunogenicity of DNA vaccines encoding multiple antigens in small animals and in nonhuman primates [8].

The transcribed 3′ and 5′ UTRs should not contain cryptic open reading frames (ORFs) since unforeseen immunogenic epitopes could be generated able to elicit a cytotoxic T lymphocyte (CTL) response [9].

A critical step of the DNA vaccine design is a careful selection and assembly of bacterial regions that provide both replication origin and selection marker necessary for propagation in different *E. coli* host strains. The compositions and orientations of the so-called relaxed origins of replication (i.e., pUC origin) can interfere with the transgene expression, manufacturing yields, and plasmid quality. Reduced expression may in part be due to the presence of TATA-containing cryptic promoter within the replication origin or the selectable marker generating spurious transcripts that triggers protein kinase R (PKR)-mediated selective translational shutdown or RNA interference (*see* Williams J.A. for a review) [10].

The use of antibiotic-resistance selection markers in DNA vaccines has safety issues and represents a key aspect for high-scale plasmid production. The European Medicines Agency (EMA) stated that "neomycin and kanamycin are of importance for veterinary and human use and that their current and potential future use cannot be classified as of no or only minor therapeutic relevance" due to current use in critical clinical settings [11]. To address these regulatory concerns, alternative non-antibiotic selection methods are being developed [10, 12]. In Chapter 6 of this book, Williams and colleagues describe the development of RNA-based antibiotic-free selection system for safer DNA vaccination.

In vector development, an important objective is plasmid backbone shortening and removal of bacterial elements. Plasmid size reduction improves pDNA structural stability and enhances the transfection efficiency, leading to increased duration of antigen expression [10, 12]. Overall, smaller plasmid size is therefore beneficial for gene delivery efficiency. Minicircle DNA technology was developed to obtain a plasmid backbone nearly devoid of any prokaryotic sequence by using site-specific recombination. The resulting miniplasmid contains almost exclusively the gene of interest and its regulating sequence motifs. The most recent developments in the field of minicircles are reviewed elsewhere [13]. Minicircle DNA-based gene therapy has been successfully exploited in mice [14] and in generation of adult human induced pluripotent stem cells (hiPSCs) [15].

To permit insertion of the sequence coding for the antigen, the plasmid should also contain a synthetic 100-base pairs DNA sequence (the multiple cloning site, MCS). The target transgene is often available in another construct but rarely presents useful restriction sites to allow directional cloning of the insert into the plasmid vector. Therefore, oligonucleotide primers are designed to contain appropriate restriction sites upstream of the ATG codon and downstream of the stop codon to amplify target transgene from the DNA template by PCR (*see* **Notes 1** and **2**). Selection of different restriction sites guarantees the insertion of the fragment in the MCS in the correct orientation for the transcription and translation.

3.2 Enhancement of Transgene Expression

Gene inserts traditionally are transferred from a genomic DNA construct directly into a DNA vaccine vectors. Notably, following the gene delivery the level of antigen expression is affected by the rate of transcription and translation. Therefore, some approaches have proved to be very helpful in enhancing antigen immunogenicity such as the effect of codon bias on rate of expression. To increase the expression of the encoded transgene, synthetic genes can be constructed such that the codons employed are those with a higher frequency of mammalian tRNA than codons present in the native sequence (i.e., codon optimization) [16]. Codon optimization matching high-use codons for the target species has been shown to dramatically increase transgene expression and immunogenicity of the DNA vaccines [17]. An optimal coding sequence is back translated from the amino acid sequence of the antigen by algorithms (*see* **Note 3**) that take into account the abundance of specific tRNAs in the cytosol of human cells and the predicted structure of the mRNA. Thereafter the synthetic gene sequence is designed and synthesized in vitro. Adverse rare codons are avoided, and secondary structures in the mRNA are minimized [18]. Thereby, the target transgene codons can benefit from codon reengineering to correspond to the available pool of tRNAs at the vaccination site leading to the induction of a specific immune response.

It is recommended, for DNA vaccines intended for potential licensure, that new gene inserts should be designed *de novo* for compatibility, regulatory compliance, and improved eukaryotic expression and then made synthetically. Gene optimization is an important factor for a successful protein expression, and synthetic transgene design is a critical step to maximize the expression of synthetic genes. Some multiparameter DNA sequence optimization procedures such as the GeneArt GeneOptimizer® software take codon usage, GC content, mRNA structure, and species-specific sequence motifs into account [17] (*see* **Note 3**). For pathogens containing various serotypes or amino acid variations, determination of amino acid sequence may involve defining a "consensus immunogen" to engineer a single broadly cross-neutralizing antigen [1, 19]. Consensus vaccines offer novel means of inducing cross-reactive cellular and humoral immune responses in humans. The ability to induce such responses would be a significant advance in the development of next-generation vaccines, especially for seasonal and pathogenic H5N1 influenza viruses, which have become endemic in many countries. These vaccines are designed using a large number of primary viral sequences and as such contain the most highly conserved characteristics of each. These synthetic constructs, with in vivo electroporation, have the ability to induce strong $CD8^+$ and $CD4^+$ cellular immune responses in small and large animal models of vaccination. Furthermore, the synthetic consensus antigens result to be more cross-reactive than their individual component antigens as showed by the ability of inhibiting divergent viruses of the H5N1 subtype [20].

A flow chart for synthetic transgene design to generate a codon-optimized antigen gene is presented in Chapter 6 of this book.

Given that plasmid DNA nuclear import is coupled to active transcription, DNA nuclear targeting sequence (DTS) could be introduced to increase the efficiency of nuclear plasmid uptake from cytoplasm and to selectively improve extrachromosomal transgene expression especially in nondividing cells, e.g., after intramuscular injection [7, 21, 22]. The sequence of Simian virus 40 (SV40) enhancer that is known to bind to distinct, ubiquitously expressed, transcription factors and to mediate plasmid nuclear entry in all cell types tested was mapped to a 372 bp region of the DNA (*see* Lam and Dean for a review) [22].

Chapter 2, "Enhancement of plasmid-mediated transgene expression," of this book elaborates on the basic concept described here.

Long-term tissue Ag expression may be achieved using scaffold/matrix attachment region (S/MAR) as *cis*-acting elements to maintain the episomal status of the circular vector [23]. Actually, Argyros et al. demonstrated prolonged expression of such vector-encoded transgenes (i.e., 6 months) in mouse tissues. Development of S/MAR minicircles led to higher and more sustained expression of transgene in vitro and in vivo [24].

3.3 Vector Modifications to Increase Innate Immunity

Recent advances provide insights into molecular and cellular mechanisms by which double-stranded structure is essential for DNA vaccine-induced immunogenicity. After transfection, DNA is sensed in the cytoplasm through nucleic acid-sensing immune machinery that uses mainly two types of immune triggers. First, immunostimulatory elements in the plasmid backbone such as

CpG motifs may play a role in the activation of immune response acting as "a built-in adjuvant" [25]. Unmethylated CpG motifs enable bacterial plasmid DNA recognition by mammalian cells as signals pathogen-associated molecular patterns (PAMPs). Second, vaccine components (such as certain adjuvants) or cell damage at injection site may induce the release of endogenous damage-associated molecular patterns (DAMPs). PAMPs and DAMPs can stimulate the innate immune system by activating conserved receptors that are referred to as pattern-recognition receptors (PRRs) [26, 27]. For example, CpG motifs allow activation of innate immune response through binding to Toll-like receptors (TLRs) and RIG-I-like receptors (RLRs) [25, 28]. However, recent studies reported that TLR-9 is dispensable for DNA vaccine-induced immune response [29]. Conversely, double-stranded DNA (i.e., the right-handed B form of structure DNA) has an essential role in triggering the nucleic acid-sensing immune machinery through a wide range of PRRs. The TRAF family member-associated NFκB activator (TANK)-binding kinase 1 (TBK1), a noncanonical IκB kinase [30], and the stimulator of interferon (IFN) genes (STING) mediate the adjuvant effect of cytoplasmic DNA. In this way, the immunogenicity of DNA vaccines is regulated by innate immune response via plasmid DNA recognition through activation of the STING/TBK1 signaling cascade. Thereby, STING/TBK1 pathway directs phosphorylation of interferon regulatory factor (IRF) 3 and IRF7 resulting in the production of type I IFNs [26, 31].

Intracellular dsDNA may also induce the absent in melanoma 2 (AIM2) that results in the activation of inflammasome and the release of biologically active interleukin-1β (IL-1β) [26]. Thus TBK1 and AIM2 pathways are critical for the induction of both innate and adaptive immune responses [32, 33].

Therefore, an alternative approach to improve DNA vaccine efficacy could be achieved through vector modifications aimed at increasing innate immunity. With the increased recognition of the impact of nucleic acid-sensing PRRs on APC function, research is well under way to directly harness these PRRs using novel adjuvants. The so-called plasmid backbone adjuvants can encode DNA- or RNA-based adjuvants. Such modifications avoid the autoimmunity concerns from expressing a human protein (i.e., immunomodulatory molecules) since the DNA or the RNA adjuvant will not be the target of adaptive immunity [10]. Addition of optimized hypomethylated CpG dinucleotide motifs may be used to increase TLR9 signaling activation that may have a role in priming CD8 T cell responses. Immunostimulatory RNA co-expressed from the vector may be utilized to activate RNA-sensing innate immune response receptors such as RIG-I-like (RLRs), namely, retinoic acid-inducible gene I (RIG-I; also known as DDX58) and melanoma differentiation-associated protein 5 (MDA5; also known as IFIH1) (for a review *see* Williams J.A. 2013) [10].

4 Notes

1. The development of new technologies provides greater opportunities to enhance efficacy of DNA vaccines. The aim of this chapter is to make the reader aware of some helpful approaches to modify basic design of plasmid DNA vaccines. Other strategies (e.g., vaccine efficacy enhancement, prime–boost, and delivery platforms) and more detailed procedures are described in other chapters of this book.
2. The sequence of interest can be amplified by PCR, or a synthetic gene sequence can be designed and generated by chemical synthesis.
3. GASCO is an algorithm developed for the optimum codon selection using genetic algorithms [34]. The software for the proposed algorithm is available on http://miracle.igib.res.in/gasco/. Entelechon has developed a proprietary software tool for gene optimization, called Leto. Leto strategically replaces synonymous codons of a given DNA sequence in order to improve a set of parameters deemed relevant for the expression yield (codon usage, homogenous GC content, mRNA secondary structure, cryptic splice sites, codon and motif repeats, restriction sites, custom motives (e.g., poly-A signals, GC islets) http://www.entelechon.com/company/synthetic-products-2/synthetic-gene/gene-optimization/). A set of free online tools (Backtranslation tool; Gene to codon usage; Codon usage table analysis) can also be accessed from this website: http://www.entelechon.com/resources/online-tools/.

References

1. Kutzler MA, Weiner DB (2008) DNA vaccines: ready for prime time? Nat Rev Genet 9:776–788
2. Fioretti D, Iurescia S, Rinaldi M (2013) Recent advances in design of immunogenic and effective naked DNA vaccines against cancer. Recent Pat Anticancer Drug Discov 9:66–82
3. Donnelly JJ, Ulmer JB, Shiver JW et al (1997) DNA vaccines. Annu Rev Immunol 15: 617–648
4. Cheng L, Ziegelhoffer PR, Yang NS (1993) In vivo promoter activity and transgene expression in mammalian somatic tissues evaluated by using particle bombardment. Proc Natl Acad Sci USA 90:4455–4459
5. Manthorpe M, Cornefert-Jensen F, Hartikka J et al (1993) Gene therapy by intramuscular injection of plasmid DNA: studies on firefly luciferase gene expression in mice. Hum Gene Ther 4:419–431
6. Cazeaux N, Bennasser Y, Vidal PL et al (2002) Comparative study of immune responses induced after immunization with plasmids encoding the HIV-1 Nef protein under the control of the CMV-IE or the muscle-specific desmin promoter. Vaccine 20:3322–3331
7. Luke JM, Vincent JM, Du SX et al (2011) Improved antibiotic-free plasmid vector design by incorporation of transient expression enhancers. Gene Ther 18:334–343
8. Barouch DH, Yang ZY, Kong WP et al (2005) A human T-cell leukemia virus type 1 regulatory element enhances the immunogenicity of human immunodeficiency virus type 1 DNA vaccines in mice and nonhuman primates. J Virol 79:8828–8834
9. Li C, Goudy K, Hirsch M et al (2009) Cellular immune response to cryptic epitopes during therapeutic gene transfer. Proc Natl Acad Sci USA 106:10770–10774

10. Williams JAV (2013) Vector design for improved DNA vaccine efficacy, safety and production. Vaccine 1:225–249
11. EMA (2007) Presence of the antibiotic resistance marker gene nptII in GM plants for food and feed uses. Vol EMEA/CVMP/56937/2007
12. Oliveira PH, Mairhofer J (2013) Marker-free plasmids for biotechnological applications: implications and perspectives. Trends Biotechnol 31:539–547
13. Schleef M (2013) Minicircle and miniplasmid DNA vectors: the future of nonviral and viral gene transfer. Wiley, Weinheim
14. Osborn MJ, McElmurry RT, Lees CJ et al (2011) Minicircle DNA-based gene therapy coupled with immune modulation permits long-term expression of alpha-L-iduronidase in mice with mucopolysaccharidosis type I. Mol Ther 19:450–460
15. Narsinh KH, Jia F, Robbins RC et al (2011) Generation of adult human induced pluripotent stem cells using nonviral minicircle DNA vectors. Nat Protoc 6:78–88
16. Andre S, Seed B, Eberle J et al (1998) Increased immune response elicited by DNA vaccination with a synthetic gp120 sequence with optimized codon usage. J Virol 72:1497–1503
17. Fath S, Bauer AP, Liss M et al (2011) Multiparameter RNA and codon optimization: a standardized tool to assess and enhance autologous mammalian gene expression. PLoS One 6:e17596
18. Weide B, Garbe C, Rammensee HG et al (2008) Plasmid DNA- and messenger RNA-based anti-cancer vaccination. Immunol Lett 115:33–42
19. Williams JA, Carnes AE, Hodgson CP (2009) Plasmid DNA vaccine vector design: impact on efficacy, safety and upstream production. Biotechnol Adv 27:353–370
20. Laddy DJ, Yan J, Kutzler M et al (2008) Heterosubtypic protection against pathogenic human and avian influenza viruses via in vivo electroporation of synthetic consensus DNA antigens. PLoS One 3:e2517
21. Rinaldi M, Fioretti D, Iurescia S et al (2008) Anti-tumor immunity induced by CDR3-based DNA vaccination in a murine B-cell lymphoma model. Biochem Biophys Res Commun 370:279–284
22. Lam AP, Dean DA (2010) Progress and prospects: nuclear import of nonviral vectors. Gene Ther 17:439–447
23. Conese M, Auriche C, Ascenzioni F (2004) Gene therapy progress and prospects: episomally maintained self-replicating systems. Gene Ther 11:1735–1741
24. Argyros O, Wong SP, Fedonidis C et al (2011) Development of S/MAR minicircles for enhanced and persistent transgene expression in the mouse liver. J Mol Med (Berl) 89:515–529
25. Coban C, Koyama S, Takeshita F et al (2008) Molecular and cellular mechanisms of DNA vaccines. Hum Vaccin 4:453–456
26. Desmet CJ, Ishii KJ (2012) Nucleic acid sensing at the interface between innate and adaptive immunity in vaccination. Nat Rev Immunol 12:479–491
27. O'Neill LA (2013) Immunology. Sensing the dark side of DNA. Science 339:763–764
28. Loo YM, Gale M Jr (2011) Immune signaling by RIG-I-like receptors. Immunity 34:680–692
29. Babiuk S, Mookherjee N, Pontarollo R et al (2004) TLR9−/− and TLR9+/+ mice display similar immune responses to a DNA vaccine. Immunology 113:114–120
30. Ishii KJ, Kawagoe T, Koyama S et al (2008) TANK-binding kinase-1 delineates innate and adaptive immune responses to DNA vaccines. Nature 451:725–729
31. Kobiyama K, Jounai N, Aoshi T, Tozuka M, Takeshita F, Coban C, Ishii KJ (2013) Innate immune signaling by, and genetic adjuvants for DNA vaccination. Vaccine 1:278–292
32. Coban C, Kobiyama K, Aoshi T et al (2011) Novel strategies to improve DNA vaccine immunogenicity. Curr Gene Ther 11:479–484
33. Li L, Saade F, Petrovsky N (2012) The future of human DNA vaccines. J Biotechnol 162:171–182
34. Sandhu KS, Pandey S, Maiti S et al (2008) GASCO: genetic algorithm simulation for codon optimization. In Silico Biol 8:187–192

Chapter 2

Enhancement of Plasmid-Mediated Transgene Expression

Daniela Fioretti, Sandra Iurescia, and Monica Rinaldi

Abstract

A large number of studies aimed at the treatment of cancer, autoimmune and metabolic diseases, neurodegenerative disorders, allergic diseases, as well as muscle disorders strengthen the fact that gene therapy could represent an alternative method to treat human diseases where conventional approaches are less effective.

To improve transgene expression from plasmid vectors, DNA nuclear targeting sequences (DTSs) can be introduced in a vector backbone to increase in vivo expression up to 20-fold using electroporation (EP) delivery in muscle tissue. The purpose of this chapter is to represent a step-by-step strategy for the construction of a plasmid vector with enhanced efficiency of nuclear plasmid uptake and the methodic for the in vivo efficiency evaluation of the obtained expression vector.

Key words DNA vaccine, Naked DNA, Plasmid, DNA nuclear targeting sequence, DTS, Electrogenetherapy, Skeletal muscle

1 Introduction

Gene transfer has been investigated over two decades as alternative method to treat human diseases where conventional approaches are less effective. Electrogenetherapy is promising for the treatment of muscle disorders as well for the systemic secretion of therapeutic proteins, DNA vaccination, immunotherapy, and cancer therapy [1–8].

Skeletal muscle is one of the most attractive tissues for gene electroporation (EP) due to its relatively easy access, long-term stable transgene expression, and excellent vascularization [9].

To achieve more efficient gene expression from nonviral vectors, DNA nuclear targeting sequences (DTSs) could be introduced in a vector backbone to increase the efficiency of nuclear plasmid uptake from cytoplasm. Such sequences would be true transient expression enhancers, as they function to enhance plasmid nuclear entry, not mRNA production or translation, which could also

Daniela Fioretti and Sandra Iurescia contributed equally to this work.

Monica Rinaldi et al. (eds.), *DNA Vaccines: Methods and Protocols*, Methods in Molecular Biology, vol. 1143,
DOI 10.1007/978-1-4939-0410-5_2, © Springer Science+Business Media New York 2014

affect chromosomal gene expression [10]. Whereas the DNA transport to the nucleus is a limiting factor for nondividing cells, e.g., skeletal muscle cells [11, 12], constructs lacking these DTSs remain confined to the cytoplasm until they are degraded [13]. In fact, the nuclear localization of plasmids in the absence of cell division requires transcription factors that bind to these specific sequences in the cytoplasm and facilitate the DNA–protein complex nuclear import.

A number of DNA transcription factor-binding sites have been identified that increase plasmid nuclear import when added to a plasmid vector backbone. The inclusion of a DNA sequence localized in a noncoding region of the SV40 virus sequence, that did not have significant sequence homology to the human genome, in a DNA vaccine vector backbone is reported to increase in vivo expression up to 20-fold using EP delivery in muscle tissue [14]. In particular the 72-bp repeats of the SV40 enhancer derived from the origin of replication and early–late promoter region facilitate maximal transport [15].

The purpose of this chapter is to represent a step-by-step strategy for the construction of a plasmid vector with enhanced efficiency of nuclear plasmid uptake and the methodic for the in vivo efficiency evaluation of the obtained expression vector.

2 Materials

2.1 Generation of the Nuclear Targeting Sequence Insert

1. pIRES vector (Clontech Laboratories Inc.) resuspended in 10 mM Tris–HCl, pH 8.5 (elution buffer) to the concentration 100 ng/μl.
2. PCR amplification reagents (Promega): *Pfu* DNA polymerase, dNTPs, Pfu DNA polymerase 10× buffer with MgSO4. PCR upstream and downstream primers.
3. Thermal cycler with hot lid.
4. Restriction enzymes.
5. Agarose gel electrophoresis equipment: For 50 ml of 1.5 % agarose gel, use 0.75 g of ultrapure agarose (electrophoresis grade) with 50 ml of 0.5× TBE. Prepare 5× stock solution in 1 l of water with 54 g of Tris base, 27.5 g of boric acid, and 20 ml of 0.5 M EDTA, pH 8.0.
6. QIAquick Gel Extraction Kit (Qiagen).

2.2 Cloning of the Nuclear Targeting Sequence Insert into the Expression Vector

1. pRC110 vector suspended in 10 mM Tris–HCl, pH 8.5 (elution buffer) to the concentration 1 μg/μl.
2. Restriction enzymes.
3. Agarose gel electrophoresis equipment.
4. QIAquick Gel Extraction Kit (Qiagen).

5. Shrimp Alkaline Phosphatase (BioLabs).
6. Ligation Kit (Takara).

2.3 Screening and Checking of the Clones

1. DH5α competent cells (Promega).
2. Miller's LB Broth for bacterial culture with ampicillin: Dissolve 5 g of peptone, 2.5 g of yeast extract, and 5 g of NaCl with 450 ml of water in 500 ml cylinder. Make up to 500 ml with water. Sterilize the solution in a Pyrex bottle for 20 min at 121 °C, cool the solution at room temperature (r.t.), and then add 500 μl of 50 mg/ml of ampicillin.
3. Miller's LB agar plates with ampicillin: Dissolve 5 g of peptone, 2.5 g of yeast extract, and 5 g of NaCl with 450 ml of water in 500 ml cylinder. Make up to 500 ml with water. Transfer the solution in a Pyrex bottle and add 7.5 g of agar. Sterilize for 20 min at 121 °C, cool the solution at 50 °C, and then add 500 μl of 50 mg/ml of ampicillin. Pour carefully into 10 mm Petri dishes avoiding bubbles. Allow gel to solidify at r.t. overnight (o.n.) and store at 4 °C for up to 2 months.
4. QIAprep Miniprep Kit (Qiagen).
5. Restriction enzymes.
6. Agarose gel electrophoresis equipment.
7. M13rev primer for sequencing (Eurofins).
8. Sequencing service at http://www.operon.com/services/dna-sequencing/getting-started.aspx (Eurofins MWG Operon).

2.4 Large-Scale DNA Preparation for In Vivo Administration

1. Miller's LB Broth for bacterial culture with ampicillin.
2. EndoFree Plasmid Mega Kit (Qiagen).
3. Endotoxin-free PBS (Sigma-Aldrich®).

2.5 In Vivo Plasmid Vector Administration

1. TX ECM 830 Pulse Generator (Genetronics, San Diego, CA) and needle-free electrode.
2. 27 G needle syringe for intraperitoneal injection.
3. 29 G needle syringe (BD) for intramuscular injection.
4. Zoletil® (Virbac) and Rompun® (Bayer).
5. Shaver or clipper.
6. Blood collection material: Sterile animal lancet, 5 mm (Goldenrod) and 1.5 ml sterile tubes for each mouse.

2.6 In Vivo Expression Assay

1. Murine IL-2 ELISA development kit (PeproTech), consisting of capture antibody, detection antibody, murine IL-2 standard, avidin–HRP conjugate.
2. ELISA flat-bottom 96-well plates.
3. ABTS Liquid Substrate Solution (Sigma–Aldrich®).
4. Dulbecco's PBS.

5. Block buffer: 1 % BSA in 1× PBS.
6. Wash buffer: 0.05 % Tween-20 in 1× PBS.

3 Methods

3.1 Generation of the Nuclear Targeting Sequence Insert

A DNA nuclear targeting sequence (DTS), consisting of two complete, contiguous copies of the functional 72-bp element of the SV40 enhancer is amplified from the pIRES vector.

1. For primers design, first generate restriction maps of the DTS and of the recipient vector (*see* **Note 1**), in this case pRC110 ([6], Fig. 1). Then choose the restriction sites on pRC110 necessary for the cloning and that do not appear in the DTS. The DNA sequence coding for these sites will be added to the region of interest as it is generated by polymerase chain reaction (PCR) so that the DTS can be cloned (*see* **Note 2**). If the recipient vector brings very few useful unique restriction sites, as showed in Fig. 1, and more vector manipulations could be needed, it's better to use only one of them for this cloning (*see* **Note 3**). For example, if *Xba*I is chosen for the cloning sites in

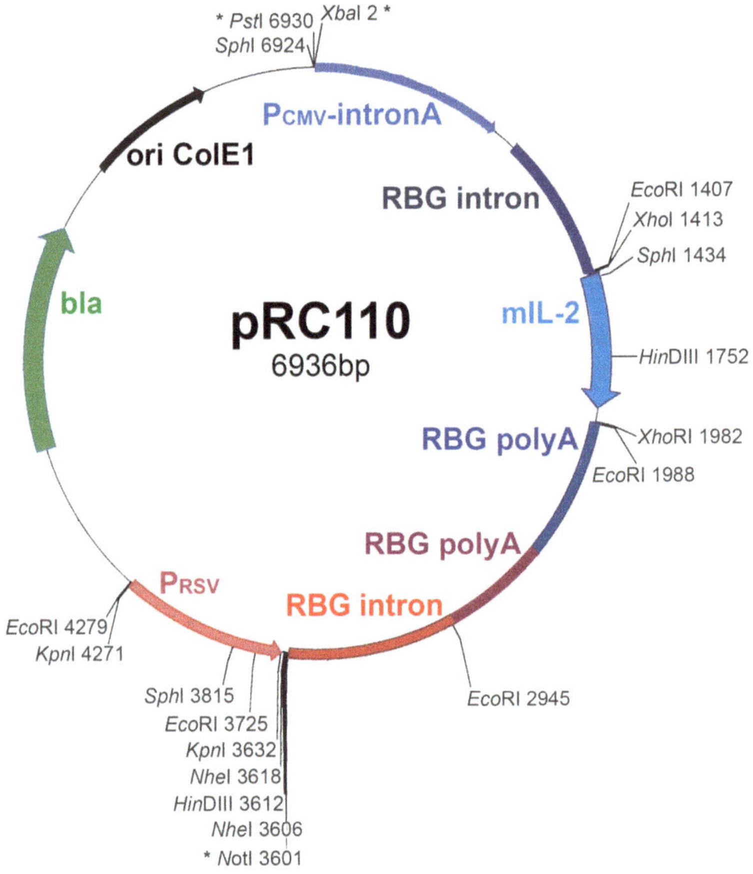

Fig. 1 pRC110 restriction map. *Note*: *Asterisk* indicates unique sites

Upstream 5' CTAGTCTAGAGGTACCTTCTGAGG 3'

Downstream 5' CTAGTCTAGAGCATGCTTTGCATA 3'

Fig. 2 Example of polymerase chain reaction (PCR) primers to generate a DNA sequence with XbaI restriction sites 5′ and 3′ of the DTS region. Extra bases are added before the restriction sites to improve cleavage at the end of PCR fragments. The XbaI restriction sites are showed in *bold*

pRC110, the *Xba*I site will be added to the 5′-end and to the 3′-end of the DTS. In Fig. 2 the two primers are showed.

2. Insert PCR amplification was carried out with *Pfu* DNA polymerase (*see* **Note 4**) following the manufacturer's protocol for standard application in 50 μl of reaction mixture (*see* **Note 5**). Preheat the thermal cycler to 95 °C, and then heat the samples at 95 °C for 1 min to ensure that the target DNA is completely denatured. Run the thermal cycling program for 30 cycles with the following setting: 45 s at 94 °C, 1 min s at 65 °C, and 2 min at 72 °C. Add 1 cycle at 72 °C for a final extension of 5 min and then soak indefinitely at 4 °C.
3. Run the PCR product by agarose gel electrophoresis, and recover the fragment using the QIAquick Gel Extraction Kit, following the manufacturer's protocol. Cut the purified fragment with *Xba*I restriction enzyme for 2–3 h, and purify the fragment another time with the same kit. Concentrate the product by ethanol precipitation adding 1/10 volume of sodium acetate (3 M, pH 5.2) (*see* **Note 6**) and 2.5× volume of at least 95 % ethanol (*see* **Note 7**). Chill at either −20 °C for 20 min or at −80 °C for 10 min, centrifuge at 4 °C for 10 min to recover the DNA, and dissolve the pellet EB.

3.2 Cloning of the Nuclear Targeting Sequence Insert into the Expression Vector

1. Digest 5 μg of the recipient plasmid with 5 U of *Xba*I, and run the linearized vector by agarose gel electrophoresis. Recover the fragment using the QIAquick Gel Extraction Kit, following the manufacturer's protocol. The restriction site is right upstream of the CMV promoter sequence.
2. Treat the linearized vector with Shrimp Alkaline Phosphatase (SAP; *see* **Note 8**). Dephosphorylated DNA termini cannot be ligated by DNA ligase so that only the insert fragment can circularize the expression vector (*see* **Note 9**). The reaction can be done in a final volume of 30 μl at 37 °C using 0.1 unit enzyme per mole DNA ends (*see* **Note 10**). Then SAP is completely and irreversibly inactivated by heating at 65 °C for 15 min.
3. Perform sticky-ended (*see* **Note 11**) DNA ligations with the standard protocol of the Takara DNA Ligation Kit—mighty mix. Combine the digested plasmid vector DNA and the DNA fragment to be inserted in a total volume of 5–10 μl. Due to the shortness of the insert fragment the recommended vector:insert molar ratio is 1:5–10. Then add one volume of

ligation mix (5–10 ml) to the DNA solution and incubate at 16 °C for 30 min (*see* **Note 12**). The ligation reaction mixture can be used directly for transformation with *E. coli* competent cells.

3.3 Screening and Checking of the Clones

1. Transform 50 μl of commercially available competent DH5α cells (*see* **Note 13**) with 50 μl of the ligation reaction.
2. Plate onto Miller's LB agar plates with ampicillin for selection, and incubate the plates at 37 °C for 16–24 h.
3. Select well-isolated colonies from the transformation plates (*see* **Note 14**), inoculate the colonies in 5 ml Miller's LB Broth for bacterial culture with ampicillin, and incubate with 250 rpm shaking for 10–16 h at 37 °C.
4. Make plasmid minipreps from the cultures, using the QIAprep Miniprep Kit.
5. Screen the clones by restriction digestion with *Pvu*II (*see* **Note 15**) followed by agarose gel electrophoresis.
6. Determine the DNA concentration in each preparation by measuring the absorbance at 260 nm using a spectrophotometer. Consider that A260 corresponds to a concentration of 50 μg/ml double-stranded DNA.
7. Clones that are identified by restriction mapping can be further verified by sequencing the insert through the sequencing service. The following commercial primer, 5′-caggaaacagctatgacca-3′ (M13rev), was chosen to map near the inserted DNA fragment upstream of XbaI site located on the recipient plasmid (*see* **Note 16**).

3.4 Large-Scale DNA Preparation for In Vivo Administration

1. Inoculate a 10 ml Miller's LB Broth plus ampicillin culture with a verified colony and incubate with 250 rpm shaking o.n. at 37 °C.
2. Use this fresh culture to inoculate 500 ml (1:500) of the same bacterial culture medium in a 1 l flask for a large-scale preparation and incubate with 250 rpm shaking o.n. at 37 °C.
3. Harvest the cells by centrifugation, and make plasmid megapreps using EndoFree Plasmid Mega Kit. Suspend plasmid DNAs in 500 μl of endotoxin-free PBS and store at −20 °C.

3.5 In Vivo Plasmid Vector Administration

The SV40 DTS+ and SV40 DTS− plasmid vectors are applied in in vivo protocols of intramuscular injection followed by electroporation.

1. Anesthetize mice (*see* **Note 17**) with a combination of Zoletil® and Rompun®, respectively, 35 and 2 mg/kg, via intraperitoneal injection using a 27 G syringe.
2. Remove the hair from both posterior legs at the site of electroporation using shaver or clipper.

3. Inject anesthetized mice in posterior muscle leg with plasmid solution, 50 μg of each plasmid diluted in 30–40 μl of sterile endotoxin-free PBS, using a 29 G syringe.
4. Immediately place the needle-free electrode over the injection site and administered electroporation using BTX ECM 830 Pulse Generator with eight pulses of 175 V, pulse duration of 20 ms, and pulse interval of 1 s.
5. Repeat injection and electroporation at the other posterior leg.
6. Collect 100 μl blood sample in 1.5 ml tube from each mouse after using the submandibular cheek bleed technique 1 week after the DNA injection. Use a lancet to puncture the vascular bundle that drains the face while the rodent is gently restrained (*see* **Note 18**).
7. Centrifuge blood samples for 15 min at 3,000 rpm using a microfuge, transfer sera in new tubes, and store at −80 °C until assayed.
8. Perform the in vivo expression assay as described in Subheading 3.6 using the diluted sera.

3.6 In Vivo Expression Assay

1. Use murine IL-2 ELISA development kit according to the manufacturer's instructions (*see* **Note 19**). Coat 96-well plates with 100 μl capture antibody (1 μg/ml) and incubate o.n. at r.t.
2. Dispense 300 μl blocking solution, 1 % BSA/1× PBS, and live at r.t. for 1 h.
3. Dilute serially the transfected cell culture supernatants (from 1:10 to 1:80) and the murine recombinant IL-2 standard (make seven dilutions from 0 to 10 ng/ml) in 1× PBS and add to the microtiter.
4. Detect the IL-2 binding by adding 100 μl of biotinylated antigen–affinity-purified anti-mIL-2 (0.2 μg/ml).
5. After 2 h at r.t. add 1:2,000 diluted avidin peroxidase.
6. After 30 min of incubation, the wells were extensively rinsed.
7. Then add the enzyme substrate ABTS.
8. Measure the absorbance after 15–30 min at 405 nm.
9. Determine the concentration of cytokine in the samples from the standard curve trendline.

4 Notes

1. There are many online software for the generation of the restriction map. One of the best is the Harry Mangalam's tacg program, version 4.3, at the site http://moo.nac.uci.edu/tacg4/form4.html.

2. Sites cannot be added to the ends of the insert region if they exist within the insert sequence, because the PCR fragment will need to be cut with those enzymes before cloning.
3. Whenever possible it is better to clone into two different sites, forcing the insert into the vector in the correct orientation for expression and preventing ligation of empty vector and, thus, minimizing the number of clones to be screened.
4. Base misinsertions that may occur infrequently during polymerization are rapidly excised by the proofreading activity of the *Pfu* DNA polymerase. Consequently, this polymerase is useful for polymerization reactions requiring high-fidelity synthesis.
5. In the reaction buffer supplied by the manufacturer, add 0.5 μl of *Pfu* (2–3 U/μl) to 50 ng of template plasmid with 50 pmol of each primer and a final dNTP concentration of 0.2 mM each in a total volume of 50 μl.
6. For ethanol DNA precipitation, the solution needs to have a high salt concentration. The best results are obtained adding sodium acetate, but NaCl could also be used.
7. In case of small DNA fragments or high dilutions o.n. incubation at 0 °C gives best results.
8. It is not necessary to remove SAP after the reaction as it is completely and irreversibly inactivated by heating phase. You must remove restriction enzymes and buffer from the digested vector, and DNA can be diluted in water or elution buffer (EB).
9. If the vector is cut with a single restriction enzyme, as in this case, chances are much higher that the vector ligates back on itself rather than on an added DNA fragment. This results in a high fraction of "empty clones" or background.
10. Linear DNA end picomoles can be calculated at the following website: http://www.promega.com/techserv/tools/biomath/calc05.htm. For example: 50 ng of pUC118 DNA (3,162 bp) corresponds to about approximately 25 fmol.
11. Blunt-ended ligations can be performed with the same protocol obtaining comparable efficiency of ligation. DNA solution should be desalted to include minimum salt prior to ligation reaction as high concentration of DNA solution lowers the ligation efficiency. The rapid ligation protocol is not recommended.
12. If low numbers of transformants are obtained, the reaction can be extended o.n. to improve ligation efficiency. Instead use the rapid protocol, 25 °C for 5 min, when the highest efficiency is not required.
13. Competent cells containing the following genetic markers, F-Φ80*lac*ZΔM15 Δ(*lac*ZYA-*arg*F) U169 *rec*A1 *end*A1 *hsd*R17 (rk−, mk+) *pho*A *sup*E44 *thi*-1 *gyr*A96 *rel*A1 λ-, result

in some benefits: the *lac*ZΔM15 enables blue/white color screening of colonies on plates containing X-gal or Bluo-gal; *rec*A1 ensures increased insert stability and prevents unwanted recombination; *end*A1 improves the yield and quality of plasmid DNA prepared from minipreps. For more information see the website http://www.invitrogen.com/site/us/en/home/Products-and-Services/Applications/Cloning/competent-cells-for-transformation/Chemically-Competent/dh5alpha-genotypes.html.

14. Typically, commercial competent cells yield >1 × 10⁶ transformants/μg control DNA per 50 μl reaction.
15. The chosen restriction enzyme allows to verify the presence of the insert, since it cuts both the vector and the insert (only one time), and to understand the orientation of the inserted fragment. This is the ideal condition; most frequently two different restriction enzymes are required.
16. Select a region for primer placement where the possibility of sequence error is low, e.g., avoid G–C-rich region. Identify potential sequencing primers that produce stable base pairing with the template DNA under conditions appropriate for cycle sequencing with the help of online software, as OligoAnalyzer 3.1 (http://eu.idtdna.com/analyzer/Applications/OligoAnalyzer/Default.aspx) and Oligo Calc: Oligonucleotide Properties Calculator (http://www.basic.northwestern.edu/biotools/oligocalc.html). Primer length should be between 18 and 30 nt with a G–C content of 40–60 %. These online software enable you to discard candidate primers that show undesirable self-dimer or hairpin. Finally, if you have several candidate primers, you might select one or a few that are more A–T rich at the 3′-end.
17. Test the anesthetic cocktail on one or two mice of your own mouse strain, prior to using the final protocol. This combination works fine on C3H/HeN mouse strain and all DBA strains derived.
18. For this technique anesthesia is not required.
19. Perform three washes after each step with 200 μl of 0.05 % Tween 20/1× PBS.

References

1. Iurescia S, Fioretti D, Fazio VM et al (2011) Design and pre-clinical development of epitope-based DNA vaccines against B-cell lymphoma. Curr Gene Ther 11:414–422
2. Iurescia S, Fioretti D, Pierimarchi P et al (2010) Genetic immunization with CDR3-based fusion vaccine confers protection and long-term tumor-free survival in a mouse model of lymphoma. J Biomed Biotechnol 2010:316069
3. Rinaldi M, Fioretti D, Iurescia S et al (2008) Antitumor immunity induced by CDR3-based DNA vaccination in a murine B-cell lymphoma model. Biochem Biophys Res Commun 370:279–284

4. Fazio VM, Ria F, Franco E et al (2004) Immune response at birth, long-term immune memory and 2 years follow-up after in-utero anti-HBV DNA immunization. Gene Ther 11:544–551
5. Rinaldi M, Catapano AL, Parrella P et al (2000) Treatment of severe hypercholesterolemia in apolipoprotein E-deficient mice by intramuscular injection of plasmid DNA. Gene Ther 7:1795–1801
6. Rinaldi M, Ria F, Parrella P et al (2001) Antibodies elicited by naked DNA vaccination against the complementary-determining region 3 hypervariable region of immunoglobulin heavy chain idiotypic determinants of B-lymphoproliferative disorders specifically react with patients' tumor cells. Cancer Res 61:1555–1562
7. Rinaldi M, Signori E, Rosati P et al (2006) Feasibility of in utero DNA vaccination following naked gene transfer into pig fetal muscle: transgene expression, immunity and safety. Vaccine 24:4586–4591
8. Fioretti D, Iurescia S, Fazio VM et al (2010) DNA vaccines: developing new strategies against cancer. J Biomed Biotechnol 2010:174378
9. Fioretti D, Iurescia S, Fazio VM et al (2013) In vivo DNA electrotransfer for immunotherapy of cancer and neurodegenerative diseases. Curr Drug Metab 14:279–290
10. Luke JM, Vincent JM, Du SX et al (2011) Improved antibiotic-free plasmid vector design by incorporation of transient expression enhancers. Gene Ther 18:334–343
11. Lam AP, Dean DA (2010) Progress and prospects: nuclear import of nonviral vectors. Gene Ther 17:439–447
12. Wagstaff KM, Jans DA (2007) Nucleocytoplasmic transport of DNA: enhancing non-viral gene transfer. Biochem J 406: 185–202
13. Miller AM, Munkonge FM, Alton EW et al (2009) Identification of protein cofactors necessary for sequence-specific plasmid DNA nuclear import. Mol Ther 17:1897–1903
14. Li S, MacLaughlin FC, Fewell JG et al (2001) Muscle-specific enhancement of gene expression by incorporation of SV40 enhancer in the expression plasmid. Gene Ther 8:494–497
15. Dean DA, Dean BS, Muller S et al (1999) Sequence requirements for plasmid nuclear import. Exp Cell Res 253:713–722

Chapter 3

Strategies for Improving DNA Vaccine Performance

Sandra Iurescia, Daniela Fioretti, and Monica Rinaldi

Abstract

The goal of active vaccination is to induce all the immune effector pathways and to establish immunological memory allowing prolonged surveillance against pathogens or cancer cells. DNA vaccination platform is an intriguing strategy owing to its ability to mobilize both branches of the immune system (i.e., innate immunity as well as adaptive immunity). Since plasmids offer several advantages for biotechnological applications due to their modular structure and easy manipulation, a wide range of strategies can be applied to improve DNA vaccine performance. This chapter discusses this topic in detail taking into account antigen/epitope selection and optimization, inclusion of intracellular targeting sequences and genetic adjuvants, and provision of T cell help.

Key words DNA vaccines, Antigen/epitope optimization, Targeting sequences, Genetic adjuvants, T cell help, Innate immunity, Adaptive immunity, Antigen targeting

1 Introduction

Nucleic acid-based DNA vaccines provide a novel class of biodrugs with great therapeutic potential as innovative approach to prevent infections or to fight infectious diseases and cancer.

DNA vaccination platform takes advantage of in vivo processes and gets the feature to harness the full power of the immune system, through engagement of multiple routes to activate both branches of the immune system (i.e., innate immunity as well as adaptive immunity).

DNA vaccines are simple vehicles for in vivo transfection and antigen delivery and production. Host cells can be transfected by a variety of routes including injection of the plasmid into the muscle or the dermis or mucosal application. Following intramuscular (i.m.) injection, the plasmid enters the nucleus of transfected cells (muscle cells or resident antigen-presenting cells (APCs), e.g., dendritic cells (DCs)). Using the host cellular machinery, the eukaryotic plasmid component initiates genes transcription and translation

Sandra Iurescia and Daniela Fioretti contributed equally to this work.

Monica Rinaldi et al. (eds.), *DNA Vaccines: Methods and Protocols*, Methods in Molecular Biology, vol. 1143, DOI 10.1007/978-1-4939-0410-5_3, © Springer Science+Business Media New York 2014

leading to foreign antigens production. These host-synthesized antigens undergo posttranscriptional modifications so that protein structures reproduce native protein conformations. Myocyte-secreted Ags may directly stimulate B cells, which in turn produce antibodies or are captured and processed within the endocytic pathway by APCs, which mediate the display of Ags on major complex of histocompatibility (MHC) class II molecules. Direct transfection of APCs leads to presentation of endogenous Ags in the context of MHC class I proteins. Cross-priming resulting from transfected somatic cells being phagocytosed by professional APCs represents the third process through which the antigen is presented to T cells. Since muscle cells are not so efficient at presenting antigenic proteins, the other two ways may be more important for DNA efficacy outcome. APCs then migrate to the proximal lymph nodes, where they present the antigenic peptides to naïve CD4$^+$ T and CD8$^+$ T cells via T cell receptor (TCR). In this way, the encoded antigens can enter the processing and presentation pathways in the context of both MHC class I and MHC class II molecules. Activated CD4$^+$ T cells trigger the differentiation of specific B cells, which can also be activated by secreted antigen that arrives to the lymph node. In theory, primed lymphocytes could be restimulated and further expanded at the immunization site via presentation of the peptide–MHC complexes displayed by transfected cells. Hence, DNA vaccination induces both humoral and cellular immune responses and provides a surveillance system [1] (*see* Fig. 1).

This chapter deals with strategies relevant to the identification/selection and optimization of a target antigen and complementary approaches to DNA vaccine design that, when combined with improved delivery devices and/or prime/boost regimens, enhance DNA vaccine performance. Inclusion of intracellular targeting sequences to ensure efficient MHC class I and MHC class II loading, incorporation of genetic adjuvants, and provision of CD4$^+$ T cell help is discussed here.

2 Materials

1. Antigen/epitope.
2. Microbial and vegetable protein for CD4$^+$ T cell help.
3. Signal sequences for intracellular compartment targeting.
4. Genetic adjuvants.
5. Immunobioinformatics tools.

3 Methods

3.1 Antigen/Epitope Selection and Optimization

A key issue in vaccine development today is what type of immune response is needed to best protect against infections or to treat established infectious diseases or cancer.

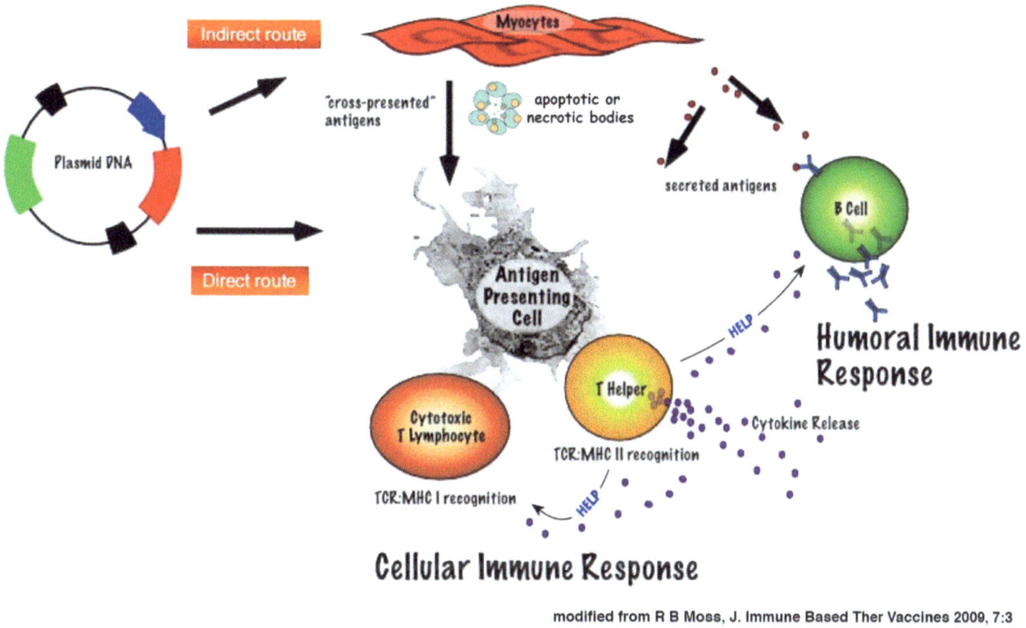

Fig. 1 Induction of cellular and humoral immunity by DNA vaccines. Using the host cellular machinery, the plasmid enters the nucleus of transfected myocytes or skin cells (indirect route) and of resident antigen-presenting cells (APCs) (direct route). Transfected muscle cells may secrete antigen that then (1) directly stimulates B cells of the immune system, which in turn produce antibodies, and (2) is captured and processed within the endocytic pathway by APCs which mediate the display of peptides on MHC II molecules. APCs have a dominant role in the induction of immunity of DNA vaccines by presenting vaccine-derived peptides also on MHC I molecules. This can follow either cross-presentation, whereby the antigen is transferred from the depot cells that are making the antigens, possibly owing to cell death, or direct transfection by the plasmid vaccine. Modified from ref. 62

Plasmid DNA vaccines are designed to encode for one single antigen as well as multiple epitopes belonging to the same antigen to trigger specific humoral and cellular responses.

A safety issue in designing anticancer vaccines for human applications is related to the use of tumor-associated antigens (TAAs) and oncoproteins. Antigen transgenes may be an exact copy of the original antigen or a modification to improve efficacy or safety. Some DNA constructs comprise mutations in the natural antigen sequences to functionally eliminate their oncogenic or catalytic activity. To this end, telomerase catalytic activity of the human telomerase reverse transcriptase (hTERT) antigen is inactivated with two site mutations as claimed in patent US8017387 [2].

The "high-risk" human papillomavirus 16 (HPV 16) E6 and E7 oncoproteins are excellent target for developing effective DNA vaccines that encompass the full epitope-encoding antigenic proteins. Nevertheless, their ability to induce degradation of the tumor suppressors p53 and retinoblastoma protein (pRb), respectively, needs to be eliminated to avoid the transforming potential of these genes before application in humans. The so-called E7(detox) (E7GGG) gene was generated through mutations in the critical Rb-binding residues, and failed to bind Rb, had no transforming activity, yet maintaining potent immunogenicity [3]. Likewise, the p53-binding site was disrupted by point mutation in the HPV E6 counterpart [4]. Recently, the more drastic approach termed gene shuffling has been applied to overcome the potential for in vivo oncogenicity of HPV E6 and E7 genes. Artificial ("shuffled") genes were generated and the scrambled genes encoded dysfunctional oncogenes that retained all immunogenic epitopes of the parental proteins [5–7].

Alternatively, the antigen may be engineered for immunogenicity using structure-based antigen design (i.e., "structural vaccinology") [8]. High-resolution structures reveal the electrostatic surfaces recognized by neutralizing antibodies and the architectures underlying these surfaces, thereby identifying which substructures must be left intact and which can be changed to optimize biochemical and immunologic performance. Indeed, small-molecule therapeutic agents often result from de novo synthesis or extensive structure-based modification of naturally derived substances. Thus, determination of structures of vaccine antigens may become a central task in vaccine optimization [9].

The process of selection of candidate immunogenic human leukocyte antigen (HLA)-restricted epitopes (i.e., T cell epitope mapping) allows identifying peptide sequences that are recognized by $CD8^+$ or $CD4^+$ T cells on antigenic proteins. Multiple applications can benefit from detection of T cell epitopes, including the design of prophylactic vaccines, diagnostics, and therapeutics [10]. Hence, the challenging task is to identify in a given antigen epitope targets able to elicit T cell responses that are intended to be modulated. The so-called T cell epitope-driven vaccine design employs immunoinformatics to identify potential targets of vaccines against cancer [11] and to develop "in silico DNA vaccines".

T cells can recognize epitope peptides that are embedded into HLA molecules expressed on the surface of APCs. The antigen processing and presentation pathways generate peptide fragments (small 8–20 amino acids long polypeptides) for loading onto MHC class I or class II molecules. The MHC–peptide complex is then presented on the cell surface, allowing recognition by epitope-specific TCRs on $CD8^+$ or $CD4^+$ T cells. The expression of CD8 or CD4 proteins dictates whether a T cell will recognize epitopes presented on the cell surface to the immune cells in the context of class I or class II MHC molecules, respectively.

The discovery of sequence motifs in peptides able to bind to various alleles of MHC I or II molecules has led to the development of algorithms for prediction of potentially immunogenic peptides within the amino acid sequence of a putative TAA or tumor-specific antigens (TSA). Predictive mapping for epitopes in full-length proteins or within a reactive peptide can then be narrowed through a search for MHC-fitting anchor motifs [12, 13]. The challenge of predicting which peptide sequences bind to which MHC molecules has been met with various computational approaches [14]. Furthermore, natural epitopes (i.e., "self" tumor-associated epitopes) that do not fit perfectly into the MHC groove can be modified, leading to the generation of epitope analogs or "heteroclitic" peptides.

The most widely used method to modify the poor immunogenicity of a "self" tumor-associated epitope is to alter the amino acid residues that interact with the MHC molecules at the peptide-binding cleft. The epitope enhancement strategy is commonly exploited in the design of cancer vaccines, allowing the change of a subdominant epitope into a dominant one by making it more competitive for specific MHC molecules [15]. Amino acid modifications in the primary and/or secondary "anchor" residues of peptides may generate potential agonist or "heteroclitic" epitopes, resulting in increased immunogenicity due to enhanced stability of the peptide–MHC complex compared with the native peptide [16]. Modification of the epitope sequence to increase affinity of the peptide–MHC complex for the TCR represents another approach to get epitope enhancement leading to more effective activation of low-avidity cytotoxic T lymphocytes (CTLs) "left" after self-tolerance mechanism in the case of the tumor Ag epitopes [15]. This strategy has been shown to be effective in increasing the in vivo expansion of T cells that recognize natural tumor epitopes [17, 18] (*see* **Note 1**).

3.2 Provision of CD4⁺ T Cell Help

Tumor antigens are often weakly immunogenic, and the immune repertoire in patients may have been tolerized. Thus, the central question is whether DNA vaccines can activate and maintain the high level of immunity required to suppress cancer cell growth. As previously discussed, DNA vaccines offer the opportunity to activate T helper (Th) cells and transform weak and ineffective immunity to a powerful antitumor attack [19]. Breakthrough knowledge of antigen processing and presentation machinery has led to the development of vaccines containing T or B cell epitopes coupled to a promiscuous $CD4^+$ Th epitope. Cognate $CD4^+$ T cell help is intrinsic to a T cell-dependent antibody response, especially when antigen is dose limiting [20], as is likely for DNA vaccines. Moreover, accumulating evidence suggests that $CD4^+$ T cells play a key role in the secondary expansion and long-term memory of $CD8^+$ T cells [21] and that fusion of immunogenic carrier sequences elicits induction of specific $CD8^+$ T cell response (see Stevenson et al. for a review [19]). The benefit of using protein-based vaccines including promiscuous MHC class II-binding epitopes, such as "universal" T helper epitopes p2 and p30 derived from tetanus toxin (TT) [22], reflected improved performance of vaccination protocols [23–25]. This minimal epitope design is also helpful at triggering immune response when delivered using a DNA vaccine formulation [26–29]. Genetic engineering allows the rational design of therapeutic DNA fusion vaccines encoding tumor antigens fused to an immunoenhancing sequence, wherein the fusion protein is able to promote a significant immune response [30]. These vaccines can be designed to activate antibody and/or T cell responses, providing focused immune attack on selected antigens. The key to bypassing immune tolerance and activating high levels of antitumor antibody or CTLs lies in inducing $CD4^+$ T cell help. Incorporation of sequences derived from tetanus toxin fragment C (TTFrC) provides for the engagement of T helper cells from a large antimicrobial repertoire to help immune responses against the tumor antigen [7, 14, 19, 26, 29, 31, 32].

Bioinformatics analyses (*see* **Note 1**) of the TTFrC polypeptide sequences can help to identify "universal" epitopes able to bind to several MHC class II haplotypes from both mouse and human [22, 33, 34]. Therefore, these "restricted" TTFrC portion retains T cell help for $CD8^+$ T cells against tumor epitopes and should lack potentially competing CTL epitopes, avoiding phenomenon of immunodominance [35, 36].

Several patents describing the use of T helper epitopes derived from tetanus toxin, *E. coli* heat-labile enterotoxin, and vegetable proteins are reviewed in Fioretti et al. [37].

3.3 Antigen Intracellular Compartment Targeting

Adaptive immune response may be improved by enhancing Ag processing and MHC class I and/or class II presentation. Due to extra- and intracellular barriers, the amount of DNA plasmid and protein antigens expressed by plasmid DNA can be reduced significantly during the delivery and presentation processes. DNA-encoded antigenic determinants can enter the antigen processing and presentation pathways and evoke all effector responses of the immune system; thus, antigen intracellular targeting represents a good example of how vaccines may be designed rationally [38]. These strategies use host protein trafficking mechanisms to target expressed proteins to particular cellular compartments or mark them for secretion, thereby facilitating antigen processing and MHC class I and/or class II presentation. The antigen processing and presentation pathways generate peptide fragments (small 8–20 amino acids long polypeptides) for loading onto MHC class I or class II molecules. The MHC–peptide complex is then presented on the cell surface, allowing recognition by epitope-specific TCRs on $CD8^+$ or $CD4^+$ T cells. Each of these pathway steps has been characterized, and their influence has been related to final presentation on the cell surface [39]. Manipulating different steps involved in the intracellular handling of protein for presentation by the class I and II molecules may lead to the enhancement of antigen immunogenicity. Generation of antigenic peptides from immunogens for MHC class I binding may be increased by tagging an antigen with ubiquitin (Ub), a molecule that plays an important role in the proteasome-dependent degradation of intracellular proteins [19, 40]. The Ub fusion vaccines were designed to accelerate turnover of the transgene antigen and increase the variety and number of peptides available for MHC binding [41].

Endoplasmic reticulum (ER) chaperones such as ER-60, tapasin (Tap), or calreticulin (CRT) may have a function in protecting antigenic peptides from cytosolic proteases and helping to deliver them to MHC class I molecules [42, 43]. The addition of ER localization and ER retention signals to the DNA vaccine backbone is sufficient to enhance immunogenicity. ER localization could increase the half-life of the vaccine-encoded antigens, thereby increasing the in vivo accumulation of the antigen pool available [44, 45].

The lysosome-associated membrane protein 1 (LAMP1) is one signal sequence widely used for targeting the expressed antigen to the endosomal compartment, thus enhancing the immunogenicity of the DNA vaccine [46].

Although proteasomal degradation of intracellular proteins represents the main route to introduce antigenic epitopes into the MHC class I presentation pathway, some candidate vaccine proteins shed from transfected cells could be taken up by non-transfected APCs. In this way, soluble proteins can enter the MHC class I processing pathway through cross-priming [41]. A strategy to introduce plasmid-encoded immunogens into the cross-priming pathways is by targeting antigenic proteins to secretory pathway. The addition of a leader sequence such as the human tissue plasminogen activator (hTPA) signal peptide [47] directs the antigen to the secretory pathway. Thus, via cross-priming mechanisms, secreted fusion proteins could induce both $CD4^+$ and $CD8^+$ T cell populations [48, 49].

The addition of leader sequences was shown to be effective in enhancing antigen presentation, bypassing proteasomal degradation and/or TAP-mediated translocation from cytoplasm to the ER. Enhanced expression of plasmid antigens has been observed using an optimized TPA signal peptide, the IgE gene leader, or a leader derived from the IgM variable heavy chain (V_H) of the B cell lymphoma BCL1 [1, 28, 50, 51].

A novel strategy to enhance the presentation of antigen through the MHC class I pathway to $CD8^+$ T cells is the exploitation of the translocation features of *Pseudomonas aeruginosa* exotoxin A (ETA) [52, 53]. Engineered chimeric proteins containing the domain II of ETA could be used to facilitate translocation from the endosomal/lysosomal compartments to the cytoplasm leading to the enhancement of MHC class I presentation of exogenous antigen.

Recent in vitro study comparing the effects of cellular targeting for different genes described the influence of signal sequences with potential interest for DNA vaccine development [54]. Chapter 4, "Enhancement of DNA vaccine efficacy by intracellular targeting," of this book constitutes a useful tool for the optimization of the design of DNA vaccines by using different intracellular compartments targeting signals.

3.4 Genetic Adjuvants

Vaccine adjuvants function through a range of mechanisms including innate immune activation, chemotaxis, antigen uptake and presentation by professional APCs, antigen depot formation, and up-regulation of co-stimulatory molecules [55].

DNA vaccines can easily be designed to co-express genetic adjuvants such as cytokines or co-stimulatory molecules so that biological adjuvants can be tailored and encoded within the same DNA vector as well (reviewed in refs. 56–58). Numerous cytokines (e.g., interleukin (IL)-2, IL-6, IL-12, granulocyte–macrophage colony-stimulating factor (GM-CSF)), chemokines (e.g., RANTES), co-stimulatory molecules (e.g., CD40, CD80, CD86), or signalling molecules (e.g., interferon regulatory factor-3 (IRF-3)) have been employed as genetic adjuvants (reviewed in refs. 37, 57, 58). Modifying the microenvironment of the vaccinated site by co-administration of genetic adjuvants improves the low immunogenicity of DNA vaccines [59]. The plasmid-mediated administration of the genetic adjuvant is thought to try to synchronize the immune stimulation with the expression of the antigen as well as to try to minimize side effects by having the cytokine or the chemokine expressed only locally. The activating cytokines are expressed and act at the site of antigen expression avoiding the toxicity of systemically administered cytokines [57]. Furthermore, the nucleic acid sequence per se may serve as an agonist for TLR-9 or other cellular sensors. In addition to cytokines and chemokines, many other classes of immune-modulatory molecules exist that target death receptors, growth factors, and adhesion molecules. Promising DNA vaccines based on the use of genetic adjuvants evaluated in preclinical and clinical trials were recently described [37].

3.5 Concluding Remarks

Advanced bioinformatics analysis such as immunoinformatics (i.e., in silico prediction of potential T cell epitopes), antigen/epitope optimization and expression, provision of $CD4^+$ T cell help, intracellular antigen targeting ensuring efficient MHC class I and class II compartment addressing, and inclusion of genetic adjuvants have been applied to improve the efficacy of DNA vaccines. Besides, epigenetic tools may be helpful to clarify DNA vaccine mechanisms as histone modifications play a crucial role in T cell differentiation from naïve to memory cells [60], and RNA interference (RNAi) technology represents a powerful new strategy for DNA vaccine enhancement. Overall, next-generation sequencing, reverse immunology, and other cutting-edge "omics" technologies [61] provide greater opportunities to design more potent DNA vaccines.

4 Notes

1. In silico prediction of putative T cell epitopes can be achieved with various computational approaches. Scoring matrices, hidden Markov models, and artificial networks are examples of algorithms that have been successful in MHC–peptide-binding prediction. The most popular web-based analytical tools used for T cell epitope prediction are listed below: SYFPEITHI (http://www.syfpeithi.de); BIMAS (http://bimas.dcrt.nih.gov/molbio/hla_bind); SMM (http://tools.immuneepitope.org/analyze/html/mhc_binding.html); NetMHCpan method is available both via the Immune Epitope Database and Analysis Resource (IEDB) website (http://www.immuneepitope.org) and at http://www.cbs.dtu.dk/services/NetMHCpan.

References

1. Kutzler MA, Weiner DB (2008) DNA vaccines: ready for prime time? Nat Rev Genet 9: 776–788
2. Scarselli E, Mennuni C, La Monica N et al. (2011) Telomerase reverse transcriptase fusion protein, nucleotides encoding it, and uses thereof. US8017387
3. Smahel M, Sima P, Ludvikova V et al (2001) Modified HPV16 E7 genes as DNA vaccine against E7-containing oncogenic cells. Virology 281:231–238
4. Cassetti MC, McElhiney SP, Shahabi V et al (2004) Antitumor efficacy of Venezuelan equine encephalitis virus replicon particles encoding mutated HPV16 E6 and E7 genes. Vaccine 22:520–527
5. Osen W, Peiler T, Ohlschlager P et al (2001) A DNA vaccine based on a shuffled E7 oncogene of the human papillomavirus type 16 (HPV 16) induces E7-specific cytotoxic T cells but lacks transforming activity. Vaccine 19: 4276–4286
6. Ohlschlager P, Pes M, Osen W et al (2006) An improved rearranged Human Papillomavirus Type 16 E7 DNA vaccine candidate (HPV-16 E7SH) induces an E7 wild type-specific T cell response. Vaccine 24:2880–2893
7. Oosterhuis K, Ohlschlager P, van den Berg JH et al (2011) Preclinical development of highly effective and safe DNA vaccines directed against HPV 16 E6 and E7. Int J Cancer 129:397–406
8. Dormitzer PR, Ulmer JB, Rappuoli R (2008) Structure-based antigen design: a strategy for next generation vaccines. Trends Biotechnol 26:659–667
9. Dormitzer PR, Grandi G, Rappuoli R (2012) Structural vaccinology starts to deliver. Nat Rev Microbiol 10:807–813
10. Kim Y, Sette A, Peters B (2011) Applications for T-cell epitope queries and tools in the Immune Epitope Database and Analysis Resource. J Immunol Methods 374:62–69
11. Rosa DS, Ribeiro SP, Cunha-Neto E (2010) CD4+ T cell epitope discovery and rational vaccine design. Arch Immunol Ther Exp (Warsz) 58:121–130
12. Rammensee HG, Friede T, Stevanoviic S (1995) MHC ligands and peptide motifs: first listing. Immunogenetics 41:178–228
13. Sette A, Buus S, Appella E et al (1989) Prediction of major histocompatibility complex binding regions of protein antigens by sequence pattern analysis. Proc Natl Acad Sci U S A 86:3296–3300
14. Iurescia S, Fioretti D, Fazio VM et al (2012) Epitope-driven DNA vaccine design employing immunoinformatics against B-cell lymphoma: a biotech's challenge. Biotechnol Adv 30:372–383
15. Berzofsky JA, Ahlers JD, Belyakov IM (2001) Strategies for designing and optimizing new generation vaccines. Nat Rev Immunol 1: 209–219
16. Borbulevych OY, Baxter TK, Yu Z et al (2005) Increased immunogenicity of an anchor-modified tumor-associated antigen is due to the enhanced stability of the peptide/MHC complex: implications for vaccine design. J Immunol 174:4812–4820
17. Slansky JE, Rattis FM, Boyd LF et al (2000) Enhanced antigen-specific antitumor immunity

with altered peptide ligands that stabilize the MHC-peptide-TCR complex. Immunity 13: 529–538
18. Buhrman JD, Slansky JE (2013) Improving T cell responses to modified peptides in tumor vaccines. Immunol Res 55:34–47
19. Stevenson FK, Rice J, Ottensmeier CH et al (2004) DNA fusion gene vaccines against cancer: from the laboratory to the clinic. Immunol Rev 199:156–180
20. Zinkernagel RM (2002) Uncertainties: discrepancies in immunology. Immunol Rev 185: 103–125
21. Janssen EM, Lemmens EE, Wolfe T et al (2003) CD4+ T cells are required for secondary expansion and memory in CD8+ T lymphocytes. Nature 421:852–856
22. Panina-Bordignon P, Tan A, Termijtelen A et al (1989) Universally immunogenic T cell epitopes: promiscuous binding to human MHC class II and promiscuous recognition by T cells. Eur J Immunol 19:2237–2242
23. Egan AF, Blackman MJ, Kaslow DC (2000) Vaccine efficacy of recombinant Plasmodium falciparum merozoite surface protein 1 in malaria-naive, -exposed, and/or -rechallenged Aotus vociferans monkeys. Infect Immun 68:1418–1427
24. Lee EA, Palmer DR, Flanagan KL et al (2002) Induction of T helper type 1 and 2 responses to 19-kilodalton merozoite surface protein 1 in vaccinated healthy volunteers and adults naturally exposed to malaria. Infect Immun 70:1417–1421
25. Valmori D, Fonteneau JF, Lizana CM et al (1998) Enhanced generation of specific tumor-reactive CTL in vitro by selected Melan-A/ MART-1 immunodominant peptide analogues. J Immunol 160:1750–1758
26. Iurescia S, Fioretti D, Pierimarchi P et al (2010) Genetic immunization with CDR3-based fusion vaccine confers protection and long-term tumor-free survival in a mouse model of lymphoma. J Biomed Biotechnol 2010:316069
27. Lund LH, Andersson K, Zuber B et al (2003) Signal sequence deletion and fusion to tetanus toxoid epitope augment antitumor immune responses to a human carcinoembryonic antigen (CEA) plasmid DNA vaccine in a murine test system. Cancer Gene Ther 10:365–376
28. Rice J, Elliott T, Buchan S et al (2001) DNA fusion vaccine designed to induce cytotoxic T cell responses against defined peptide motifs: implications for cancer vaccines. J Immunol 167:1558–1565
29. Iurescia S, Fioretti D, Fazio VM et al (2011) Design and pre-clinical development of epitope-based DNA vaccines against B-cell lymphoma. Curr Gene Ther 11:414–422
30. Rice J, Ottensmeier CH, Stevenson FK (2008) DNA vaccines: precision tools for activating effective immunity against cancer. Nat Rev Cancer 8:108–120
31. Chaise C, Buchan SL, Rice J et al (2008) DNA vaccination induces WT1-specific T-cell responses with potential clinical relevance. Blood 112:2956–2964
32. Signori E, Iurescia S, Massi E et al (2010) DNA vaccination strategies for anti-tumour effective gene therapy protocols. Cancer Immunol Immunother 59:1583–1591
33. Demotz S, Barbey C, Corradin G et al (1993) The set of naturally processed peptides displayed by DR molecules is tuned by polymorphism of residue 86. Eur J Immunol 23: 425–432
34. James EA, Bui J, Berger D et al (2007) Tetramer-guided epitope mapping reveals broad, individualized repertoires of tetanus toxin-specific CD4+ T cells and suggests HLA-based differences in epitope recognition. Int Immunol 19:1291–1301
35. Yewdell JW, Bennink JR (1999) Immunodominance in major histocompatibility complex class I-restricted T lymphocyte responses. Annu Rev Immunol 17:51–88
36. Chen W, McCluskey J (2006) Immunodominance and immunodomination: critical factors in developing effective CD8+ T-cell-based cancer vaccines. Adv Cancer Res 95:203–247
37. Fioretti D, Iurescia S, Rinaldi M (2013) Recent advances in design of immunogenic and effective naked DNA vaccines against cancer. Recent Pat Anticancer Drug Discov 9:66–82
38. Wang G, Pan L, Zhang Y (2011) Approaches to improved targeting of DNA vaccines. Hum Vaccin 7:1271–1281
39. Jensen PE (2007) Recent advances in antigen processing and presentation. Nat Immunol 8:1041–1048
40. Yewdell JW, Bennink JR (2001) Cut and trim: generating MHC class I peptide ligands. Curr Opin Immunol 13:13–18
41. Leifert JA, Rodriguez-Carreno MP, Rodriguez F et al (2004) Targeting plasmid-encoded proteins to the antigen presentation pathways. Immunol Rev 199:40–53
42. Cheng WF, Hung CF, Chai CY et al (2001) Tumor-specific immunity and antiangiogenesis generated by a DNA vaccine encoding

calreticulin linked to a tumor antigen. J Clin Invest 108:669–678
43. Lin CT, Chang TC, Chao A et al (2005) Enhancement of DNA vaccine potency through linkage of antigen gene to ER chaperone molecules, ER-60, tapasin, and calnexin. J Biomed Sci 12:279–287
44. Bins AD, Wolkers MC, van den Boom MD et al (2007) In vivo antigen stability affects DNA vaccine immunogenicity. J Immunol 179:2126–2133
45. Oosterhuis K, Aleyd E, Vrijland K et al (2012) Rational design of DNA vaccines for the induction of human papillomavirus type 16 E6- and E7-specific cytotoxic T-cell responses. Hum Gene Ther 23:1301–1312
46. Ji H, Wang TL, Chen CH et al (1999) Targeting human papillomavirus type 16 E7 to the endosomal/lysosomal compartment enhances the antitumor immunity of DNA vaccines against murine human papillomavirus type 16 E7-expressing tumors. Hum Gene Ther 10:2727–2740
47. Delogu G, Li A, Repique C et al (2002) DNA vaccine combinations expressing either tissue plasminogen activator signal sequence fusion proteins or ubiquitin-conjugated antigens induce sustained protective immunity in a mouse model of pulmonary tuberculosis. Infect Immun 70:292–302
48. Norbury CC, Princiotta MF, Bacik I et al (2001) Multiple antigen-specific processing pathways for activating naive CD8+ T cells in vivo. J Immunol 166:4355–4362
49. Watts C (1997) Capture and processing of exogenous antigens for presentation on MHC molecules. Annu Rev Immunol 15:821–850
50. Wang S, Hackett A, Jia N et al (2011) Polyvalent DNA vaccines expressing HA antigens of H5N1 influenza viruses with an optimized leader sequence elicit cross-protective antibody responses. PLoS One 6:e28757
51. Wang JY, Song WT, Li Y et al (2011) Improved expression of secretory and trimeric proteins in mammalian cells via the introduction of a new trimer motif and a mutant of the tPA signal sequence. Appl Microbiol Biotechnol 91:731–740
52. Hung CF, Cheng WF, Hsu KF et al (2001) Cancer immunotherapy using a DNA vaccine encoding the translocation domain of a bacterial toxin linked to a tumor antigen. Cancer Res 61:3698–3703
53. Mohammed AF, Abdul-Wahid A, Huang EH et al (2012) The Pseudomonas aeruginosa exotoxin A translocation domain facilitates the routing of CPP-protein cargos to the cytosol of eukaryotic cells. J Control Release 164:58–64
54. Carvalho JA, Azzoni AR, Prazeres DM et al (2010) Comparative analysis of antigen-targeting sequences used in DNA vaccines. Mol Biotechnol 44:204–212
55. Li L, Saade F, Petrovsky N (2012) The future of human DNA vaccines. J Biotechnol 162:171–182
56. Fioretti D, Iurescia S, Fazio VM et al (2010) DNA vaccines: developing new strategies against cancer. J Biomed Biotechnol 2010:174378. doi:10.1155/2010/174378
57. Saade F, Petrovsky N (2012) Technologies for enhanced efficacy of DNA vaccines. Expert Rev Vaccines 11:189–209
58. Tovey MG, Lallemand C (2010) Adjuvant activity of cytokines. Methods Mol Biol 626:287–309
59. Anderson RJ, Schneider J (2007) Plasmid DNA and viral vector-based vaccines for the treatment of cancer. Vaccine 25(Suppl 2):B24–B34
60. Cuddapah S, Barski A, Zhao K (2010) Epigenomics of T cell activation, differentiation, and memory. Curr Opin Immunol 22:341–347
61. Kennedy RB, Poland GA (2011) The top five "game changers" in vaccinology: toward rational and directed vaccine development. OMICS 15:533–537
62. Moss RB (2009) Prospects for control of emerging infectious diseases with plasmid DNA vaccines. J Immune Based Ther Vaccines 7:3

Chapter 4

Enhancement of DNA Vaccine Efficacy by Intracellular Targeting Strategies

Elisabete Borges Freitas, Ana Margarida Henriques, Miguel Fevereiro, Duarte Miguel Prazeres, and Gabriel Amaro Monteiro

Abstract

Immune response against an encoded antigenic protein can be elicited by including targeting sequences to DNA vaccines that promote protein sorting to processing pathways, related with antigen presentation by major histocompatibility complexes (MHC). Candidate DNA vaccines coding for neuraminidase 3 of the avian influenza virus were designed to encode different sequences that direct the protein to specific cellular compartments such as endoplasmic reticulum (i.e., adenovirus E1A), lysosomes (i.e., LAMP), and the combination of protein targeting to the endoplasmic reticulum and lysosome (i.e., E1A-LAMP). The DNA vaccine prototypes were engineered by biomolecular techniques and subsequently produced in *E. coli* cells. The biological activity of the vaccines was tested firstly in vitro, in Chinese hamster ovary cells, through flow cytometry and real-time polymerase chain reaction analysis. Then, an essential in vivo study was performed in chickens, in order to evaluate the efficacy of DNA prototype vaccines, by measuring the antibody production by enzyme-linked immunosorbent assay.

Key words DNA vaccines, Targeting sequences, Antigenic protein, Neuraminidase, MHC, E1A, LAMP, Flow cytometry, RT-PCR, ELISA

1 Introduction

DNA vaccines are purified antigen-encoding bacterial plasmids, used as preventive or therapeutic vaccines, designed to mimic infections safely [1]. After administration into the host, a functional antigen is synthesized, which is able to induce a long-lasting immune response against a pathogen. The immune response can be modulated by controlling both the expression levels of the antigenic protein and the way it is presented to the immune system, generating a humoral and cellular response [2]. To improve the immunity of DNA vaccines, some targeting sequences may be added to the DNA vector to direct the antigenic protein to specific cell compartments. This approach directs the antigen peptides to the major histocompatibility complexes I and II (MHCI and MHCII)

Monica Rinaldi et al. (eds.), *DNA Vaccines: Methods and Protocols*, Methods in Molecular Biology, vol. 1143,
DOI 10.1007/978-1-4939-0410-5_4, © Springer Science+Business Media New York 2014

pathways, increasing the number of peptides available for presentation by antigen-presenting cells (APCs) and somatic cells. Examples of targeting sequences [3–5] include (1) a secretion signal (Sc) composed of the first 21 amino acids of the tissue plasminogen activator (tPA) signal, which promotes protein secretion to the extracellular space [6]; (2) the lysosomal associated membrane protein (LAMP) that targets the protein through a vesicular pathway directly to the lysosomes [7]; (3) the adenovirus early region 1A (E1A), the first viral transcription unit expressed after viral infection, that is a targeting sequence to the endoplasmic reticulum [8]; and (4) the combination of E1A-LAMP that also promotes the transport to the ER and to the lysosome. Other antigen-targeting strategies and sequences are used in experimental studies, such as the membrane anchoring using human alkaline phosphatase and proteasome targeting through murine ubiquitin A76 [1, 9] and the bovine herpes virus-1 glycoprotein D signal [10]. The targeting sequences LAMP, Sc, E1A, and E1A-LAMP are considered in this work. The transport to the lysosome promotes the degradation of proteins and loading of the peptides to MHCII molecules that display them at the cell surface [11]. This generates a more pronounced humoral response due to the activation of T helper lymphocytes ($CD4^+$ T cells) that secrete cytokines and activate B lymphocytes, which start the production of identical copies of an antibody, and differentiate into memory cells [12]. The secretion of cytokines also leads to the recruitment of macrophages, granulocytes, and other lymphocytes as well as to the activation of T cytotoxic lymphocytes ($CD8^+$ T cells) and differentiation of more T lymphocytes with production of long-term memory cells [13]. Thus, the indirect development of a cellular response is one of the main advantages of DNA vaccines, stimulating a long-lasting immune response. The trafficking to the extracellular space mediated by Sc sequence increases the probability of the antigen to be phagocytized/endocytosed, digested at the lysosome/endosome, and displayed by MHCII molecules, promoting the $CD4^+$ T cell activation that also leads to the development of a cellular response [14]. The endoplasmic reticulum (ER) promotes the association of MHCI molecules to the peptides digested at proteasome. Thus, the transport to the ER by E1A can favor the MHCI pathway and a cellular response, with activation of $CD8^+$ T cells that lead to cytokine cascade and gene activation, in order to stimulate the apoptosis of cells expressing the recombinant protein and the expansion of more T lymphocytes. The MHCII pathway and a humoral response can also be promoted since the ER is also responsible for the trafficking of MHCII molecules through the Golgi to the endocytic route. A possible cross-presentation of the antigens by transfected cells to naïve APCs that acquire the released exogenous antigens and promote protein degradation at endocytic vesicles can also occur. Then, the peptide is displayed by MHCII molecules, which promotes the activation of $CD4^+$ T cells [4, 12, 15, 16].

In this way, the transport of antigens to specific cell compartments, related with MHCI/MHCII pathways, can promote cellular and humoral immune responses with a more efficient production of memory cells. The localization of the expressed protein significantly influences the efficacy of DNA vaccines [2].

DNA vaccines encoding for an antigen, such as a neuraminidase (NA) from avian influenza virus together with targeting sequences can improve the immune response. As an example, this chapter describes the engineering of DNA vaccines in order to encode for the influenza virus glycoprotein N3 (GenBank: HM849031.1) in association with different targeting sequences—Sc, E1A, LAMP, and E1A-LAMP—to understand if an enhanced immune response is achieved, clarifying their importance in the context of DNA vaccination. The experimental designs described in this chapter include the following steps in Subheadings 1.1–1.3: design of the plasmids, in vitro studies, and in vivo studies.

1.1 Design of the Plasmids

Plasmids are engineered by directional cloning, using two restriction enzymes that generate cohesive ends. Double digestions, with different enzymes, allow the isolation and generation of compatible cohesive ends, essential to clone the targeting sequences in the N3 vector (in frame with the N3 gene sequence). Beyond the antigen-coding sequence, DNA vaccines contain essential elements such as a prokaryotic origin, an antibiotic selection marker for kanamycin for example, multiple cloning site (MCS), an eukaryotic promoter such as the cytomegalovirus (CMV) promoter and enhancers, as well as transcription termination/polyadenylation (polyA) signal sequence to promote gene expression in vaccine recipients [1]. The green fluorescent protein (GFP) reporter gene when fused to the antigenic protein gene allows the assessment of its expression and cellular localization.

1.2 In Vitro Studies

Chinese hamster ovary (CHO) cell line can be used to test plasmids and their biological activity and integrity. This is done by flow cytometry, a semiquantitative analysis, essential to verify if the plasmids are correctly constructed for expression of the N3 recombined protein with the targeting sequences. Protein expression is analyzed in terms of GFP reporter gene expression, measured by transfection efficiency and mean fluorescence levels of cultured CHO cells, clarifying the differences among the N3–GFP fusion proteins that should be directed for processing in different compartments, depending on the targeting sequence used.

Plasmids are also tested in vitro by quantitative real-time polymerase chain reaction (qPCR), which is a highly sensitive method that quantifies the plasmid copy number in CHO cells through the amplification of the GFP sequence present in the plasmids. qPCR is also essential to quantify mRNA content per ng of total RNA, after the synthesis of a first strand of cDNA, to allow the assessment of the gene expression from different plasmid constructs.

1.3 In Vivo Studies

In vivo experiments are performed to focus both on the evaluation of the antibody response in Leghorn chickens and on the efficacy of the plasmids as DNA vaccines, after vaccination with avian influenza N3 DNA prototype vaccines. A heterologous prime boost strategy is done, and the evaluation of the amount of antibodies present in sera collected during in vivo assays is measured by enzyme-linked immunosorbent assay (ELISA).

2 Materials

2.1 Culture Medium for Recombinant E. coli Cells

1. Kanamycin (30 μg/mL) from a stock solution (30 mg/mL), 0.2 μm filtered.
2. Luria Broth—Miller's modified medium (LB) (25 g/L) with kanamycin: Prepare LB medium with yeast extract 5 g/L, sodium chloride 10 g/L, and casein enzymatic hydrolysate 10 g/L in distilled water by distributing 30 mL of LB medium to 100 mL Erlenmeyers or 200 mL of LB medium to 2 L Erlenmeyers or 5 mL of LB medium to 15 mL conical centrifuge tubes. Autoclave at 120 °C for 20 min. Store the media at 4 °C. Aseptically (in a flow chamber) add kanamycin (30 μg/mL) before the addition of the cells.
3. Miller's LB agar (40 g/L) plates: For plates with solid medium prepare tryptone 10 g/L, yeast extract 5 g/L, NaCl 10 g/L, and agar 15 g/L in 200 mL of distilled water. Autoclave at 120 °C for 20 min. Add 200 μL of kanamycin (30 μg/mL) after cooling and before the solidification of the medium. Distribute and store the plates at 4 °C.
4. Orbital shaker, 37 °C.

2.2 Transformation of E. coli DH5α Cells

1. Competent *E. coli* DH5α cells.
2. Plasmids: pVAX1-N3-GFP (N3 vector), pVAX1-E1A-N3-LAMP (E1A-N3-LAMP), and pVAX1-ISG-GFP-LAMP (ISG-LAMP), and pVAX1-Sc-ISG-GFP (Sc-ISG) encoding for the invariant surface glycoprotein (*see* **Note 1**).
3. Sterile liquid LB medium (25 g/L).
4. LB-agar plates (35 g/L) supplemented with kanamycin (30 μg/mL).
5. Water bath.
6. Incubator, 37 °C.
7. Orbital shaker, 37 °C.
8. Cryovials.
9. Glycerol for molecular biology (99 %).

2.3 Small-Scale Plasmid Purification for Molecular Cloning

1. Small-scale mini kit High Pure Plasmid Isolation Kit (Roche) (*see* **Note 2**).
2. Spectrophotometer.

2.4 Large-Scale Plasmid Purification for Cell Transfection and In Vivo Expression

1. Plasmid Purification Midi and HiSpeed Maxi Kit (QIAGEN) (*see* **Note 3**).
2. 100 % (v/v) isopropanol.
3. 70 % (v/v) ethanol.
4. Phosphate-buffered saline (PBS): 0.9 % NaCl, 10 mM sodium phosphate pH 7.2.
5. Spectrophotometer.

2.5 Cloning of the Targeting Sequence Inserts into the N3 Vector

1. Control plasmid: pVAX1-GFP [17] suspended in 10 mM Tris–HCl pH 8.5 (elution buffer).
2. Antigen-encoding plasmids: pVAX1-N3-GFP (N3 vector), pVAX1-E1A-N3-LAMP (E1A-N3-LAMP), pVAX1-ISG-GFP-LAMP (ISG-LAMP), and pVAX1-Sc-ISG-GFP (Sc-ISG) suspended in 10 mM Tris–HCl pH 8.5 (elution buffer).
3. ApE© software.
4. Restriction enzymes and respective buffers.
5. MilliQ sterilized water.
6. Agarose gel electrophoresis equipment.
7. Agarose for routine analysis of nucleic acids.
8. Loading buffer (6×): 60 mM Tris–HCl pH 8.6, 100 mM NaCl, 2 mM β-mercaptoethanol. Weigh 16 g of sucrose and 100 mg of bromophenol blue and add to 40 mL of MilliQ water. Add 1.5 mL of loading buffer per microcentrifuge tube and store at 4 °C.
9. Tris-acetate–EDTA (TAE) buffer: 40 mM Tris base, 20 mM acetic acid, 1 mM EDTA pH 8 from a 50× stock solution.
10. Ethidium bromide (0.5 mg/mL): Add 40 μL of ethidium bromide (0.5 mg/mL) to 1 L of MilliQ water in a plastic box and protect from sunlight.
11. Molecular weight marker.
12. UV protection mask.
13. Scalpel tip.
14. Transilluminator.
15. QIAquick Gel Extraction Kit (QIAGEN).
16. 80–100 % (v/v) isopropanol.
17. Preheated sterilized MilliQ water at 65 °C (*see* **Note 4**).
18. Spectrophotometer.

2.6 Ligation and Bacterial Cell Transformation Processes

1. Fragment corresponding to the linearized vector pVAX1-N3-GFP and the inserts (E1A, LAMP, or Sc).
2. T4 DNA ligase 3 U/μL.
3. T4 DNA ligase buffer: 300 mM Tris–HCl pH 7.8, 100 mM $MgCl_2$, 100 mM DTT, 10 mM ATP.
4. MilliQ sterilized water.
5. DNA SpeedVac® Concentrator.
6. Competent *E. coli* DH5α cells.
7. Sterile liquid LB medium (25 g/L).
8. LB-agar plates (35 g/L) supplemented with kanamycin (30 μg/mL).
9. Water bath.
10. Incubator 37 °C.
11. Cryovials.
12. Glycerol for molecular biology (99 %).

2.7 Culture of Chinese Hamster Ovary Cells

1. 75 cm^2 T flasks.
2. Culture medium (Invitrogen): 22.5 mL of F12 (HAM) nutrient mixture 1×, 1 % of antibiotic–antimycotic 100×, 1 % of modified Eagle medium—nonessential amino acids (MEM—NEAA) 100×, 1 % of 100 mM sodium pyruvate, 0.1 % of gentamicin (50 mg/mL), 1.5 mL of inactivated foetal bovine serum (FBS).
3. One vial of CHO cells (with approximately 4×10^6 cells) resuspended in 1 mL of FBS and 10 % DMSO.
4. Incubator at 37 °C and 5 % of CO_2 humidified environment.
5. 50 mL conical centrifuge tubes.

2.8 Transfection of CHO Cells

1. Lipofectamine™ 2000 (Invitrogen).
2. Plasmid DNA.
3. Incomplete F12 (HAM) medium (without antibiotics).
4. Sterile 24-well plates.
5. Complete F12 (HAM) medium with antibiotics and 10 % of FBS.
6. Incubator at 37 °C and 5 % of CO_2 humidified environment.

2.9 Analysis of Protein Expression in CHO Cells by Flow Cytometry

1. Sterilized PBS: 0.9 % NaCl, 10 mM sodium phosphate pH 7.2.
2. Trypsin versene solution: 8 g/L NaCl, 0.4 g/L KCl, 1 g/L glucose, 0.58 g/L $NaHCO_3$, 0.5 g trypsin (1:250), 0.2 g/L EDTA pH 7.2–7.4.
3. Incubator at 37 °C and 5 % of CO_2 humidified environment.

4. 15 mL conical centrifuge tubes.
5. 2 % (w/v) paraformaldehyde (PFA).
6. Conical centrifuge tubes and aluminum foil to protect samples from light.
7. CellQuest Pro Software©.

2.10 Analysis of Plasmid Copy Number in CHO Cells by Quantitative Real-Time Polymerase Chain Reaction

1. Sterile PBS: 0.9 % NaCl, 10 mM sodium phosphate pH 7.2.
2. Trypsin versene solution: 8 g/L NaCl, 0.4 g/L KCl, 1 g/L glucose, 0.58 g/L $NaHCO_3$, 0.5 g trypsin (1:250), 0.2 g/L EDTA pH 7.2–7.4.
3. 15 mL conical centrifuge tubes.
4. Incubator at 37 °C and 5 % of CO_2 humidified environment.
5. Neubauer chamber.
6. FastStart DNA Master SYBR Green I Kit LightCycler (Roche).
7. Thermal cycler with hot lid such as LightCycler detection system.
8. Forward and reverse GFP primers (10 μM) (*see* **Note 5**).

2.11 Analysis of mRNA Transcription (via cDNA) by Quantitative Real-Time Polymerase Chain Reaction

1. Sterilized PBS: 0.9 % NaCl, 10 mM sodium phosphate pH 7.2.
2. Trypsin versene solution: 8 g/L NaCl, 0.4 g/L KCl, 1 g/L glucose, 0.58 g/L $NaHCO_3$, 0.5 g trypsin (1:250), 0.2 g/L EDTA pH 7.2–7.4.
3. Incubator at 37 °C and 5 % of CO_2 humidified environment.
4. 15 mL conical centrifuge tubes.
5. Pure RNA Isolation Kit (Roche).
6. Spectrophotometer.
7. First Strand cDNA Synthesis Kit for real-time polymerase chain reaction (RT-PCR) (AMV) (Roche).
8. FastStart DNA Master SYBR Green I Kit LightCycler (Roche).
9. Thermal cycler with hot lid such as LightCycler detection system.
10. Forward and reverse GFP primers (10 μM) (*see* **Note 5**).

2.12 In Vivo Expression Assay

1. Plasmid DNA.
2. N3 purified protein.
3. Sterile PBS.
4. Lipofectamine™ 2000 (Invitrogen).
5. Three-week-old Leghorn chickens.
6. Microneedles (27G × 1/2," 0.4 × 12 mm, 1 mL).
7. 96-well ELISA plates.

8. ELISA plate washing machine.
9. Pool of chicken sera.
10. Serum anti-H5N3.
11. Serum anti-H5N1.
12. Carbonate–bicarbonate buffer: 15 mM Na_2CO_3, 35 mM $NaHCO_3$ pH 9.6.
13. Blocking buffer: 2 % of milk powder, 5 % of Newborn Calf Serum (Invitrogen), 5 % *E. coli* lysate, sterilized PBS (*see* **Note 6**).
14. PBS/Tween 0.05 %: 950 mL H_2O, 50 μL PBS (20×), 500 μL of polyoxyethylene sorbitan monolaurate.
15. Anti-chicken serum IgG (Fc):HPr (Abd Serotec).
16. Peroxidase substrate: Tetramethylbenzidine (TMB).
17. Stop solution: H_2SO_4 (95–97 % w/v).
18. Absorbance microplate reader.

3 Methods

Carry out the work in a sterile environment (e.g., in a flow chamber) mainly when working with media and bacterial and animal cell cultures. The use of a clean room is also critical when working with DNA and RNA. Ensure that plastics (pipette tips, centrifuge tubes, 24-well plates and 96-well plates, and T flasks) and solutions (culture media, phosphate saline buffer, lipofectamine, trypsin versene, newborn calf serum, and foetal bovine serum) used in cell cultures and in in vivo experiments are sterilized. Sterilize the appropriate amount of MilliQ and distilled water required to perform the experiments.

3.1 Heat-Shock Transformation of *E. coli* DH5α Competent Cells

1. Transform 50 μL of *E. coli* DH5α competent cells with 1 μL of plasmid DNA of interest: pVAX1-GFP (control plasmid), N3 vector, E1A-N3-LAMP, ISG-LAMP, and Sc-ISG [3, 4, 18].
2. Incubate on ice for 30 min.
3. Heat the mixture for 1 min at 42 °C (in a water bath) and subsequently cool on ice for 2 min.
4. Add 1 mL of sterile liquid LB medium and incubate at 37 °C without agitation for 1 h.
5. Centrifuge the cells at 5,000 ×*g* for 3 min.
6. Resuspend the cells in 100 μL of the supernatant and then plate in LB-agar supplemented with kanamycin.
7. Incubate overnight at 37 °C.
8. Prepare a cell bank as follows. Inoculate one single colony in 5 mL of LB medium (25 g/L) with 5 μL of kanamycin (30 μg/mL) at 37 °C and 250 rpm in an orbital shaker. Use inoculum to

start cell growth with an initial O.D.$_{600\ nm}$ = 0.1 in a 100 mL Erlenmeyer containing 30 mL of LB medium (25 g/L), supplemented with 30 μL of kanamycin (μg/mL), at 37 °C with orbital agitation at 250 rpm. When cells reach an O.D.$_{600\ nm}$ ≈ 1 distribute 160 μl of culture into cryovials containing 40 μl of glycerol, mix well, and store at −80 °C.

3.2 Small-Scale Plasmid Purification for Molecular Cloning

1. Add 20 μL of each cryopreserved aliquot of cells harboring the plasmids of interest to 15 mL conical centrifuge tubes with 5 mL of liquid LB medium (25 g/L) and 5 μL of kanamycin (30 μg/mL). Incubate tubes overnight in an orbital shaker at 37 °C and 250 rpm.
2. Harvest cells at an approximate optical density (O.D.) at 600 nm of ≈ 2 to ensure that exponential phase is reached.
3. Centrifuge cells at 3,000 × *g* for 10 min at 4 °C.
4. Purify the plasmids with the High Pure Plasmid Isolation kit (Roche), following the manufacturer's instructions.
5. The number of aliquots with purified plasmid DNA is dependent on the number of tubes used to perform the growth of the cells. Therefore, store one aliquot with the purified plasmid DNA at 4 °C, in order to be used afterwards, and store the remaining ones at −20 °C, until needed.
6. Measure DNA concentration by reading the absorbance at 260 nm (*see* **Note 7**).

3.3 Large-Scale Plasmid Purification for Cell Transfection and In Vivo Expression

1. Inoculate *E. coli* DH5α cells harboring the plasmids of interest in 200 mL of LB medium (25 g/L), with 200 μL of kanamycin (30 μg/mL), in 2 L Erlenmeyer flasks.
2. Incubate overnight in an orbital shaker at 37 °C and 250 rpm until O.D.$_{600\ nm}$ ≈ 3.
3. After growth, centrifuge the culture at 3,000 × *g* for 20 min.
4. Perform the plasmid purification according to the instructions of the manufacturer of the Plasmid Purification Midi and HiSpeed Maxi Kit (QIAGEN).
5. Elute the plasmid DNA in PBS. After purification, freeze the samples at −20 °C or keep at 4 °C for long-term and short-term storage.
6. Determine the DNA concentration by reading absorbance at 260 nm (*see* **Note 7**).

3.4 Cloning of the Targeting Sequence Inserts into the N3 Vector

1. Clone the targeting sequences in the 5,097 bp backbone pVAX1-N3-GFP (N3 vector), in frame with the influenza A virus glycoprotein N3-encoding gene.
2. Clone the Sc and E1A sequences at the 5′ end upstream the N3 sequence and LAMP at the 3′ end downstream of the N3–GFP sequence.

Table 1
Pairs of restriction enzymes used to double digest N3 vector and ISG-LAMP, E1A-N3-LAMP, and Sc-ISG in order to isolate fragments containing the targeting sequences

Plasmids	Double digestions	Fragment size (kb)
pVAX1-N3-GFP (**vector**)	BsrGI/XbaI MluI/NheI	**5,082**[a]/15 **4,430**[a]/667
pVAX1-ISG-GFP-**LAMP**[b] (ISG-LAMP) (**insert**)	BsrGI/XbaI	5,223/**126**[b]
pVAX1-**E1A**[b]-N3-GFP-LAMP (E1A-N3-LAMP) (**insert**)	MluI/NheI	4,542/**416**[b]
pVAX1-**Sc**[b]-ISG-GFP (Sc-ISG) (**insert**)	MluI/NheI	4,063/**729**[b]/511

[a]Fragment of interest corresponding to the N3 vector where targeting sequences are inserted
[b]Fragments of interest containing targeting nucleotide sequences to be cloned in the N3 vector

3. Use the ApE© software to analyze the restriction sites present in plasmids and to design the desired clones in silico (*see* **Note 8**).
4. Perform restriction reactions using 0.5 μL of enzyme (10 units/μL), tenfold enzyme buffers, 3 μg of plasmid DNA, and MilliQ sterilized water (*see* **Note 9**).
5. Double digest the ISG-LAMP and N3 vector with the BsrGI and XbaI enzymes in order to clone the LAMP sequence on the N3 vector. Incubate the mixture for 3 h at 37 °C.
6. To clone E1A sequence, double restrict the E1A-N3-LAMP and N3 vector with MluI and NheI, allowing the isolation of the E1A sequence and its subsequent cloning. Add NheI only after 3 h of incubation at 37 °C with MluI. Perform an additional incubation at 37 °C with both enzymes for 2 h.
7. To clone the Sc sequence use again MluI and NheI for a double digestion of the Sc-ISG and N3 vector.
8. Check the size of the expected fragments (*see* Table 1), resulting from the double digestion of the N3 vector, ISG-LAMP, E1A-N3-LAMP, and Sc-ISG in order to isolate fragments containing the targeting sequences.
9. Set up a 1 % agarose gel by weighing 1 g of agarose and adding 100 mL of TAE 1×.
10. Melt the agarose, transfer the mixture to a tray, and allow the gel to polymerize for 30 min (*see* **Note 10**).
11. Place the tray with the gel inside the electrophoresis chamber and apply 6 μL of the molecular weight marker in the first well of the gel.
12. Add 5 μL of sixfold loading buffer to the 25 μL digestion mixtures.
13. After addition of the loading buffer, apply 2 μL of each digestion mixture in the first well next to the molecular weight

ladder. Leave an interval of two wells, and apply the remainder 28 μL of the digestion mixture to well number five.

14. Run the 1 % agarose gel electrophoresis at 120 V for 1 h and 20 min.
15. After electrophoresis, remove the gel from the electrophoresis chamber and cut it longitudinally between wells two and five. The gel portion containing the molecular weight ladder and the 2 μL sample is then removed from the tray and stained for 20 min in an ethidium bromide solution. Protect the second portion of the gel by placing aluminum foil under it.
16. Put on the UV protection mask, and with a scalpel tip, mark the bands corresponding to the vector pVAX1-N3-GFP and to the different insert targeting sequences on the stained portion of the gel.
17. Turn off the UV light, and place the unstained portion of the gel next to the stained portion.
18. By comparison, cut the bands containing the vector and targeting sequences (*see* Fig.1) in the non-stained gel portions that were not exposed to ethidium bromide and UV light (*see* **Note 11**).

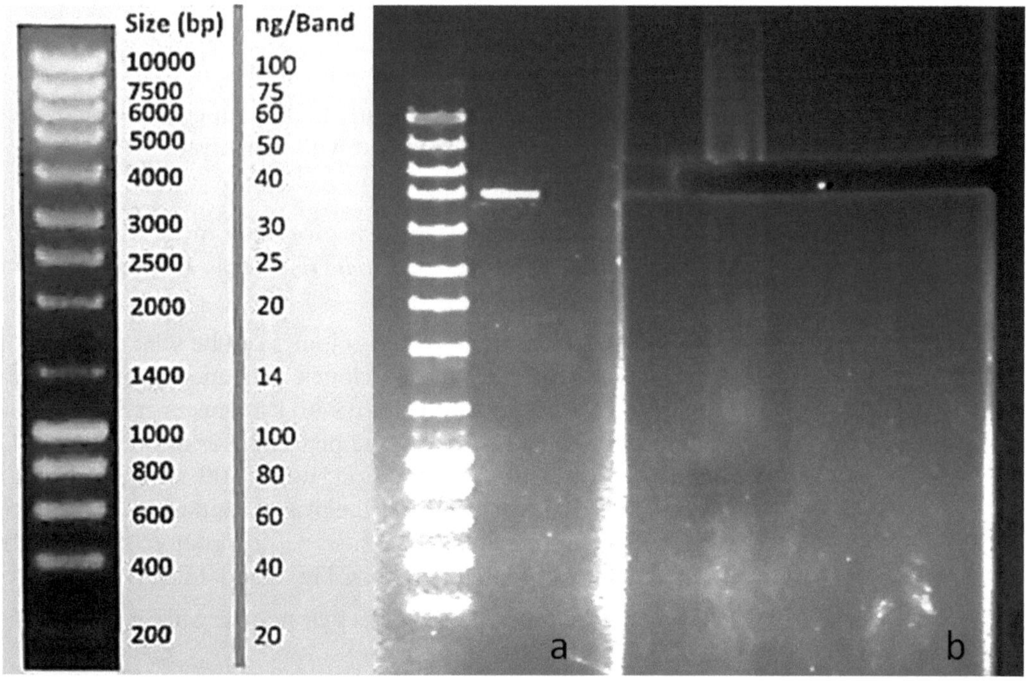

Fig. 1 Example of the method used to remove the desired gel band containing linear DNA without exposure to ethidium bromide and UV light. (**a**) Band with 2 μL of a double digestion of the pVAX1-N3-GFP (N3 vector) with BsrGI/XbaI. (**b**) Removal of the desired N3 vector band containing the linear plasmid DNA. Molecular weight marker used was NZYDNA ladder III

19. Purify the DNA from the agarose gel slices with the QIAquick Gel Extraction Kit (QIAGEN) following the manufacturer's instructions, taking care to elute the plasmid DNA with pre-heated MilliQ water.
20. Measure DNA concentration of the desired purified vector and target sequences by reading absorbance at 260 nm (*see* **Note 7**). Confirm (optional) concentration by checking the band intensity in a 1 % agarose gel electrophoresis (*see* **Note 12**).

3.5 Ligation and Bacterial Cell Transformation Processes

1. Perform a ligation mixture with 2 μL of T4 DNA ligase (3 units/μL) and 2 μL of the tenfold T4 ligase buffer, for a final volume of 20 μL, considering the mass of insert and vector needed for the reaction (*see* **Note 13**).
2. Incubate the mixture for 3 h at room temperature.
3. Transform 50 μL of competent *E. coli* DH5α cells with 10 μL of ligation mixture, and store the remaining 10 μL of the mixture overnight at 4 °C.
4. Incubate the cells with the 10 μL of the ligation mixture for 30 min on ice.
5. Heat the mixture for 1 min at 42 °C (in a water bath), and subsequently cool it on ice for 2 min.
6. Add immediately 1 mL of sterile liquid LB medium, and incubate the cells at 37 °C without agitation for 1 h.
7. Centrifuge the cells at 3,000 × *g* for 3 min.
8. Resuspend the cells in 100 μL of the supernatant and then plate in LB-agar supplemented with kanamycin.
9. Incubate overnight at 37 °C.
10. Next day, repeat the transformation step incubating another 50 μL aliquot of competent *E. coli* DH5α cells with the remaining 10 μL of the ligation mixture.
11. After incubation, transformed colonies can be seen. To confirm the presence of the expected clones, pick and inoculate several colonies in 5 mL of LB medium with kanamycin at 37 °C, overnight with agitation. Purify the plasmids (*see* Subheading 3.2), and perform a double digestion, using 500 ng of each, with KpnI and XbaI, which cleaves the plasmid and the N3 sequence. Run the mixtures in 1 % agarose gel in order to confirm the correct size of the fragments (*see* Fig. 2 and Table 2).
12. Prepare cell banks of the correct clones (*see* Subheading 3.1).

3.6 Culture and Transfection of CHO Cells

1. Grow a starting culture by adding one vial of frozen CHO cells (with approximately 4×10^6 cells resuspended in 1 mL of FBS + 10 % DMSO) to a 75 cm^2 T flask with 22.5 mL of F12 (HAM) nutrient mixture supplemented with 1 % of

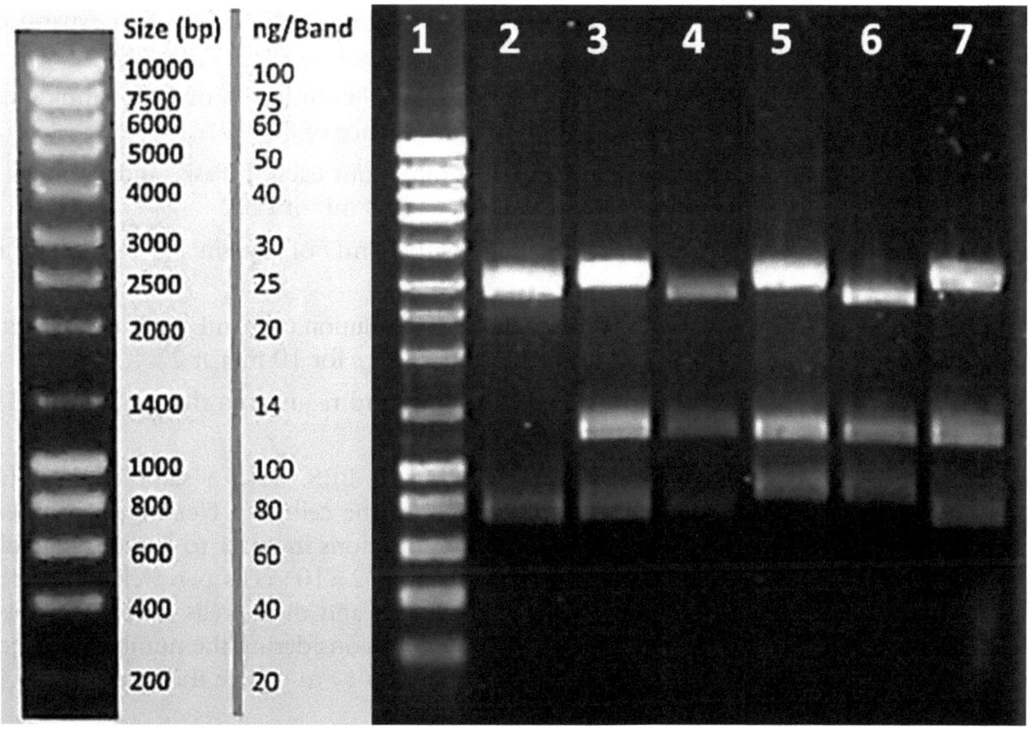

Fig. 2 Agarose gel with the DNA plasmids double digested with KpnI/XbaI. pVAX1-GFP (*2*); N3 vector (*3*), E1A-N3 (*4*), N3-LAMP (*5*), E1A-N3-LAMP (*6*), Sc-N3 (*7*). Molecular weight marker (*1*) NZYDNA ladder III

Table 2
DNA plasmid sizes (bp) and expected fragments for double digestion with KpnI/XbaI

DNA plasmid vectors	Size (bp)	KpnI/XbaI-expected fragments
pVAX1-GFP	3,697	2925/772
pVAX1-N3-GFP (N3)	5,097	3070/1257/770
pVAX1-E1A-N3-GFP (E1A-N3)	4,850	2820/1257/773
pVAX1-N3-GFP-LAMP (N3-LAMP)	5,208	3070/1257/881
pVAX1-E1A-N3-GFP-LAMP (E1A-N3-LAMP)	4,958	2820/1257/881
pVAX1-Sc-N3-GFP (Sc-N3)	5,162	3132/1257/773

antibiotic–antimycotic 100×, 1 % of MEM—NEAA 100×, 1 % of 100 mM sodium pyruvate, 0.1 % of gentamicin (50 mg/mL), and 10 % of inactivated FBS.

2. Incubate the T flasks at 37 °C in a 5 % of CO_2 humidified environment.
3. After a confluence of 80–90 % is reached, split CHO cells to two 75 cm^2 T flasks using the conditions described above to

ensure that enough cells are present for subsequent expansion in 24-well plates.

4. Incubate the T flasks at 37 °C in a 5 % of CO_2 humidified environment up to a confluence of 80–90 %.
5. Discard the F12 medium from each T flask, and perform a quick wash of the cells with 8 mL of PBS.
6. Trypsinize the cells with 8 mL of trypsin for 5–10 min at 37 °C.
7. Transfer cells on the trypsin solution to 50 mL conical centrifuge tubes and centrifuge at 230 × *g* for 10 min at 22 °C.
8. Discard the supernatant, and resuspend the pellet in 5 mL of PBS.
9. Perform a 1:10 dilution in PBS. Take a sample from the diluted solution and count the cells in a Neubauer chamber. Then make appropriate dilutions in order to have a final volume of 500 μL containing 2×10^5 cells per well. Consider three wells per each plasmid and three wells for the negative control. Do the calculations considering the number of effective wells plus one (n° wells + 1) to ensure that the mixture is enough to fill all wells.
10. Centrifuge a second time, and resuspend the pellet in a final volume, per well, of 450 μL of F12 medium without antibiotics (referred along the chapter as incomplete medium) and 50 μL of inactivated FBS. Thus, multiply the n° wells + 1 per 450 μL of F12 medium and per 50 μL of inactivated FBS and resuspend the pellet in this total volume.
11. Distribute 500 μL of the mixture (F12 incomplete medium with FBS) per well.
12. Incubate the plate(s) at 37 °C in a 5 % of CO_2 humidified environment for approximately 24 h in order to reach a confluence of 80–90 %.
13. For each well prepare separately one mixture containing 2 μL of lipofectamine and 48 μL of incomplete F12 medium and one mixture containing 1 μg of plasmid DNA and incomplete F12 medium up to a final volume of 50 μL. For the negative controls prepare a mixture of 2 μL lipofectamine and 98 μL of incomplete F12 medium. Consider performing triplicates for both plasmid DNAs and negative controls.
14. Prepare a transfection mixture with a final volume of 100 μL by mixing the two solutions. Incubate at room temperature for 20 min.
15. Meanwhile, remove the 24-well plate(s) with growing CHO cells from the CO_2 chamber and discard 250 μL of medium from each well.

16. Add 100 μL of the prepared transfection mixtures to each one of the wells in the plate. Add 100 μL of the plasmid-free transfection mixture to each of the negative control wells.
17. Incubate the 24-well plate(s) at 37 °C in a 5 % of CO_2 humidified environment for 4–6 h.
18. Remove the medium and transfection mixture from each well and add fresh complete F12 medium with antibiotics and 10 % FBS up to a final volume of 500 μL per well.
19. Incubate cells for 48 h at 37 °C in a 5 % of CO_2 humidified environment.
20. After incubation, harvest the cells and perform flow cytometry and quantitative real-time polymerase chain reaction (qPCR) analysis.

3.7 Analysis of Protein Expression in CHO Cells by Flow Cytometry

1. Harvest transfected CHO cells after 48 h of incubation.
2. Discard the complete F12 medium, and wash each well with 800 μL of PBS.
3. Add 200 μL of trypsin to each well, and incubate the plate(s) for 5 min at 37 °C.
4. Centrifuge the cells from each well in 15 mL conical centrifuge tubes.
5. Add 1 mL of PBS to each well in order to recover cell leftovers and transfer to the corresponding centrifuge tube.
6. Centrifuge at 22 °C and 230 × *g* for 10 min.
7. Discard the supernatant, and resuspend the resulting pellet in 800 μL of 2 % PFA, fixing cells and allowing flow cytometry analysis to be postponed.
8. Cover the tubes with aluminum foil and keep at 4 °C until flow cytometry analysis.
9. Use the CellQuest Pro Software© (Becton-Dickinson, NJ, USA) to analyze the green fluorescence intensity corresponding to GFP expression level and generate histograms and dot plots that lead to the determination of transfection efficiency and mean fluorescence (*see* Fig. 3) (*see* **Note 14**).

3.8 Analysis of Plasmid Copy Number in CHO Cells by Quantitative Real-Time Polymerase Chain Reaction

1. Harvest transfected CHO cells after 48 h of incubation.
2. Discard complete F12 medium, and wash the cells with 800 μL of PBS.
3. Discard the PBS, trypsinize with 200 μL of trypsin, and incubate for 5 min at 37 °C.
4. Transfer to 15 mL conical centrifuge tubes.
5. Centrifuge the samples at 22 °C at 230 × *g* for 10 min.

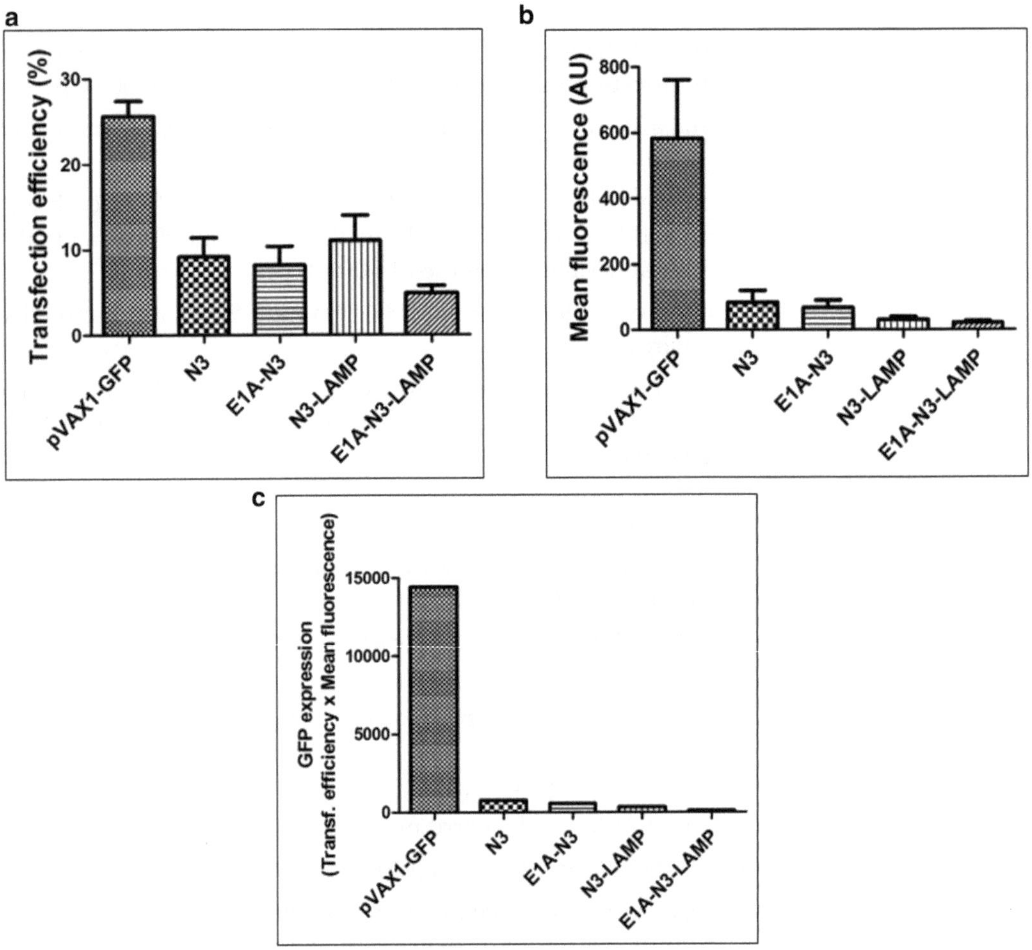

Fig. 3 Analysis of transfection efficiency (**a**), as a measure of percentage of cells that show GFP fluorescence; (**b**) as a measure of intensity of fluorescent cells (mean fluorescence); (**c**) GFP expression levels by the product between transfection efficiency and mean fluorescence levels, assessed by flow cytometry for CHO cells harvested 48 h after transfection with pVAX1-GFP, N3 vector, E1A-N3, N3-LAMP, and E1A-N3-LAMP. Data represent two independent assays with three replicates each. Vertical bars show standard deviation. One-way ANOVA, Tukey's multiple comparison test, was performed with a level of confidence of $p < 0.05$

6. Discard the supernatant, and resuspend the pellet in 500 μL of PBS.
7. Take a sample and count the cells in a Neubauer chamber.
8. Perform a second centrifugation at 22 °C and 230×*g* for 10 min, and resuspend the cell pellets in sterile RNase-free water to a final concentration of 6,250 cells/μL [19].
9. Store samples at −20 °C until analysis.
10. Perform the qPCR analysis in capillaries using the LightCycler FastStart DNA Master SYBR Green I kit (Roche) and the

LightCycler® detection system (Roche), following the recommended protocol (*see* **Note 15**).

11. Proceed with several dilutions of the purified plasmid DNA with the High Pure Plasmid Purification kit (Roche), in sterilized and filtered MilliQ water, in order to obtain different concentrations (2.5, 25, 250, 2,500, and 25,000 pg/μL). These samples are needed for the construction of the calibration curves required to determine plasmid copy number per transfected CHO cell, for each plasmid.
12. Prepare a reaction mixture with 2 μL of SYBR Green, 0.8 μL of both forward and reverse GFP primers (0.4 μM final), 1.6 μL of $MgCl_2$ (3 mM final), and 10.8 μL of PCR-grade water, for a final volume of 16 μL.
13. Prepare several capillaries for a final volume of 20 μL per capillary. First, prepare a negative control, adding 4 μL of sterilized and filtered MilliQ water to 16 μL of the above reaction mixture.
14. Set up a second negative control with 2 μL of non-transfected cells (12,500 cells), 2 μL of sterilized and filtered MilliQ water, and 16 μL of reaction mixture.
15. Add 2 μL of transfected CHO cells (12,500 cells, for each plasmid) to 2 μL of sterile and filtered MilliQ water and to 16 μL of reaction mixture.
16. Finally, to obtain the calibration curves, mix 2 μL of non-transfected cells (12,500 cells) with 2 μL of each pDNA diluted solution (5, 50, 500, 5,000, 50,000 pg) and 16 μL of reaction mixture.
17. Perform the amplification in a LightCycler detection system. Start the program with the incubation of the samples at 95 °C for 10 min. Perform 30 cycles of RT-PCR with a denaturation step of 10 s at 95 °C, followed by 5 s at 55 °C for annealing and finalized with an extension at 72 °C for 7 s. At the end perform an extension step at 70 °C for 30 s, followed by a denaturation step based on a temperature gradient, with an increase of 0.1 °C per second, from 70 to 95 °C, and with a final cooling at 40 °C for 30 s (*see* **Note 16**).
18. If desirable, after RT-PCR, analyze the samples in a 1 % agarose gel electrophoresis in order to visualize DNA fragments corresponding to the 108 bp of the amplified GFP gene.

3.9 Analysis of mRNA Transcription (via cDNA) by Quantitative Real-Time Polymerase Chain Reaction

1. Harvest transfected CHO cells after 48 h of incubation.
2. Discard the complete F12 medium, and wash each well with 800 μL of PBS.
3. Add 200 μL of trypsin to each well, and incubate the plate(s) for 5 min at 37 °C.

4. Transfer to 15 mL conical centrifuge tubes.
5. Add 1 mL of PBS to each well to recover leftover cell and transfer to conical centrifuge tubes.
6. Centrifuge at 22 °C and 230 × *g* for 10 min.
7. Discard the supernatant, and resuspend cell pellet in 200 μL of PBS.
8. Extract the total RNA using the Pure RNA Isolation Kit (Roche), following the manufacturer's instructions and taking care to elute the RNA in 50 μL of PCR-grade water.
9. Quantify the RNA by absorbance at 260 nm (*see* **Note 7**).
10. Use the First Strand cDNA Synthesis Kit for RT-PCR (AMV) (Roche), following the recommended protocol, to synthesize a cDNA chain using the Oligo-p(dT)15 primer, which anneals to the 3′ end of the polyA mRNA tail and 1 μg of RNA per sample.
11. Perform an RT-PCR following the same protocol previously described (*see* Subheading 3.8) to quantify the mRNA content in CHO cells (*see* **Note 17**).
12. Prepare a reaction mixture with 2 μL of SYBR Green, 0.8 μL of both forward and reverse GFP primers (0.4 μM final), 1.6 μL of $MgCl_2$ (3 mM final), and 10.8 μL of PCR-grade water, for a final volume of 16 μL.
13. Construct calibration curves using 2 μL of each plasmid DNA-diluted solution (5, 50, 500, 5,000, 50,000 pg), 16 μL of reaction mixture, and 2 μL of sterile and filtered MilliQ water.
14. Prepare the samples using 2 μL of cDNA, 2 μL of MilliQ water, and 16 μL of reaction mixture.
15. Set up a negative control using 4 μL of MilliQ water and 16 μL of reaction mixture.
16. Prepare a second control of 2 μL of MilliQ water with 2 μL of cDNA from non-transfected cells and 16 μL of reaction mixture.
17. Perform the amplification in a LightCycler detection system (*see* Subheading 3.8).

3.10 *In Vivo* Expression Assay

1. Start the experiment with 3-week-old Leghorn chickens. Divide the animals in groups (at least five animals per group). Immunize groups with the plasmid DNAs encoding N3 antigen fused to the different targeting sequences. Prepare adequate negative controls by injecting one group with a pVAX1-GFP plasmid control and one group with PBS.
2. Prepare a mixture of 20 μL of lipofectamine with 100 μL of sterile PBS and incubate at room temperature for 30 min.
3. During incubation time, set up a mixture of 100 μg of pDNA with sterile PBS, to a final volume of 280 μL.

4. Mix both samples, and incubate the 400 μL of lipofectamine mixture with pDNA at room temperature for 15 min.
5. Immunize each chicken with a total volume of 400 μL of the pDNA mixture with lipofectamine, by two intramuscular injections with 200 μL in each side of the breast.
6. Perform two boosts with 2 weeks of interval after prime administration, at days 14 and 28.
7. Finally, 24 days after the last boost, at day 52, do a heterologous boost with administration of 100 μL of the purified N3 protein (*see* **Note 18**).
8. Take a blood sample from the wing of each chicken every week, for the first 3 weeks, and before each immunization. Collect samples at days 42, 52, 66, 78, and 90.
9. Centrifuge blood samples at 2,400 × *g* for 10 min, discharge the cells, and store the serum at 4 °C.
10. In the next day, perform the ELISA in 96-well ELISA plate(s) (*see* **Note 19**).

Day one of the ELISA experiment:

1. Coat ELISA plate(s) with N3-purified protein diluted 1:500 in carbonate–bicarbonate buffer (pH 9.0).
2. Add 100 μL to each well of the ELISA plate and incubate overnight at 4 °C.
3. Mix the serum samples previously stored at 4 °C, using 20 μL from each chicken serum inside of each group, in order to form a pool of sera for every group of study.

Day two of the ELISA experiment:

4. Dilute 100-fold in blocking buffer the pools of chicken serum and both anti-H5N3 and anti-H5N1 controls.
5. Incubate at 37 °C for 30 min (*see* **Note 20**).
6. During incubation time, wash the ELISA plate(s) four times in a washing machine with a solution of PBS/Tween 0.05 %.
7. Transfer 100 μL of diluted solutions per well of the ELISA plate(s).
8. Incubate at room temperature for 1 h.
9. During incubation time, prepare 1:10,000 dilution of anti-chicken serum IgG (Fc):HPr (Abd Serotec) in blocking buffer and store at 37 °C until used.
10. After 1 h of incubation, wash the plate(s) four times with the same washing buffer.
11. Add 100 μL per well of anti-chicken IgG serum.
12. Incubate the plate(s) at 37 °C for 30 min and wash again.

13. Add 100 μL of TMB to each well in order to generate a signal.
14. 30 min after the addition of TMB, add 100 μL of sulfuric acid to the wells in order to stop the reaction.
15. Place the plate in a microplate reader, and measure the absorbance at 450 nm (*see* **Note 21**).

4 Notes

1. The targeting sequences Sc, E1A, and LAMP were cloned in pVAX1-Sc-ISG-GFP (Sc-ISG), pVAX1-E1A-N3-GFP-LAMP (E1A-N3-LAMP), and pVAX1-ISG-GFP-LAMP (ISG-LAMP) in previous studies [3, 4, 18]. Alternatively, the targeting sequences could also be obtained from synthesized oligonucleotides. The desired targeting sequence (*see* Table 3) can be ordered from a DNA synthesis company. Order two complementary oligonucleotides (forward and reverse) that have to be subsequently annealed, following the Thermo Scientific—Anneal Complementary Pairs of Oligonucleotides protocol, and cloned in the N3 vector. The lyophilized oligonucleotides are resuspended in annealing buffer (10 mM Tris, 1 mM EDTA, 50 mM NaCl, pH 8.0) to a concentration of 100 pmol/μL. An efficient annealing was obtained when 10 μM of oligonucleotides were used. A thermocycler is used for annealing, of the oligonucleotides, for 5 min at 95 °C, followed by a decrease of 2 °C per minute, until 25 °C. After annealing, the oligonucleotides must be stored at 4 °C.
2. In this experiment and using these plasmids, it is frequent to obtain purification yields between 100 and 230 ng/μL.

Table 3
Nucleotide sequence of targeting sequences and localization of the targeted protein

Targeting sequences	Nucleotide sequence	Targeting compartment
E1A	ATGCGCTATATGATTCTGGGCCTGCTGGCGCTGGCGGCGGTG TGCAGC (Protein: **MRYMILGLLALAAVCSA**)	Endoplasmic reticulum
LAMP	TTGATCCCCATTGCTGTGGGCGGTGCCCTGGCAGGGCTGGTCC TCATCGTCCTCATTGCCTACCTCATTGGCAGGAAGAGGAGTC ACGCCGGCTATCAGACCATCTAG (Protein: **LIPIAVGGALAGLVLIVLIAYLIGRKRSHAGYQTI**)	Lysosomes
Sc	ATGGACGCCATGAAGCGCGGCCTGTGCTGCGTGCTGCTGCTGT GCGGCGCCGTGTTCGTGAGCGCCCGC (Protein: **MDAMKRGLCCVLLLCGAVFVSAR**)	Extracellular space

Table 4
Sequence of the GFP gene forward and reverse primers

GFP forward primer	GFP reverse primer
5′-TCGAGCTGGACGGCGACGTAAA-3′	5′-TGCCGGTGGTGCAGATGAAC-3′

3. It is frequent to obtain yields from 100 to 1,000 ng/μL, depending on the plasmid being purified. The constructs with N3 and targeting sequences showed lower yields than the plasmid pVAX1-GFP and N3 vector.
4. The elution step using preheated sterile MilliQ water promotes higher yields. However, these values of DNA recovery are frequently low—approximately 7 and 20 ng/μL, respectively, for targeting sequences and for vectors.
5. The sequences of the forward and reverse primers used to amplify the GFP gene in RT-PCR are displayed in Table 4.
6. The blocking buffer should be left to dissolve overnight in order to ensure a proper dissolution of the components. To perform the several washing steps of the 96-well ELISA plate use a solution of PBS/Tween 0.05 % and drain the plate in paper towels.
7. Perform the calibration of the spectrophotometer using the same solution used to elute the DNA or the RNA in order to obtain a more reliable concentration value. Make a 1:50 dilution using 2 μL of sample and 98 μL of MilliQ water. The $Abs_{260\ nm}$ readings are converted into mass concentrations by the following correlations: 1 U $Abs_{260\ nm}$ = 50 μg/mL for DNA and 1 U $Abs_{260\ nm}$ = 40 μg/mL for RNA.
8. The user-friendly ApE© free software is a sequence editor program (http://biologylabs.utah.edu/jorgensen/wayned/ape/) used to analyze linear and circular DNA. It has numerous features that are described on the website, which can be explored. After copying and pasting the DNA sequence of interest go to the "Features" and select "Annotate features using Library." It is also possible to find a sequence that was not identified by the program by selecting "Edit → Find." Paste the nucleotide sequence, and select "Find next." This will highlight the sequence. Then, press the right button of the mouse and choose "New Feature" in order to save and identify the sequence of interest. To analyze the restriction sites and simulate digestions with different enzymes, select "Enzymes" and "Enzyme selector." Choose the enzyme, and select the "Highlight" button. To visualize the graphic map, select the icon beneath "Help," with a circular map. This will show a

linear or a circular map, according to the feature selected in the "circular" or the "linear" button, in the upper right corner of the program window. To analyze the resulting fragments, press the button right next to the frag icon. This software reads DNA Strider, Fasta, Genbank, and EMBL files and allows exporting or saving DNA maps.

9. The restriction mixture should be started by adding plasmid DNA and then the MilliQ sterilized water, followed by the enzyme(s) buffer and, in the end, the restriction enzymes, which should be stored immediately after the procedure. The reaction volume depends on the concentration of the DNA, though it is advised to maintain the reaction volume up to a maximum of 25 μL. For double digestions select the buffer that gives maximum activity of both enzymes.

10. It is advised to mix vigorously and perform microwave heating without boiling, until all agarose is melted. Use a pipette tip to remove bubbles before agarose jellifies in the tray.

11. Split the agarose gel. Stain the part of the gel where the small sample was run with ethidium bromide and expose to UV light. Try to remove excess of liquid from the gel surface with a paper tissue. Hand the gel steadily, and mark the band with a scalpel tip. Turn off the UV light, and place the non-stained part of the gel, containing the remainder sample, right next to the stained gel. Then, slice the desired non-stained sample of the linear DNA, by comparison, with the help of a ruler or a similar object. Finally, stain and expose the whole gel to UV light in order to ensure that the desired DNA fragment was removed.

12. Visualization of bands with low concentration (<10 ng/μL) of targeting sequences is difficult.

13. For the reaction mixture, consider a 3:1 insert/vector molar ratio and a mass of insert of approximately 200–300 ng. Do the calculations according to the following equation:

$$\frac{\text{ng of vector} \times \text{kbsize of insert}}{\text{kb size of vector}} \times \text{molar ratio of} \frac{\text{insert}}{\text{vector}} = \text{ng of insert}$$

A final volume of 20 μL for the ligation mixture should be used. If needed, DNA can be concentrated using a DNA SpeedVac® Concentrator. After DNA drying add the adequate volume of MilliQ sterile water. Then start the ligation by adding DNA followed by the T4 buffer and T4 ligase. Mix carefully with a pipette and incubate under the desired conditions.

14. Perform flow cytometry up to 4 days after cell harvesting. Always use the same time for each set of experiments. Use the cytometer to record the forward scatter (FSC), side scatter (SSC), and green fluorescence (FL1) in each run. For each

sample, the cytometer discriminates cells from the debris due to their characteristics of FSC versus SSC, distinguishing cells from debris that are outside the gate. Background autofluorescence of non-transfected cells is taken into account considering the difference between total cell population inside the gate and the background autofluorescence of non-transfected cells indicated by FL1 parameter [19]. This establishes the M1 and M2 parameters, corresponding to non-transfected and transfected cells with green fluorescence, respectively. Transfection efficiency is defined as the percentage of cells that express fluorescence above a threshold level, defined by autofluorescence of non-transfected cells. CHO cells transfected with the plasmids N3-LAMP or E1A-N3-LAMP tend to show less fluorescence than cells transfected with E1A-N3, because sorting to lysosomes and subsequent degradation are expected. Low fluorescence values are also anticipated for CHO cells transfected with plasmid Sc-N3, because this protein is likely to be secreted to the extracellular space. Nevertheless, the N3 protein has a transmembrane domain that could retain the protein at the membrane level, increasing the measured fluorescence. Finally, CHO cells expressing the non-fused GFP protein are expected to exhibit higher levels of fluorescence due to the expression of GFP molecules instead of a fusion protein [4]. The intensity of fluorescent cells is also measured by mean fluorescence, which is expected to be higher in the case of the E1A-N3 fused proteins and probably lower for cells transfected with Sc-N3. The intensity of fluorescence of cells transfected with the plasmid N3-LAMP is expected to be lowest due to the degradation of the protein in the lysosome, and even lower in the case of cells transfected with the plasmid E1A-N3-LAMP, where the protein is processed at ER and can be more efficiently targeted for degradation at lysosomes. Once again, the intensity of the GFP protein encoded by the control plasmid must be higher when compared with the other plasmid vectors. Protein expression can also be analyzed in terms of GFP reporter gene expression, by multiplying transfection efficiency and mean fluorescence levels, clarifying the relationship between the different plasmids.

15. This is a very sensitive method that requires a very rigorous handling. It is advisable to avoid exhalation near the samples, to use pipette tips with filter, and, if possible, to use the same pipettes in every experiment due to the fact that a small variation can affect the results. It is important to minimize the variations that occur during the work, such as the carryover that occurs during the pipetting of the samples, which has to be taken into account. It is essential to ensure that the exterior of the pipette tip is as clean as possible during the transfer of the samples.

16. Start the amplification program with the incubation of the samples at 95 °C for 10 min to activate the DNA polymerase and to promote the cell lysis [19]. Consider each one of the 30 cycles with incubation periods of 10 s at 95 °C, for denaturation step, followed by 5 s at 55 °C, to allow the annealing of the primers and the annealing of the dye SYBR Green I, and finalize at 72 °C for 7 s to terminate the DNA strand amplification. The presence of the plasmids inside CHO cells does not necessarily mean that protein expression is occurring. RT-PCR measures the total plasmid DNA content in CHO cells independently of its localization inside the cells [17]. It is expected that the difference in size, due to the addition of the N3 and targeting sequences, influences plasmid uptake. Therefore, it is probable that CHO cells transfected with the smaller plasmid pVAX1-GFP show an increase in plasmid copy number comparing with CHO cells transfected with the N3 vector and the plasmids Sc-N3, E1A-N3, N3-LAMP, and E1A-N3-LAMP. Conversely, the differences between cells transfected with N3 vector and with plasmids encoding the N3 protein and targeting sequences should not be significant since the difference between these vectors is of approximately 100 base pairs. Plasmid degradation by nucleases in lysosomes or in cytoplasm decreases the number of intact plasmids that reach the nucleus and are available to be expressed [19]. Thus, RT-PCR may detect nicked or fragmented plasmid DNA molecules that are not able to drive the expression of the target antigen.
17. A positive correlation is anticipated between the high copy number of plasmids and the levels of mRNA content. This does not mean that high levels of gene transcription are being achieved, as this is perceived by the ratio between mRNA and plasmid DNA. After transfection, plasmids are present in CHO cells at different copy numbers due to the different plasmid sizes, that affect plasmid uptake, and also due to different plasmid susceptibilities to nuclease degradation. Therefore, mRNA will vary according to the plasmid copy number inside cells as well as its intracellular localization. Thus, it is important to establish a relationship between the plasmid copy number, previously determined, and the mRNA content in CHO cells.
18. After a prime immunization with plasmid DNA at day 0, give two subsequent boosts of plasmid DNA in every 2 weeks and, 24 days after the last boost, perform an injection of the purified N3 protein, performing a heterologous prime boost strategy, in order to enhance the humoral response. During prime and subsequent boosts, there is an activation of T cell responses against N3 recombinant protein encoded by the DNA vectors. However, with the administration of the N3 antigenic protein, the immune response is potentiated, probably because memory

lymphocytes developed before heterologous prime are stimulated, developing a focused immune response against N3 protein [3, 20]. Hence, it is probable that, after administration of the N3 purified protein, a dramatic increase in the antibody production could be observed.

19. The use of a multichannel pipette is very useful, since this is a very sensitive method that involves a large number of samples. It is advised to use always the same pipettes and to do the measurements rigorously in order to minimize the variations that are usually to occur among different assays (due to differences in the plates in the performance of the several dilutions as well as due to the variations between blocking buffers done in different days) and achieve reproducible and reasonable results inside a certain limit of variation.
20. Since it is difficult to separate all the proteins during N3 protein purification, which is produced in *E. coli*, the natural antibodies produced in chickens against *E. coli* proteins will bind to those molecules. In this way, the blocking buffer is essential to minimize these ligations that are very frequent to occur and really undesirable, minimizing the background.
21. The production of antibodies is expected to be higher for chickens immunized with N3-LAMP and with E1A-N3-LAMP. The protein is targeted for degradation in the lysosome, where MHCII molecules are loaded with the peptides that are displayed at cell surface. The probability of activation of $CD4^+$ T cells is increased as well as a direct humoral response. In the second case, it is also probable that a targeting to the ER promotes a more efficient targeting to the lysosome that is, in this way, reinforced, inducing a more pronounced antibody production. On the other hand, the production of antibodies in chickens immunized with E1A-N3 is expected to be lower. The targeting of the protein to the ER promotes the MHCI pathway. The MHCII pathway and the development of a humoral response can be indirectly generated due to the cross-presentation of antigenic particles, released by transfected cells that follow the endocytic pathway where MHCII molecules are aggregated with the peptides and displayed at cell surface, activating $CD4^+$ T cells and a humoral response [16]. Antibody production in chickens immunized with Sc-N3 or N3 vector can also be high due to the fact that the expressed protein can probably be retained in the cell membrane, due to the presence of the transmembrane domain of the N3 protein, that was predicted with the web tool TMHMM [4, 18, 21], reinforcing the fact that the neuraminidase is divided in a small cytoplasmic tail and in an extracellular domain [22]. A humoral response could be generated by a direct presentation at cell surface, to $CD4^+$ and $CD8^+$ T cells. Otherwise, if not retained in the cell

membrane, the protein expressed by the N3 vector could be free in the cytoplasm and intracellulary digested at the endosome by autophagy, following the endocytic pathway and being displayed by MHCII molecules. In the case of the plasmid Sc-N3, the expressed protein could be secreted to the extracellular space and be phagocytized or endocytosed by other APCs, following the endocytic pathway, correlated with MHCII presentation and development of a humoral response.

Acknowledgment

This work was supported by Fundação para a Ciência e a Tecnologia (project PTDC/EBB-EBI/113650/2009).

References

1. Williams JA, Carnes AE, Hodgson CP (2009) Plasmid DNA vaccine vector design: impact on efficacy, safety and upstream production. Biotechnol Adv 27(4):353–370
2. Carvalho JA, Prazeres DM, Monteiro GA (2009) Bringing DNA vaccines closer to commercial use. IDrugs 12:642–647
3. Henriques AM, Fevereiro M, Prazeres DM et al (2007) Development of a candidate DNA vaccine against Maedi-Visna virus. Vet Immunol Immunopathol 119(3–4):222–232
4. Carvalho JA, Azzoni AR, Prazeres DM et al (2010) Comparative analysis of antigen-targeting sequences used in DNA vaccines. Mol Biotechnol 44(3):204–212
5. Berhane S, Areste C, Ablack JN et al (2011) Adenovirus E1A interacts directly with, and regulates the level of expression of, the immunoproteasome component MECL1. Virology 421(2):149–158
6. Qiu JT, Liu B, Tian C et al (2000) Enhancement of primary and secondary cellular immune responses against human immunodeficiency virus type 1 gag by using DNA expression vectors that target gag antigen to the secretory pathway. J Virol 74(13):5997–6005
7. Ruff AL, Guarnieri FG, Staveley-O'Carroll K et al (1997) The enhanced immune response to the HIV gp160/LAMP chimeric gene product targeted to the lysosome membrane protein trafficking pathway. J Biol Chem 272(13): 8671–8678
8. Persson H, Jornvall H, Zabielski J (1980) Multiple mRNA species for the precursor to an adenovirus-encoded glycoprotein: Identification and structure of the signal sequence. Proc Natl Acad Sci U S A 77(11):6349–6353
9. Leifert JA, Rodriguez-Carreno MP, Rodriguez F et al (2004) Targeting plasmid-encoded proteins to the antigen presentation pathways. Immunol Rev 199:40–53
10. Liang R, van den Hurk JV, Zheng C et al (2005) Immunization with plasmid DNA encoding a truncated, secreted form of the bovine viral diarrhea virus E2 protein elicits strong humoral and cellular immune responses. Vaccine 23(45):5252–5262
11. Braulke T, Bonifacino JS (2009) Sorting of lysosomal proteins. Biochim Biophys Acta 1793(4):605–614
12. Storni T, Kundig TM, Senti G et al (2005) Immunity in response to particulate antigen-delivery systems. Adv Drug Deliv Rev 57(3): 333–355
13. Wiesel M, Oxenius A (2012) From crucial to negligible: functional CD8(+) T-cell responses and their dependence on CD4(+) T-cell help. Eur J Immunol 42(5):1080–1088
14. Kaur M, Rai A, Bhatnagar R (2009) Rabies DNA vaccine: no impact of MHC class I and class II targeting sequences on immune response and protection against lethal challenge. Vaccine 27(15):2128–2137
15. Villadangos J, Schnorrer P (2007) Intrinsic and cooperative antigen-presenting functions of dendritic-cell subsets in vivo. Nat Rev Immunol 7:543–555
16. Kutzler MA, Weiner DB (2008) DNA vaccines: ready for prime time? Nat Rev Genet 9(10):776–788
17. Azzoni AR, Ribeiro SC, Monteiro GA et al (2007) The impact of polyadenylation signals on plasmid nuclease-resistance and transgene expression. J Gene Med 9(5):392–402

18. Gil JB (2011) Development of DNA vaccines prototypes against avian influenza viruses. Master thesis
19. Carapuça E, Azzoni AR, Prazeres DM et al (2007) Time-course determination of plasmid content in eukaryotic and prokaryotic cells using Real-Time PCR. Mol Biotechnol 37:120–126
20. Bridge SH, Sharpe SA, Dennis MJ et al (2011) Heterologous prime-boost-boost immunisation of Chinese cynomolgus macaques using DNA and recombinant poxvirus vectors expressing HIV-1 virus-like particles. Virol J 8:429
21. Freitas EB (2012) DNA vaccines against avian influenza virus: enhancing immune response by protein targeting. Master thesis
22. Kamps B, Hoffmann C, Preiser W (2006) Influenza Report. Flying Publisher, Wuppertal

Chapter 5

Progresses in DNA-Based Heterologous Prime-Boost Immunization Strategies

Ronald J. Jackson, David B. Boyle, and Charani Ranasinghe

Abstract

Although recombinant DNA and recombinant viral vectors expressing HIV antigens have yielded positive outcomes in animal models, these vaccines have not been effectively translated to humans. Despite this, there is still a high level of optimism that poxviral-based vaccine strategies could offer the best hope for developing an effective vaccine against not only HIV-1 but also other chronic diseases where good-quality T and B cell immunity is needed for protection. In this chapter we discuss step by step (1) how recombinant poxviral vectors co-expressing HIV antigens and promising mucosal/systemic adjuvants (e.g., IL-13Rα2) are constructed, (2) how these vectors can be used in alternative heterologous prime-boost immunization strategies, (3) how systemic and mucosal samples are prepared for analysis, followed by (4) two immunological assays: multicolor intracellular cytokine staining and tetramer/homing maker analysis that are used to evaluate effective systemic and mucosal T cell immunity.

Key words Poxvirus vectors, Fowlpox virus, Vaccinia virus, DNA vaccines, Prime-boost immunization, Mucosal/systemic vaccination, Intracellular cytokine staining, Tetramer analysis, Homing marker analysis, T cell immunity, Mucosal/systemic cell preparation

1 Introduction

Recombinant plasmid DNA-based prime-boost immunization strategies that were developed in the 1990s emerged as exciting, safe, and cost-effective methods to induce T and B cell immunity to encoded HIV-1 vaccine antigens [1–7]. However, following the failures of many of these recombinant DNA prime-boost vaccination strategies as well as the single modality vaccines to translate effectively to clinical trials [8–11], the focus has now largely shifted towards the use of viral/viral or viral/protein prime-boost vaccination strategies to induce effective protective immunity against HIV-1 [12–14]. Among the large range of vectors currently being tested, poxviral vectors offer good optimism for the future [14–16]. Our studies have clearly demonstrated that vaccine vector combination used (which vector is used in the prime or the booster), route

Monica Rinaldi et al. (eds.), *DNA Vaccines: Methods and Protocols*, Methods in Molecular Biology, vol. 1143, DOI 10.1007/978-1-4939-0410-5_5, © Springer Science+Business Media New York 2014

of delivery, and the cytokine cell milieu at the vaccination site play a critical role in governing the induction of effective protective immunity [12, 17–21]. We have now established that intranasal (i.n.) recombinant fowlpox (FPV-HIV) prime followed by intramuscular (i.m.) recombinant vaccinia (VV-HIV) vaccination strategy induces not only good magnitude of mucosal and systemic T cell immunity but also high avidity CD8 T cell immunity with broader cytokine profiles compared to rDNA prime boost immunization strategies or purely systemic rFPV/rVV vaccination strategies [12]. Many studies have now demonstrated that not only the magnitude of immunity measured by IFN-γ production, but also multi-functionality and quality or avidity of T and B cells play a crucial role in inducing protective immunity. In this chapter we describe the design of recombinant poxviral-based HIV-1 vaccines co-expressing an IL-13 inhibitor together with HIV antigens that offer good prospects for the future. Specifically, some of the vaccine strategies that can induce long lasting mucosal and systemic immunity and the immunological assays used in evaluating efficacy of these novel vaccines [21]. Immunological assays, namely, (1) multiparameter intracellular cytokine analysis that has been successfully used to evaluate HIV-specific multifunctional CD8 T cell immune responses following vaccination, which is known to be a hallmark of protective immunity and (2) MHC-I specific $K^dGag_{197-205}$-tetramer staining to evaluate the number of HIV-1 specific systemic (i.e., spleen) and mucosal (iliac nodes or Peyer's patch) CD8 T cells, together with the use of gut-specific mucosal homing marker analysis (α4B7 and CCR9) are described [21].

2 Materials

The standard molecular biology procedures used are essentially as described in many molecular biology protocol manuals such as Green and Sambrook (2012) "Molecular Cloning: A laboratory manual (Fourth Edition)" and are not discussed in detail [22]. To simplify the procedures we use commercially prepared kits and reagents for common molecular techniques such as nucleic acid isolation and analysis, PCR and DNA cloning with reliable results.

2.1 Plasmid DNA Construction, Production, and Purification

2.1.1 Isolation of Mouse IL-13Rα2 cDNA and Subcloning into Poxvirus Vectors

1. Total RNA or polyA+RNA isolation kit, e.g., RNeasy Protect Mini Kit (QIAGEN) includes RNAlater RNA Stabilization Reagent.
2. Micro-pestles suitable for 1.5 mL microcentrifuge tubes.
3. RT-PCR kit, e.g., OneStep RT-PCR Kit (QIAGEN).

4. Oligonucleotide primers (*see* **Note 1**).
   - Forward AGATCTGAAATGGCTTTTGTGCATATCAG ATGCTTGTG
   - Reverse GAGCTCTTAACAGAGGGTATCTTCATAAGC
5. PCR machine with heated lid suitable for 200 μL thin walled PCR tubes.
6. PCR cloning vector kit, e.g., PCR Cloning^{plus} Kit (QIAGEN) contains pDrive and EZ Competent Cells.
7. Agarose, DNA electrophoresis grade.
8. DNA molecular size standards.
9. Mini-gel DNA electrophoresis apparatus and power supply.
10. DNA stain.
11. UV or blue-light transilluminator.
12. 50× TAE electrophoresis buffer : 2 M Tris base, 1 M acetic acid, 50 mM EDTA pH 8.0.
13. Agarose Gel DNA purification kit.
14. Transfection-competent *E. coli* strain.
15. X-Gal (20 mg/mL), IPTG (100 mM) stocks for blue-white colony color selection.
16. DNA restriction enzymes, modifying enzymes, T4 DNA ligase for subcloning DNA fragments.
17. L-broth and L-agar plates.
18. 50 mg/mL Ampicillin (1,000×) dissolved in water and filter-sterilized.
19. Plasmid DNA mini-prep kit.
20. cDNA (IL-13Rα2) for subcloning into poxvirus vectors (*see* **Note 2**).
21. Poxvirus plasmid vectors pTK7.5A [23] for VV and pAF09 [24] for FPV (*see* **Notes 3** and **4**).

2.2 Cell Culture and Recombinant Viruses

1. Recombinant poxviruses VV336 and FPV086 [25].
2. Minimal Essential Media with l-glutamine (MEM).
3. Sterile Fetal Bovine Sera (FBS), heat-inactivated 56 °C, 30 min.
4. (100×) Penicillin (5 mg/mL)–Streptomycin (5 mg/mL)–Neomycin (10 mg/mL) (PSN) antibiotic mixture.
5. Complete MEM contains 5 % (v/v) FBS and 1× antibiotics (PSN).
6. 2× Temin's Modified Eagle's Media without phenol red (TMEM).
7. 1 M HEPES, use at 10 mM to buffer complete MEM for use with roller bottles without CO_2 buffering.

8. Hank's balanced salt solution (HBSS).
9. Dulbecco's Phosphate-Buffered Saline, with calcium and magnesium, without phenol red (DPBS) for washing cells prior to dissociation.
10. 1× Trypsin–EDTA cell dissociation reagent.
11. HuTK- 143B (ATCC: #CRL8303) human osteosarcoma cells for both positive and negative selection of thymidine kinase expression by vaccinia virus.
12. 2 % (w/v) Low gelling temperature agarose prepared in water, cell culture tested.
13. Lipofectamine 2000 Transfection Reagent (Invitrogen).
14. X-gal (20 mg/mL).
15. HAT (hypoxanthine–aminopterin–thymidine) supplement (50×): 5 mM hypoxanthine, 20 μM aminopterin, 0.8 mM thymidine.
16. 10 mg/mL mycophenolic acid (400×) dissolved in ethanol.
17. 10 mg/mL xanthine (40×) dissolved in 0.1 M NaOH.
18. 6× well culture dishes TC surface treated and CellBIND surface (Corning) for chicken embryo skin cells.
19. A range of T25, T75, T175 flasks and multi-well plates for tissue culture.
20. 850 cm^2 roller bottles with CellBIND surface (Corning).
21. Roller bottle incubator, 37 °C.
22. Standard Tissue culture incubator, 37 °C, 5 % CO_2, humidified.
23. SPF fertile chicken eggs.
24. Egg incubator with wet bulb thermometer.
25. Small battery powered torch.
26. Small dissection scissors, medium and large forceps, sterilize by autoclaving.
27. 5 mg/mL Collagenase from *Clostridium histolyticum* Type VIII dissolved in HBSS, filtered.
28. 1 M $CaCl_2$, filtered.
29. Four layers of gauze loosely attached across the top of a glass funnel, sterilize by autoclaving.
30. Sterile 1-L bottles and flasks for chick embryo skin cell isolation.
31. Sterile disposable 250 mL centrifuge bottles and 50 mL tubes.
32. 1 % (w/v) Virkon S disinfectant freshly prepared.
33. 0.1 % (w/v) crystal violet in 20 % (v/v) ethanol.

2.2.1 PCR Confirmation of Recombinant Viruses

1. InstaGene matrix (Bio-Rad) (*see* **Note 5**).
2. HotStar Taq Master Mix Kit (QIAGEN).
3. Oligonucleotide primer pairs, 10 μM each.
   - IL-13Rα2 primers as above for RT-PCR
   - FPVTK-F 5′-GGTAATAGGTATAGACGAGGCTC
   - FPVTK-R 5′-CTAATATCGAACTCCATTCCGTG
   - VV-F 5′-GAGCAGCTTCGTCCACGTACACCGC
   - VV-R5′-GGTTCTTATAATTGTAACATCCTTCTCTTCC
4. PCR machine with heated lid suitable for 200 μL thin walled tubes.
5. Mini-gel DNA electrophoresis apparatus, TAE, agarose, DNA stain, DNA size standards.

2.2.2 Immuno-blotting

1. Equipment and culture media for tissue culture and viruses
2. 0.1 μM filters.
3. Mini-gel protein electrophoresis tank and power supply.
4. Mini-gel western transfer tank and cassettes.
5. Precast polyacrylamide gels, e.g., 4–20 % LongLife Tris–HEPES gels (NuSep).
6. 2× loading buffer: 4 % (w/v) SDS, 20 % (v/v) glycerol, 0.01 % (w/v) bromophenol blue, 125 mM Tris–HCl pH 6.8, 2 % (v/v) 2-mercaptoethanol.
7. 10× Protein electrophoresis buffer: 1 M Tris-base, 1 M HEPES, 1 % (w/v) SDS.
8. Western Transfer buffer: 25 mM Tris base, 192 mM glycine, 10 % (v/v) methanol.
9. Phosphate buffered saline (PBS): 10 mM Na-phosphate pH 7.2, 0.9 % (w/v) NaCl.
10. Complete Protease Inhibitor EDTA free tablets (Roche).
11. Precut 0.45 μm PVDF membranes.
12. Blocking buffer: PBS containing, 0.5 % (w/v) Skim Milk Powder, 0.05 % (v/v) Tween-20.
13. Primary antibody, goat anti-mouse IL13Rα2 (R&D Systems)
14. Secondary antibody, anti-goat biotin conjugate (Sigma).
15. Streptavidin–horseradish peroxidase conjugate (Amersham).
16. Western Lightning™ Chemiluminescence Reagent Plus (PerkinElmer).
17. LAS 1000 Luminescence Image Analyzer (Fujifilm).

2.3 Heterologous Prime-Boost Immunization

1. BALB/c (H-2d) (female). Control and treatment groups (n = 5–8).
2. Isoflurane anesthetic and machine.
3. Concentrated plasmid preparations of pHis-64 and pHis-HIV-B [26], prepared using a midi or maxi-scale plasmid isolation kit.
4. Wild-type vaccine strain FPV-M3 control and recombinant FPV-HIVgag/pol (FPV-086) [25], FPV-HIV-IL13Rα2 [21], VV-HIVgag/pol (VV336) [25], and VV-HIV-IL13Rα2 [21] stocks prepared as described above.
5. Bath sonicator.
6. 1 mL insulin syringes with 10 μL graduations.
7. Phosphate buffered saline.
8. P40 air displacement pipette and 20 μL tips for intranasal inoculations.

2.4 Sample Collection and Lymphocyte Preparation

1. Biological Safety Cabinet Class II.
2. PIPETBOY.
3. Air displacement pipettes and associated tips.
4. Centrifuge and adaptors to hold 50 mL centrifuge tubes.
5. 1 mL insulin syringes with 10 μL graduations fine needle attached.
6. Scissors and forceps sterilize by autoclaving.
7. Sterile stainless steel sieves or disposable cell strainers (Falcon).
8. Plastic plunger inserts from a 1 mL syringe, rubber stopper removed. Sterilize by autoclaving.
9. Sterile gauze.
10. Sterile 50 mL tubes.
11. Freshly prepared complete RPMI-1640 supplemented with 10 % (v/v) heat-inactivated FBS, 2 mM sodium pyruvate, 2 mM l-glutamine , 57 μM 2-mercaptoethanol, 25 mM HEPES, 1× PSN antibiotics. Store at 4 °C.
12. TRIS Red Blood Cell Lysis Buffer: 144 mM NH_4Cl, 17 mM Tris–HCl pH 7.6. Sterilize through a 0.22 μm filter and store at 4 °C. Check pH after filtering as filtering may increase the pH.
13. Sterile petri dishes.
14. Sterile transfer pipettes.
15. Sterile 10 and 25 mL pipettes.
16. Hemocytometer.

2.5 Immunological Assays

1. Fluorescent labelled cytokine of interest (IFN-γ-FITC, TNF-α-PE, IL-2-Pacific Blue) BD Pharmingen, BioLegend, or e-Biosciences (*see* **Notes 6** and 7).

2. Fluorescent labelled CD8 and CD4 (CD8-APC, CD4-PerCP), obtained same suppliers as above (*see* **Notes 6** and 7).
3. Fluorescent labelled homing markers CCR9, α4B7 with appropriate fluorochromes, obtained same suppliers as above (*see* **Notes 6** and 7).
4. APC or PE labelled MHC-1 $K^dGag_{197-205}$ tetramer or any tetramer of interest (*see* **Note 8**).
5. Isotype controls (BD Pharmingen, BioLegend, or e-Biosciences).
6. Cells of interest (from spleen, lymph nodes, Peyer's patch).
7. Complete RPMI-1640 Sigma or Invitrogen.
8. Phosphate buffered saline.
9. FACS buffer: PBS, 1 % (v/v) heat-inactivated FBS.
10. Paraformaldehyde PBS buffer: PBS, 0.5 % (v/v) paraformaldehyde.
11. Cell activators (e.g., 1–5 μg /mL ConA as positive control, HIV Peptides, HIV Proteins).
12. IC Block = 2 mM Monensin or 1× Brefeldin A, from BD Pharmingen, BioLegend, or e-Biosciences.
13. IC Fix from BioLegend.
14. IC Perm from BioLegend.
15. 96-well round bottom culture plates.
16. Pipettes and tips.
17. Small FACs tubes/and rack.
18. Ice bucket.
19. Vortex.
20. Centrifuge with 96-well plate adapter.

3 Methods

We will expand here the main methods used to construct and characterize recombinant poxviruses expressing vaccine antigens and immune-modulator products such as cytokines or soluble cytokine inhibitors as recently described in Ranasinghe et al. [21]. The construction of the recombinant DNA vaccine is described briefly in this article, as pHIS HIV-B vaccine construction was discussed in detail in a previous edition of Molecular Methods [26].

3.1 Vaccine Construction

3.1.1 DNA Vaccines

The DNA vaccine construct contains approximately 65 % of the HIV genome, without compromising the safety of the vaccine. The HIV-1 genome was modified from the B subtype pNL(AD8) provirus and inserted into the plasmid DNA vaccine vector pHIS-64 (Coley Pharmaceuticals) containing kanamycin-selection,

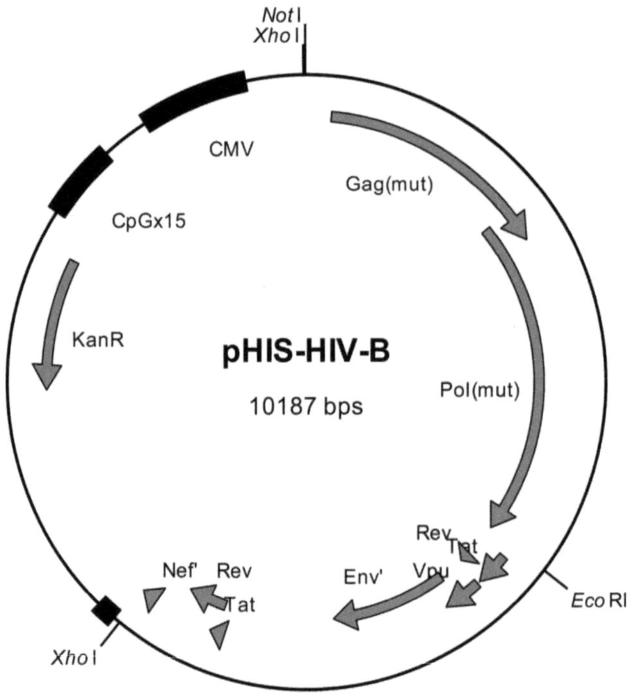

Fig. 1 Schematic map of pHIS-HIV-B. The DNA vaccine contains approximately 65 % of the HIV genome (clade B) from pNL(AD8) provirus including specific mutation and removal of the RNA-binding zinc finger motifs from the nucleocapsid region of Gag, abolishing reverse transcriptase (RT) activity, RNase H activity, and integrase activity from Pol, truncation of Env and Nef and the full sequences for Vpu, Tat, and Rev. Deletion of the coding regions for Vpr, Vif, and the LTRs

a CMV promoter, the bovine growth hormone polyA signal, and 15 primate-optimized CpG immunostimulatory sequences [26]. The final DNA vaccine pHIS-HIV-B (Fig. 1) contains sequences expressing modified Gag, modified RT (reverse transcriptase), Protease, Rev, Tat, Vpu, truncated Nef (the first 31 codons), and truncated Env (the first 275 codons only). HIV-1 genes for Integrase, LTRs, Vif, Vpr that are known to be of risk were deleted. The pHIS-HIV-B insets were fully sequenced in both directions using ABI Sequencing (Biomolecular Resource Facility, The John Curtin School of Medical Research, ANU, Canberra). The plasmid was amplified and manufactured under GMP conditions (Qiagen, Germany). These rDNA vaccines were prepared as described previously [26].

3.1.2 Isolation of Mouse IL-13Rα2 cDNA

Mouse spleen total RNA was isolated using the RNeasy Protect Mini Kit (QIAGEN) using methods for extraction of total RNA samples from animal tissues as recommended by the manufacturer. Reverse transcription PCR (RT-PCR) was conducted using the QIAGEN Onestep RT-PCR kit. The choice of oligonucleotide primers to copy and amplify the mouse IL13Rα2 cDNA [21] was based on the sequences available in Genbank. The primers were designed to match the natural sequence of the mouse gene (*see* **Note 1**).

1. Remove mouse spleens from freshly euthanized mice and store in RNAlater at −20 °C.
2. Homogenize 10 mg of tissue in a 1.5 mL microcentrifuge tube using a micro-pestle in 600 μL Buffer RLT contained in the RNeasy Protect Mini Kit, followed by DNA shearing using a 20 gauge syringe. Purify total RNA using the supplied columns and solutions according to the manufacturer's instructions, store RNA samples at −20 °C.
3. Mouse IL13Rα2 cDNA is isolated by RT-PCR using the gene-specific forward and reverse primers and the QIAGEN OneStep RT-PCR Kit as recommended by the manufacturer. Reverse transcription 50 °C 30 min; PCR amplification, initial activation 95 °C 15 min; 40 cycles 95 °C for 1 min, 55 °C for 1 min, 72 °C for 1 min; final extension 72 °C 10 min.
4. Separate the PCR cDNA fragments by gel electrophoresis using 1 % (w/v) agarose, in 1× TAE buffer and a mini-gel apparatus. Include DNA size standards in a separate well.
5. Visualize the PCR cDNA fragments using a transilluminator. The mouse IL-13Rα2 cDNA will migrate as two species; 1.1 kb and a 1.0 kb (splice variant deleted for exon 10). Excise and gel fragments containing the desired PCR cDNA bands using a sterile scalpel blade and place in a 1.5 mL tube.
6. Purify the PCR DNA fragments using an agarose gel DNA purification kit and ligate into the PCR cloning vector pDrive. The ligated plasmid DNA is used to transfect competent *E. coli* and plated onto LB agar plates containing 50 μg/mL ampicillin. Use 40 μl each of X-gal and IPTG stock solutions per plate mixed and spread with the bacterial suspension to allow selection using blue-white screening.
7. Screen "white" *E. coli* colonies for cloned DNA by isolation of the plasmid DNA using a plasmid isolation kit. Clones containing the correct sized DNA fragments are identified by restriction enzyme digestion with *Bgl*II and *Sac*I followed by agarose gel electrophoresis. The DNA fragment will insert into the PCR vector in either orientation.
8. Confirm cloned cDNAs for sequence identity and integrity by DNA sequencing.

3.1.3 Subcloning IL-13Rα2 cDNA into Poxvirus Vectors

For the construction of recombinant vaccinia viruses co-expressing HIV gag/pol and immune modulator genes we have used the vaccinia virus transfer vector pTK7.5A [23]. The vector TK7.5A contains part of the vaccinia virus *Hin*dIII-F fragment to facilitate

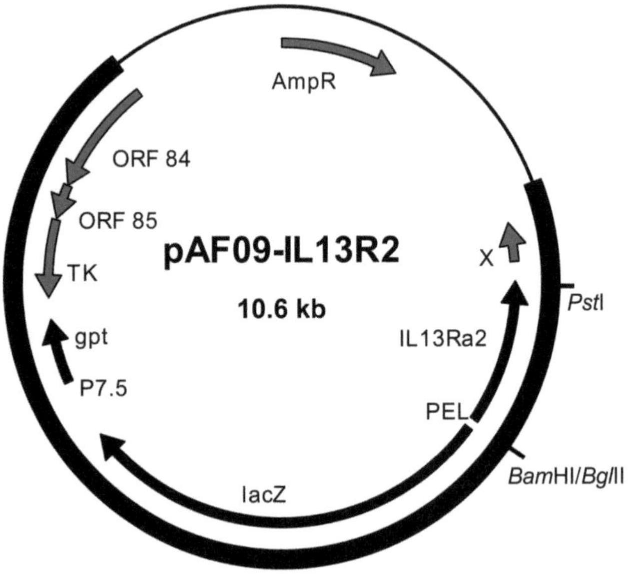

Fig. 2 Schematic map of pAF09-IL13Rα2. The FPV vector pAF09 contains a region of the viral genome encoding the viral *tk* gene. The *E. coli gpt* and *lacZ* genes and a multiple cloning site are inserted between the ORFs 86 (*tk*) and 87 (X ORF). The mouse IL13Rα2 cDNA is ligated between the *Bam*HI and *Pst*I sites of the vector, so it is under the transcriptional control of the bidirectional FPV early/late promoter (PEL) in the recombinant virus

homologous recombination with the vaccinia virus genome and the selectable Herpes simplex *tk* gene (*see* **Note 3**). For construction of recombinant fowlpox viruses, the vector pAF09 (Fig. 2) [24] which contains a region of the FPV genome encoding the viral *tk* gene for homologous recombination (Fig. 3) is used. This vector relies on the insertion of a foreign gene adjacent to the *tk* gene without disrupting expression and selection for the bacterial *gpt* (xanthine-guanine phosphoribosyltransferase) and *lacZ* (β-galactosidase) gene expression (*see* **Note 4**).

1. Digest the pDrive-IL13Rα2 plasmid prepared above with *Sac*I at 37 °C for 1 h, followed by inactivation at 65 °C for 15 min.
2. Treat the *Sac*I digested plasmid DNA with Klenow polymerase and 33 μM each dNTPs at 25 °C for 15 min to digest the protruding 3′ end to produce a blunt end. Heat-inactivate the Klenow polymerase at 75 °C for 20 min.
3. Digest the Klenow treated DNA with *Bgl*II to excise the IL-13Rα2 cDNA from the pDrive vector. Separate the DNA fragments by agarose gel electrophoresis and purify the IL-13Rα2 DNA fragment.
4. Ligate the purified DNA fragment into pTK7.5A previously digested with *Bam*HI and *Hinc*II so that the inserted IL13Rα2

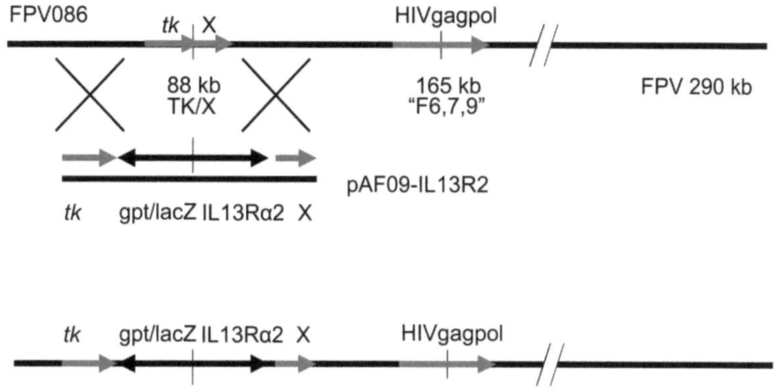

Fig. 3 Depiction of homologous recombination between FPV genome and *tk*, X genes contained in the vector pAF09 resulting in the insertion of *gpt*, *lacZ*, and IL13Rα2 genes into recombinant virus. rFPV isolated by selection for *gpt* and *lacZ* expression

gene is in the correct orientation under the transcriptional control of the VV P7.5 early/late promoter.

5. To subclone the IL-13Rα2 cDNA into pAF09 digest pDrive-IL13Rα2 with *Bgl*II and *Pst*I (located in pDrive MCS, orientation dependent) and ligated between the *Bam*HI and *Pst*I sites of pAF09 so that the gene is under the transcriptional regulation of the FPV early/late promoter.
6. Transfect competent *E. coli* cells with the ligated plasmid DNA and plate onto LB agar plates containing 50 μg/mL ampicillin. Isolate and characterize clones for the correct DNA insert.
7. Prepare plasmid mini-prep DNA starting with a 5 mL overnight bacterial culture since the plasmids are large and low copy number.

3.2 Cell Culture and Recombinant Virus Isolation

While vaccinia virus has a broad host range and will grow on a range of continuous tissue culture cells derived from multiple species, FPV is restricted and will only grow in cells of avian origin. FPV preferentially grows on epithelial cells rather than fibroblasts; we describe below a method for preparation of primary chicken embryo skin cells using collagenase to release the cells from the embryos.

3.2.1 Isolation of Chick Embryo Skin Cells

1. Incubate 30 fertile chicken eggs in an egg incubator at 55–60 % humidity, 37.5 °C, 31 °C wet bulb thermometer with 2 hourly rotation for 13 days.
2. Visualize the airspace (candling) by shining a torch light through the egg in a darkened room. Using a pencil make a circle on the outside of the egg indicating the air space. Discard any unfertilized, cracked or contaminated eggs.

3. The eggs are washed in a warm solution of 1 % (w/v) Virkon S to remove any debris and disinfect the outer egg surface.
4. Inside a Biological Safety Cabinet Class II, aseptically remove the eggshell covering the air space with sterile scissors and forceps, preferably leaving the inner membrane intact, removing any stray pieces of shell.
5. Embryos are aseptically removed from the eggs through the air space inner membrane using a large pair of sterile forceps and placed into a sterile 1-L bottle. Leave the yolk, liquids and membranes within the egg and discard.
6. The embryos (usually 20–25) are rinsed twice with 250 mL sterile pre-warmed HBSS to remove transferred yolk and blood cells.
7. The chick embryos are then incubated in 500 mL pre-warmed HBSS containing 10 mL 5 mg/mL Collagenase and 2.5 mL 1 M $CaCl_2$ with gentle agitation at 37 °C for 40 min to release skin cells. Extended incubations will liberate a greater percentage of fibroblast cells relative to epithelial cells.
8. The embryo remains are removed from the cell suspension by passage through 4 layers of sterile gauze in a glass funnel and the filtered cell suspension collected into a sterile flask. Discard the remains of the chick embryos.
9. Transfer the cell suspension to sterile disposable 250 mL centrifuge bottles and gently centrifuge at $230 \times g$ for 10 min at room temperature to recover the cells.
10. Gently suspended the cells in 100 mL of pre-warmed HEPES buffered complete MEM and dispense into four 850 cm^2 roller bottles with CellBIND surface at approximately 1×10^8 cells per bottle. Adjust media volume to 200 mL with HEPES buffered complete MEM with antibiotics.
11. After overnight incubation at 37 °C in a roller incubator, replace the media to remove non-adhered cells, debris and blood cells. The CES cells should be confluent within 48–72 h incubation. The CES cells should be mainly epithelial cells upon isolation; however, there will be a significant percentage of fast growing fibroblasts.
12. When confluent the CES cells can be seeded into smaller flasks (~5×10^7 cells per T175) or 6× well plates (~1×10^6 cells per well). Avoid seeding at low densities as fibroblasts will quickly dominate the cell population.
13. Aseptically decant the media from the roller bottle into a waste beaker. Wash the monolayer with 10 mL pre-warmed DPBS by gently rotating the bottle and remove the DPBS using a pipette. Dislodged the cells in 10 mL Trypsin–EDTA in a 37 °C roller incubator until the cells are no longer clumped, 5–10 min.

Suspend the cells in complete MEM, count and seed new TC flasks. A confluent roller bottle will contain between $2–3 \times 10^8$ cells.

14. CES cells can be maintained in HEPES buffered complete MEM in roller bottles with CellBIND surface at 37 °C for up to a week, although should be further subdivided and media replenished if cell rounding occurs. We maintain CES cells for 2 months in CellBIND roller bottles with regular weekly splitting (1:2 or 1:3) back into the 850 cm^2 roller bottle and feeding with fresh media.

3.2.2 Construction of Recombinant Poxviruses

Since the 1980s Boyle et al. have constructed a number of vaccinia virus and fowlpox virus specific vectors and developed methods for the generation and isolation of recombinant poxviruses co-expressing multiple antigens or immunomodulators [23–25, 27] (*see* **Note 9**). Recently, we have made extensive use of these vectors and methods for the construction of recombinant poxviruses expressing HIV genes. The recombinant vaccinia virus VV336 contains a disruption of the vaccinia virus *tk* gene due to the insertion of the HIV gag/pol (AE clade) gene sequences using the vector pJmcs [25] and is therefore TK^- (*see* **Note 10**). We have used the recombinant FPV co-expressing the HIV gag/pol and/or env genes from either the "F6,7,9" or "REV" sites [25] while still maintaining the wild-type TK/X insertion site enabling the use the vector pAF09 [24] for subsequent manipulations. In Ranasinghe et al. [21] we used FPV086 as the parent virus which contains the insertion of the HIV gag/pol (B clade) gene into the "F6, 7, 9" insertion site.

1. All cell culture and live viral procedures described should be conducted within a Biological Safety Cabinet Class II (*see* **Note 11**). Similar methods are used to construct recombinant vaccinia and fowlpox viruses, although different cells and selection methods are required to isolate the recombinant viruses.
2. Start by preparing HuTK– 143B cells for VV, or CES cells for FPV, by seeding T25 flasks with 5×10^6 cells and grow to near confluence by incubation at 37 °C, 5 % CO_2 in complete MEM.
3. The confluent cells are infected with poxvirus (VV336 or FPV086) at a multiplicity of infection (MOI) of 0.01 plaque forming units (PFU) per cell diluted in 500 μL MEM and incubated for 1 h at 37 °C with regular gentle mixing to ensure the monolayer does not dry out. Add 3.5 mL of complete MEM (without antibiotics).
4. While the cells are being infected prepare: (1) 15–20 μL (2 μg) of plasmid mini-prep DNA mixed with 500 μL MEM (without serum and antibiotics); (2) 20 μL Lipofectamine 2000 reagent mixed with 500 μL MEM (without serum and antibiotics);

(3) After 5 min incubation at room temperature mix the two samples and incubated for a further 20 min to allow liposome-DNA complexes to form.

5. The liposome mixture is added to the cells previously infected with poxvirus and gently mixed, total volume is 5 mL. Incubate at 37 °C, 5 % CO_2. CES cells are sensitive to Lipofectamine 2000, so replenish the media after overnight incubation with complete MEM.
6. After 48–72 h post-transfection, when the cells have reached near 100 % cytopathic effect (CPE). The culture cells are dislodged using a cell scraper, frozen and thawed, and aliquots disrupted by sonication for 15 s to release virus particles.

3.2.3 Isolation of Recombinant Vaccinia Virus

1. Confluent monolayers of HuTK– 143B cells in 6-well culture dishes are separately infected with several small aliquots, 10–100 μL, of sonicated virus suspension in a total of 500 μL MEM per well for 1 h at 37 °C. Remove virus inoculum.
2. The infected cells are overlaid with 2 mL MEM containing 1× HAT supplement to select for growth of recombinant viruses co-expressing the HSV*tk* gene. When the infected cells have reached near 100 % CPE they are again frozen, thawed, and sonicated to release progeny virus.
3. The above selection process is repeated until a stock containing visible viral plaques growing on HuTK– 143B cells is produced.
4. To plaque purify the recombinant VV the infected cultures are frozen and thawed and an aliquot treated with an equal volume of trypsin dissociation reagent and incubated at 37 °C for 30 min prior to sonication for 15 s.
5. Serial 10-fold dilutions (400 μL) of virus suspension in MEM are used to infect confluent HuTK– 143B cells in 6-well dishes for 1 h and the inoculum removed.
6. Each well is overlaid with 2 mL of a prepared media consisting of 1 mL of 2× TMEM (warmed to 37 °C), 1× HAT supplement, and 1 mL 2 % (w/v) low gelling temperature agarose (at 37 °C).
7. Allow the agarose overlay to set at room temperature and then incubated at 37 °C, 5 % CO_2 until plaques are visible (24–48 h).
8. Well-separated individual plaques are picked using a sterile 1,000 μL tip attached to a pipette set at 100 μL, by stabbing through the agarose, scraping the plaque and immediate cells, and gently sucking back into the tip. The agarose plug is resuspended in 200 μL complete MEM and stored at −20 °C.
9. The agarose plug/MEM is sonicated for 15 s and several 5–50 μL aliquots plated as above to obtain individual plaques.

This process is repeated at least three times to obtain plaque purified virus.

10. Confirm recombinant virus by PCR for the desired insert and absence of wild-type virus insertion site.

3.2.4 Isolation of Recombinant Fowlpox Viruses

1. Several small aliquots of the culture sonicate from the transfected cells are used to infected confluent monolayers of CES cells in either T25 flasks or 6-well plates with MEM containing 25 μg/mL mycophenolic acid, 250 μg/mL xanthine, 1× HAT supplement to select for recombinant FPV co-expression of the *E. coli gpt* gene.
2. The selection procedure should be repeated several times to amplify *gpt*+ recombinant FPV.
3. To plaque purify the recombinant FPV the infected cultures are frozen and thawed, an aliquot sonicated for 15 s and serial tenfold dilutions prepared in complete MEM.
4. 400 μL of the dilutions are used to infect confluent monolayers of CES cells in 6× well culture dishes for 1 h at 37 °C. The inoculum is removed.
5. The cells are overlaid with 2 mL 1× TMEM, 0.2 % FBS, 1 % (w/v) low gelling temperature agarose (at 37 °C). Allow the overlay to set at room temperature and incubate at 37 °C for 4–5 days to allow the plaques to develop.
6. The plaques containing recombinant viruses are identified by adding a second 2 mL 1 % agarose, 1× TMEM overlay containing 200 μg/mL X-gal on top of the first overlay. Incubate for a further 24–48 h to visualize blue FPV plaques due to co-expression of the *lacZ* gene.
7. Pick plaques as described above and plaque purify three times.

3.2.5 PCR Confirmation of Recombinant Viruses

To determine if the putative recombinant viruses contain insertion of the desired foreign gene and were not contaminated with parent virus, two PCR reactions were performed: first, using gene specific primers to detect the presence of the inserted gene, and second, using a primer set which flanks the insertion site on the parent virus to detect wt viral DNA sequences. If the plaque purified (at least three times) virus stocks are positive for the insert but negative for the wt viral sequences they are considered to be plaque purified. The IL13Rα2 primer pair used is the same as for RT-PCR giving PCR product of 1.0 kb for IL13Rα2Δ10. Using the suggested primers the wt FPV genome will give PCR a product of 396 bp, while a purified recombinant FPV will be unlikely to generate a product due to size of the insert. The VV-F region the primers will generate a PCR product of 402 bp for wt VV.

1. Plaque purified (three times) recombinant viruses are used to infect confluent CES or 143B cells in either 6× well plates or T25 culture flasks until near 100 % CPE was achieved.
2. Recover infected cells into the culture media using a cell scraper or a plunger from a 1 mL syringe and transfer to a sterile tube.
3. Cells are mixed using a vortex mixer to disperse the clumps.
4. 100 μL of cell/virus suspension is transferred to sterile 0.5 mL microcentrifuge tube and the cell debris recovered by centrifugation and the media carefully removed without disturbing the pellet.
5. Flick the tube to suspend the pellet in the small amount of media left behind.
6. Add 100 μL of InstaGene matrix suspension using a 1,000 μL tip and mix.
7. Incubate at 56 °C for 15 min, and 100 °C for 8 min with mixing before and after each incubation.
8. Centrifuged for 5 min to pellet the InstaGene resin.
9. Use 1–2 μL of the cleared solution for PCR reactions. Store the remainder of the cleared solution and InstaGene matrix pellet at −20 °C.
10. At room temperature prepare:

    12.5 μL 2× HotStar Taq Master Mix

    1 μL Forward Primer diluted to 10 μM (0.4 μM final)

    1 μL Reverse Primer diluted to 10 μM (0.4 μM final)

    RNase free water to 25 μL

    1–2 μL virus DNA prepared using InstaGene matrix
11. PCR program: 15 min 95 °C; 35 cycles 30 s 94 °C, 30 s 55 °C, 1 min 72 °C, 10 min extension 72 °C.
12. Analyze 10 μL PCR reaction by DNA agarose gel electrophoresis.

3.2.6 Immuno-blotting

The recombinant poxviruses were further characterized for expression of the desired antigen or immunomodulator by immuno-blotting to ensure no mutation or deletion occurred during the recombination events, which prevent expression.

1. Recombinant poxviruses are cultivated by infecting confluent CES or 143B cells using 6× well plates at a MOI of 1 PFU/cell. Incubate with serum-free MEM to reduce serum albumin in the final samples.
2. When 100 % CPE is observed, VV 24–48 h or FPV 4–6 days, the infected cells are recovered by scraping the cells into the culture media using the plunger from a sterile 1 mL syringe.

3. The cells and media are transferred to a centrifuge tube and clarified at 930 × *g* for 5 min at room temperature.
4. Media is retained by gentle decanting without disturbing the cell pellet and filtered through a sterile 0.1 μm filter to remove cell debris and virus particles.
5. Cell pellets and filtered media are stored at −20 °C.
6. The cell pellets (approximately 2×10^6 cells from a 6-well plate) are suspended in 100 μL of PBS containing 2× Complete protease inhibitor. An equal volume of 2× LongLife Gel loading buffer is added and mixed by vortexing. The viscosity of the suspension can be reduced by shearing the DNA using a 26G syringe needle.
7. 20 μL of the filtered media samples is mixed with 20 μL 2× loading buffer.
8. Protein and loading buffer samples, 5–10 μL cell pellets or 40 μL filtered media, are heated at 100 °C for 5 min to denature proteins. Spin in a microcentrifuge for 2 min to collect solution in the bottom and pellet any insoluble material.
9. Proteins are separated using a protein mini-gel electrophoresis apparatus and precast 4–20 % LongLife Tris–HEPES gels with 10 wells. Electrophoresis conditions: 150 V for 30–45 min using a Tris–HEPES–SDS buffer.
10. Proteins are transferred to PVDF membranes using a standard wet tank transfer apparatus and Western transfer buffer at 8 V/cm inter-electrode distance (40 V) for 1–2 h.
11. Completely dry the membranes at room temperature.
12. Air-dried PVDF membranes (do not pre-wet with methanol or blocking buffer) containing transferred proteins are incubated for 1 h with primary antibody (0.2 μg/mL) diluted in blocking buffer.
13. Wash membranes twice in PBS and then incubate for 30–60 min with secondary antibody anti-goat biotin conjugate at 1:1,000 dilution in blocking buffer.
14. Wash membranes twice in PBS and incubate with 1:1,000 dilution of streptavidin–horseradish peroxidase conjugate diluted in PBS for 30 min.
15. Wash membranes twice in PBS, and develop using a chemiluminescence reagent and capture images using a Luminescence Image Analyzer.

3.2.7 Poxvirus Vaccine Stock Preparation

The aim of preparing high-titer poxvirus stock is to harvest the cell-associated forms of the virus, rather than the dilute forms released into the media. Stocks can be prepared in either T175 flasks or roller bottles. We have found that higher-titer viral stocks are usually recovered by growing cells on a flat surface constantly submerged in culture media (*see* **Note 12**).

1. Confluent monolayers of HuTK- 143B (for VV) or CES (FPV or VV) cells are prepared by seeding 3×10^7 cells per layer of a single T175 or 3-layer (525 cm^2) flasks. Cells should be confluent and ready to infect following overnight incubation at 37 °C.
2. Infect cell monolayers at an MOI of approximately 1.0 PFU/cell. Mix an equal volume of virus stock with trypsin reagent and incubate at 37 °C for 30 min, then sonicate for 15 s to disperse viral clumps.
3. Use approximately 5–10 mL of prepared virus mixed with complete MEM per flask layer to evenly cover the cell monolayer. Gently rock the flasks every 15 min to disperse the inoculum across the monolayer.
4. Incubate at 37 °C for 1 h.
5. Add complete MEM to 40–50 mL per flask layer.
6. Incubate at 37 °C, 5 % CO_2 until 100 % CPE is observed.
7. FPV infected CES cells, are ready to harvest when cells detached as a sheet 4–5 days post infection.
8. Vaccinia virus the cells are ready to harvest 48–72 h post infection, using a cell scraper to dislodge cells into the culture media.
9. Transfer the cells and media to disposable centrifuge bottles or tubes and pellet the infected cells at $2{,}000 \times g$ for 10 min.
10. Discard the culture media and resuspend the cell pellet in DPBS. For FPV resuspend the cells from a triple-layer flask in 1 mL DPBS. For VV resuspend in 1 mL per T175 flask.
11. Sonicate the virus suspension three times for 15 s and vortex, until cell clumps are no longer visible.
12. Dispense aliquots in cryovials and store at −70 °C.
13. We routinely achieve FPV stock titers between 5×10^8 to 1.0×10^9 PFU/mL using this method. Vaccinia virus stock titers are usually greater than 10^9 PFU/mL. (These high-titer stocks can be used for intranasal vaccinations).

3.2.8 Titration of Poxviruses

We have found that CES cells adhere poorly to plates and flasks with standard TC treated surfaces. Specifically cells can detach after the virus inoculum or overlay is applied. In contrast, multiwell plates with CellBIND surface (Corning) have shown consistent results with the cells being healthier and adhering for a week or longer of culture, which is necessary to observe FPV plaque formation. Regular TC coated surfaces suffice when using HuTK- 143B cells for vaccinia virus titrations.

1. CES (for FPV or VV) or HuTK– 143B (for VV) are subcultured at 1×10^6 cells per well of a 6× well tissue culture plate (use CellBIND surface for CES cells). The cells should be confluent and ready for plaque titrations after overnight incubation at 37 °C, 5 % CO_2.
2. To prepare virus stocks for titration, a 50 μL aliquot is mixed with an equal volume of trypsin reagent and incubated at 37 °C for 30 min.
3. The virus suspension is further disrupted by sonication for 10–15 s to disperse viral clumps.
4. The volume is adjusted to 500 μL (10^{-1} dilution) with complete MEM. Serial tenfold dilutions of the virus stock are prepared in complete MEM.
5. Remove the media from the confluent cells in the 6× well plates.
6. A standard 400 μL volume of 10^{-6} to 10^{-8} dilutions for FPV, or 10^{-8} to 10^{-10} for VV, is applied to the cell monolayers in duplicate and incubated for 1 h at 37 °C. This is enough liquid to evenly cover the cells without drying during the incubation. Every 10–15 min gently rock the plates to disperse the virus suspension across the wells.
7. After 1 h the viral inoculum is removed and 2 mL of complete MEM applied.
8. Incubate the plates at 37 °C, 5 % CO_2 without moving them so as not to disperse virus which will result in inaccurate titers.
9. For VV, plates are incubated for 48 h and then fixed/stained by carefully removing the media and applying approximately 0.5 mL/well of 0.1 % (w/v) crystal violet in 20 % ethanol to cover the cells for 30–60 s.
10. The stain is removed and the plates inverted and air-dried inside a Class II Biological Safety Cabinet.
11. The vaccinia virus plaques should be visible as cleared areas in the stained monolayer.
12. For FPV, after 5–7 days incubation at 37 °C, foci of infection should be visible when stained as above. FPV plaques will appear as irregular spots on the stained monolayer.
13. Count plaques on wells containing between 20 and 100 plaques and calculate viral titer using the formula: 2.5 × average plaque count/well × 1/virus dilution = PFU/mL [e.g., $2.5 \times 40 \times 1/10^{-7} = 1.0 \times 10^9$ PFU/mL].

3.3 Heterologous Prime-Boost Immunization

Studies have clearly established that the route and the vector combination can induce varying immune outcomes. Here we describe vaccine strategies that can induce long lasting mucosal and systemic immunity.

Table 1
Prime boost vaccine strategies that induce mucosal immunity

	Prime	Boost
1	i.m. pHIS-HIV	i.n. FPV-HIV
2	i.m. pHIS-empty	i.n. FPV-empty
3	i.n. FPV-HIV-13Rα2	i.m. VV-HIV-IL-13Rα2
4	i.n. FPV-HIV	i.m. VV-HIV

The parent rFPV and rVV constructs encoding HIV-1 gag/pol antigens originated from (FPV-HIV 086 and VV-HIV 336, respectively) [25]. i.n. = intranasal, i.m. = intramuscular. The immune outcomes observed with the strategies 1 and 2 are described in [17, 20]. The immune outcomes observed following IL-13Rα2 adjuvanted vaccines are described in Ranasinghe et al. [21]

3.3.1 Recombinant DNA/FPV Prime-Boost Immunization Strategy to Induce Mucosal and Systemic Immunity

1. BALB/c (H-2d) female mice aged 6–8 weeks are divided between control and treatment groups (5–8 per group).
2. Mice in the treatment group are primed by injecting pHIS-HIV-B plasmid expressing HIV gag/pol genes (50 μg in 100 μL PBS) 50 μL into each quadriceps muscle (i.m.) under mild isoflurane anesthetic on weeks 0 and 2.
3. Immediately prior to immunization the rFPV is sonicated 3× 20 s to obtain a homogeneous viral suspension. Sonication is critical to obtain effective virus uptake (*see* **Note 13**).
4. Mice are boosted with 1×10^7 PFU of FPV-gag/pol vaccine at weeks 4–6 weeks intranasally (i.n.) with 10 μL per nostril using a pipette (Table 1—strategy 1).
5. Mice in the control group are immunized in the same manner with plasmid DNA and FPV not containing any HIV-1 gene inserts (Table 1—strategy 2) [20] (*see* **Note 14**).

3.3.2 Recombinant FPV/VV Prime-Boost Immunization Strategy to Induce Mucosal and Systemic Immunity

1. BALB/c (H-2d) mice aged 6–8 weeks are divided between control and treatment groups (n = 5–8 per group).
2. Immediately prior to delivery rFPV and rVV are sonicated 3× 20 s to obtain a homogeneous viral suspension (*see* **Note 13**).
3. Mice in the treatment group are primed intranasally with FPV co-expressing HIV gag/pol and IL-13Rα2 (1×10^7 PFU in PBS) 10 μL per nostril under mild isoflurane anesthetic and 2 weeks later boost i.m. with 1×10^7 PFU of VV-gag/pol–IL-13Rα2 (Table 1—strategy 3).
4. Mice in the control group were immunized in the same manner with FPV-HIV and VV-HIV control vaccines (Table 1—strategy 4) [20, 21] (*see* **Note 15**).

3.4 Sample Collection and Lymphocyte Preparation

To evaluate effector and memory CD8 T cell immunity mice will be sacrificed at 2, 4, or 8 weeks post final booster immunization. To assess systemic immunity spleens and to assess mucosal immunity lymph nodes and Peyer's patch in will be collected.

1. Label two 15 mL tubes for the treatment and control groups and for each tissue to be removed. Add 5 mL of complete RPMI-1640 aseptically to one of the tubes for each of the tissues.
2. Use a fresh set of sterile scissors and forceps per group, and a sterile sieve for each set of tissues is required.
3. Euthanize each mouse just prior to tissue collection by CO_2 asphyxiation in a Class II Biological Safety Cabinet (at time points indicated above).
4. Swab the abdomen of the mouse with 70 % ethanol. Pinch up the skin on the abdomen and make a small cut in the skin using scissors. Holding each side of the cut with your fingers, pull back the skin. Using sterile scissors and forceps, cut open the body cavity to expose the organs. Locate the required tissues, e.g., spleen, caudal/lumbar lymph nodes or mesenteric lymph nodes and remove using curved forceps.
5. Place each tissue into the labelled 15 mL tube containing 5 mL complete RPMI-1640 and place on ice. Spleens can be harvested individually; however, pool the mucosal samples to obtain sufficient cell numbers to perform the assays.
6. Carefully collect the Peyer's patches from intestine, place in a petri dish, and remove all intestinal debris by washing in complete medium before placing in sample collection tube. Mucosal samples should be collected in the tubes with 5 mL complete RPMI-1640.
7. Dispose of mouse carcasses by incineration.
8. Use a Class II Biological safety cabinet for cell preparation and setting up immune assays.
9. Tip tissue and medium from the 15 mL tube onto a sterile stainless steel sieve (or disposable cell strainer). Gently push the tissue through the sieve with the flat end of a plastic plunger from a 1 mL syringe into a sterile 50 mL tube.
10. Pellet the cells by centrifugation at $140 \times g$ at 4 °C for 5 min in a bench centrifuge.
11. Discard the supernatant into a 2 L beaker containing a small amount of bleach.
12. Wash cells by resuspending the pellet in 5 mL of complete RPMI-1640 by gently tapping the side of the tube. Do not vortex the sample (spleen cells can be washed in normal RPMI-1640 without FBS).
13. Centrifuge at $140 \times g$, 4 °C for 5 min.

14. For spleen cells, resuspend pellet in 10 mL of red blood cells lysis buffer and leave on ice for 5 min. Add 20–30 mL of RPMI-1640 to dilute lysis buffer before centrifugation.
15. Both spleen and lymph node cell preparations should be washed twice by resuspending in complete RPMI-1640 and centrifuged as above.
16. The Peyer's patch samples should be passed through 3 layers of sterile gauze into a 50 mL tube to remove debris, prior to washing twice with complete RPMI-1640.
17. Resuspend the pellets in 5 mL of complete RPMI-1640 for spleens and 0.5–1 mL for lymph nodes. Perform a viable leukocyte count using a hemocytometer. Place cells on ice until required.
18. Set up T cell assays using the mouse specific antibodies as described below.

3.5 Immunological Assays

3.5.1 Intracellular Cytokine Staining (ICS)

Intracellular cytokine analysis is performed to evaluate the HIV-specific CD4 and/or CD8 T cell immune responses following vaccination. Single cytokine or multiple cytokine staining can be performed [28]. Multi-color flow cytometry analysis is normally performed to evaluate poly-functionality of the immune T cells, which is known to be a hallmark of protective immunity.

1. Day 1 Cell culture and stimulation. Add between 5×10^5–2×10^6 cells per well in 96-well culture plate in complete RPMI-1640, in a 100 μL volume.
2. Add 100 μL medium containing the activator (1–5 μg/mL of ConA, the HIV peptide or protein), mix cells by gently shaking the plate.
3. Culture cells for 15–16 h at 37 °C, 5 % CO_2 if only IFN-γ and TNF-α is measured.
4. Dilute the 2 mM Monensin 1:100 in complete medium RPMI-1640 and add 10 μL per well. (Alternatively 1× Brefeldin A can be used).
5. Incubate cells for further 4–5 h at 37 °C, 5 % CO_2, to block the cell transport.
6. When measuring IFN-γ, TNF-α, and IL-2. Culture the cells for 2 h at 37 °C and then as in step below in Monensin or Brefeldin A and incubate for further 4 h to block the cell transporters. Do not leave overnight (*see* **Note 16**).
7. Centrifuge cells at 1,200–1,400 rpm at 4 °C for 2–3 min to pellet the cells.
8. Remove supernatant—by flicking plate carefully once.
9. Wash cells with 200 μL FACs buffer. Centrifuge and pellet the cells as above and remove supernatant. Store cells on ice.

10. Surface Staining: Diluted CD8-APC and CD4-PerCP antibodies as titrated to obtain optimal staining (i.e., 1/200) in FACs buffer. Prepare enough diluted antibody for entire assay (*see* **Notes 6** and 7).
11. Add 50 μL diluted antibody per well and incubate cells on ice (4 °C), in the dark cover the ice bucket with foil for 10 min.
12. Add 100 μL FACs buffer and centrifuge as before for 2–3 min at 4 °C, remove supernatant as above.
13. Fixing cells: Add 100 μL IC-FIX to each well and incubate cells on ice (4 °C), in the dark for 10 min.
14. Intracellular cytokine staining: Spin as before for 5 min at 4 °C, remove supernatant, add 100 μL 1× IC-Perm and incubate at room temperature for 10 min. Spin as before and remove supernatant.
15. Dilute IFN-γ-FITC and TNF-α-PE as follows: Add 0.25–0.5 μL of each antibody (*see* **Notes 6** and **7**) into 25 μL 1× IC Perm mix and add 25 μL per well and incubate cells on ice (4 °C), in the dark for 30 min.
16. Add 100 μL 1× IC-Perm, spin as before, and remove supernatant.
17. Add 100 μL PBS to wash the cells, spin and remove supernatant, repeat wash.
18. Resuspend the final pellet in 100 μL paraformaldehyde PBS buffer. Transfer into FAC cluster tubes using a multichannel pipette. Store in the dark at 4 °C. Acquire data immediately or within 1–2 days.
19. For each fluorochrome used an isotype control and single color control need to be prepared alongside which will help calibrate the FACs machine
20. FACs analysis: Start the machine according to the manufacturer's instructions.
21. Pre-run the machine with PBS for 5 min to remove contaminants.
22. Set up acquisition data folder. (Always use the DATE and save first tube as 1).
23. Run unstained cells and gate the lymphocyte population on forward scatter (FSC-s) and side scatter (SCC-s) plots.
24. Run the single color controls and adjust the compensation to remove any overlap of antibodies during acquisition.
25. Once the populations are compensated. Start acquiring the data (100,000–200,000 gated events). Start from the control tubes.
26. Record the name of each saved data file in lab book if required.
27. Shut down machine according to the manufacturer's instructions.

3.5.2 MHC-I Tetramer Staining and Homing Marker Analysis

Generally MHC-I tetramer staining is performed to evaluate the number of CD8 T cells that are HIV-specific in systemic compartment spleen or mucosal compartments (iliac nodes, lung, or Peyer's patch). Here we also describe the use of homing markers α4B7 and CCR9 to assess HIV-specific CD8 T cells in the gut mucosae.

1. Aliquot systemic or mucosal cells 1×10^6–4×10^6 (prepared as in Subheading 3.4) per well in 96-well round bottom plates make up the volume to in 100 μL volume with FACs buffer.
2. Centrifuge cells at 1,200–1,400 rpm at 10 °C for 2 min to pellet the cells in a bench centrifuge.
3. Immediately remove supernatant by quickly flicking as explained before
4. Wash cells with 100 μL FACS buffer. Centrifuge and pellet the cells as above, remove supernatant. Store cells on ice.
5. *Tetramer preparation:* freshly dilute tetramer 1: 50 or 1: 100 in FACS buffer. Add 20 μL of diluted tetramer per well (*see* **Note 17**).
6. Dilute the CD8-FITC 1:100 or 1:200 in FACS buffer. Add 20 μL diluted CD8-FITC per well and incubate cells at room temperature in the dark wrapped in foil (make sure the samples are kept flat) for maximum 40 min (*see* **Note 17**).
7. If homing marker analysis is performed, first stain with tetramer for 30 min at room temperature as in **step 6**. Then add the diluted CD8-FITC and homing markers (CCR9, α4B7) in a final volume of 40 μL and incubate on ice for 30 min. Do not stain tetramer and homing markers at room temperature for 40 min this will prevent effective tetramer biding.
8. Wash cells twice with 100 μL FACS buffer.
9. Prepare isotype controls and single color controls as for Subheading 3.
10. Resuspend in 100 μL paraformaldehyde PBS buffer.
11. Acquire data using FAC machine of choice according to the fluorochromes used. Collect minimum 100,000 gated events per sample for spleen. If using mucosal samples collect at least 200,000 gated events per sample or run the whole sample.

4 Notes

1. The forward primer includes a short sequence upstream of the ATG codon, so the ORF will be in-frame with the poxvirus late promoter transcription/translation site (TAAATG) contained in the pAF09 [24] vector. In the example described, the IL13Rα2 cDNA is inserted at the *Bam*HI site of pAF09

generating the sequence TAAATG GAC Gga tct gaa **ATG** so that the IL13Rα2 methionine codon (bold) is in-frame with the poxvirus late promoter ATG (underlined). Terminal restriction enzyme sites are included in the primers to facilitate subcloning into the poxvirus vectors. A shorter forward primer was initially used; however, all cDNAs isolated contained a single base frameshift immediately adjacent to the 3′ end of the primer. A longer primer is described to cover the region of the single base deletion. A consensus Kozak sequence (GCCRCCaugG) [29] could be incorporated surrounding the ATG codon to improve translation efficiency in the final recombinant viruses.

2. Alternatively, a preexisting cDNA clone or custom synthetic DNA could be used. Characterized cDNA clones for mouse and humans are readily available from commercial suppliers. Synthetic custom DNA sequences can be made and subcloned into the poxvirus transfer vectors. The advantages of these DNA services are as follows: (1) the costs and time compared to traditional molecular methods in isolation of a base perfect cDNA clone ready for amplification and subcloning; (2) an original source of animal tissue is not required; (3) the synthetic DNA sequences can be customized to remove unwanted restriction enzyme sites or add unique sequences at the ends to enable easier subcloning of the DNA fragments; (4) the DNA sequences can be optimized to the target cell/host species to improve stability and expression; (5) specific regulatory signals (e.g., poxvirus promoters and early transcription terminators) can be co-synthesized as part of the gene cassettes; (6) the removal of cryptic poxvirus early transcription termination sequences (T_5NT) from the coding regions of genes, as this improves early gene expression which is important to stimulate T cell responses [30, 31].
3. pTK7.5A [23] utilizes a site located at the start of the "F7L" ORF to insert foreign DNA sequences. The vector contains the Herpes simplex I thymidine kinase (*tk*) gene to allow positive selection of recombinant viruses with a TK^+ phenotype in the presence of HAT (hypoxanthine–aminopterin–thymidine) supplement on TK^- cells. Aminopterin inhibits de novo synthesis of purine and pyrimidine precursors. Free hypoxanthine is salvaged by the cells to synthesize purines while free thymidine is salvaged by the viral expressed thymidine kinase to form TMP. The vaccinia virus P7.5 early/late promoter is included to drive expression of desired foreign genes (e.g., cytokine) ligated into the multiple cloning site. The short multiple cloning site contains unique sequences for common restriction endonucleases: *Bam*HI, *Sal*I/*Hinc*II, *Pst*I, *Hind*III, and *Eco*RI.

4. pAF09 [24] vector contains the *E. coli gpt* and *lacZ* genes to enable selection of the recombinant viruses. The selectable marker gene, *gpt*, encodes the enzyme xanthine-guanine phosphoribosyltransferase, which gives the viral infected cell the ability to salvage free xanthine to form xanthine monophosphate, a precursor of guanosine monophosphate synthesis. In the presence of mycophenolic acid, an inhibitor of inosine monophosphate dehydrogenase, cellular XMP and GMP become limiting inhibiting viral and cell growth. The recombinant *gpt*+virus infected cells salvage free xanthine to make XMP which is then converted by GMP synthetase to GMP enabling viral replication. HAT supplement contains aminopterin which inhibits both purine and pyrimidine synthesis improving the mycophenolic acid selection by inhibiting precursors for de novo GMP synthesis. Free thymidine and hypoxanthine are salvaged by the cells to form TMP and AMP and used for nucleic acid synthesis. The *E. coli lacZ* gene is included to enable visualization of recombinant viral plaques expressing β-galactosidase using the colorless substrate X-gal which is hydrolyzed to give an insoluble blue product. pAF09 contains a FPV early/late promoter with a consensus TAA<u>ATG</u> late gene transcription initiation site. Therefore, for maximal poxvirus late gene expression of the inserted gene, it should be in-frame and as close as possible with the methionine codon (underlined). pAF09 contains a short MCS with common restriction endonuclease sites *Bam*HI, *Sma*I, *Sal*I, *Pst*I, and *Hin*dIII.
5. InstaGene matrix is a prepared mixture of proteinase-K and Chelex resin. Protease digestion and boiling releases the viral and cellular nucleic acids, while the Chelex resin absorbs inhibitors that can interfere with PCR reactions. This quick method avoids the use and disposal of hazardous organic solvents and ethanol DNA precipitations for isolation of viral DNA.
6. During multicolor analysis fluorochromes must be selected carefully to avoid overlap of colors, this information can be obtained by visiting a Web site such as http://www.biolegend.com/spectraanalyzer.
7. When using new batches of fluorescent antibodies, first titrate each antibody to find out the best staining concentration (1/50–1/800 dilution). For best results and cost efficacy, perform all staining in minimum volumes (25–50 μL). At any time do not reuse old diluted staining material. Always use freshly diluted antibody. Sometimes adding several fluorochromes at the same time can cause problems, and this should be kept in mind when performing these assays.
8. Allophycocyanin-conjugated $K^dGag_{197-205}$ tetramers were synthesized at the Bio-Molecular Resource Facility at The John Curtin School of Medical Research, The Australian National

University. Tetramers can be made to specification from different commercial companies.

9. A detailed discussion of the generation of recombinant vaccinia viruses using alternative plasmid vectors and methods can be found in Earl et al. and companion articles [32–34]. The description of a number of alternative FPV vectors with different insertion sites and transient dominant selection methods for the construction of recombinant FPVs can be found in Boyle et al. [27]. Recombinant FPV expressing one or more HIV antigen genes from different sites, "F6, 7, 9", "REV", or TK/X sites, was described by Coupar et al. [25].
10. TK– VVs are highly attenuated and have limited replication in the mammalian host while growing normally in tissue culture cells [35]. Recombinant viruses generated using pTK7.5A express HSVtk using the PF early promoter and have an intermediate but still attenuated virulence relative to wild-type vaccinia virus [36].
11. The Australian Office of the Gene Technology Regulator (OGTR) recommends vaccination of laboratory works handling recombinant vaccinia viruses expressing immune modulators. GMO dealings with recombinant poxviruses should be conducted using certified Physical Containment 2 (PC2) facilities and work practices.
12. For FPV grown on CES cells, infected cell recovery is easily accomplished since the heavily infected cells detach from the flasks or roller bottle as a sheet into the media and can be recovered by centrifugation. For FPV we have found using 3 or 5 layer flasks convenient with the space between the layers allowing cell recovery. Attempts to use flasks with more layers have resulted in loss of infected cells due to the cell sheets becoming lodged between the layers preventing recovery. For vaccinia virus, although a significant quantity of infected cells detach, many remain attached to the culture vessel and need to be recovered using a cell scraper. We do not recommend using multi-layer flasks for vaccinia virus stock preparation due to this reason. Dissociation agents can be used to dislodge VV infected cells; however, this may result in lower recovery due to release of virus from the infected cells.
13. When sonicating virus, necessary steps should be taken not to over-sonicate the samples, which will result in reduced viral titers.
14. To induce optimum immunity during heterologous rDNA/viral prime-boost immunization, rDNA should be delivered in the priming immunization not the booster immunization. To induce effective immunity rDNA should be delivered systemically (e.g., i.m.). The i.n. delivery of rDNA does not induce

good immunity mainly due to poor uptake of rDNA by the mucosae [17, 20].

15. In a viral/viral prime-boost immunization setting, to induce effective high magnitude and high quality T cell immunity, rFPV should be given in the prime and rVV should be delivered in the booster vaccination [18, 20]. The above two points clearly demonstrate that careful selection of vector combinations plays a crucial role in inducing effective immunity to vaccine antigens. Our studies have shown that rFPV is an excellent mucosal delivery vector [17, 20]. The recombinant poxviral vectors such as rFPV and recombinant Modified vaccinia Ankara (a safer vector compared to rVV) co-expressing molecules such as IL-13Rα2 can be used to induce effective HIV-specific mucosal and systemic immunity [21].
16. When performing cytokine or chemokine analysis using intracellular cytokine staining or any other assay (antibody arrays, ELISA), the expression kinetics of each cytokines can be vastly different, and this needs to be carefully taken into consideration [12, 21, 37–39].
17. During tetramer staining, do not use previously diluted old tetramer preparations. FACS buffer used in tetramer assays should not contain azide at any time. The staining should be done at room temperature.

Acknowledgements

This work was supported by the Australian National Health and Medical Research Council project grant award 525431 (CR), Australian Centre for Hepatitis and HIV Virology EOI 2010 (CR) & 2012 grant (CR&RJJ), and Bill and Melinda Gates Foundation GCE Phase I grant OPP1015149 (CR).

References

1. Leong KH, Ramsay AJ, Boyle DB et al (1994) Selective induction of immune responses by cytokines coexpressed in recombinant fowlpox virus. J Virol 68:8125–8130
2. Ramsay AJ, Husband AJ, Ramshaw IA et al (1994) The role of interleukin-6 in mucosal IgA antibody responses in vivo. Science 264: 561–563
3. Kent SJ, Zhao A, Best SJ et al (1998) Enhanced T-cell immunogenicity and protective efficacy of a human immunodeficiency virus type 1 vaccine regimen consisting of consecutive priming with DNA and boosting with recombinant fowlpox virus. J Virol 72:10180–10188
4. Hanke T, Samuel RV, Blanchard TJ et al (1999) Effective induction of simian immunodeficiency virus-specific cytotoxic T lymphocytes in macaques by using a multiepitope gene and DNA prime-modified vaccinia virus Ankara boost vaccination regimen. J Virol 73:7524–7532
5. Amara RR, Villinger F, Altman JD et al (2001) Control of a mucosal challenge and prevention of AIDS by a multiprotein DNA/MVA vaccine. Science 292:69–74
6. Stambas J, Brown SA, Gutierrez A et al (2005) Long lived multi-isotype anti-HIV antibody responses following a prime-double boost immunization strategy. Vaccine 23:2454–2464

7. Metcalf D, Begley CG, Johnson GR et al (1986) Effects of purified bacterially synthesized murine multi-CSF (IL-3) on hematopoiesis in normal adult mice. Blood 68:46–57
8. Kelleher AD, Puls RL, Bebbington M et al (2006) A randomized, placebo-controlled phase I trial of DNA prime, recombinant fowlpox virus boost prophylactic vaccine for HIV-1. AIDS 20:294–297
9. Hemachandra A, Puls RL, Sirivichayakul S et al (2010) An HIV-1 clade A/E DNA prime, recombinant fowlpox virus boost vaccine is safe, but non-immunogenic in a randomized phase I/IIa trial in Thai volunteers at low risk of HIV infection. Hum Vaccin 6:835–840
10. McCormack S, Stohr W, Barber T et al (2008) EV02: a Phase I trial to compare the safety and immunogenicity of HIV DNA-C prime-NYVAC-C boost to NYVAC-C alone. Vaccine 26:3162–3174
11. Sekaly RP (2008) The failed HIV Merck vaccine study: a step back or a launching point for future vaccine development? J Exp Med 205: 7–12
12. Ranasinghe C, Turner SJ, McArthur C et al (2007) Mucosal HIV-1 pox virus prime-boost immunization induces high-avidity CD8+ T cells with regime-dependent cytokine/granzyme B profiles. J Immunol 178:2370–2379
13. Gomez CE, Najera JL, Jimenez EP et al (2007) Head-to-head comparison on the immunogenicity of two HIV/AIDS vaccine candidates based on the attenuated poxvirus strains MVA and NYVAC co-expressing in a single locus the HIV-1BX08 gp120 and HIV-1(IIIB) Gag-Pol-Nef proteins of clade B. Vaccine 25:2863–2885
14. Esteban M (2009) Attenuated poxvirus vectors MVA and NYVAC as promising vaccine candidates against HIV/AIDS. Hum Vaccin 5: 867–871
15. Rerks-Ngarm S, Pitisuttithum P, Nitayaphan S et al (2009) Vaccination with ALVAC and AIDSVAX to prevent HIV-1 infection in Thailand. N Engl J Med 361:2209–2220
16. Yuan Z, Chen W, Zhang J et al (2012) Development of an immunoassay for differentiating human immunodeficiency virus infections–from vaccine-induced immune response in Tiantan vaccine trials in China. Clin Biochem 45:1219–1224
17. Kent SJ, Dale CJ, Ranasinghe C et al (2005) Mucosally-administered human-simian immunodeficiency virus DNA and fowlpoxvirus-based recombinant vaccines reduce acute phase viral replication in macaques following vaginal challenge with CCR5-tropic SHIVSF162P3. Vaccine 23:5009–5021
18. Ranasinghe C, Medveczky JC, Woltring D et al (2006) Evaluation of fowlpox-vaccinia virus prime-boost vaccine strategies for high-level mucosal and systemic immunity against HIV-1. Vaccine 24:5881–5895
19. Ranasinghe C, Ramshaw IA (2009) Immunisation route-dependent expression of IL-4/IL-13 can modulate HIV-specific CD8(+) CTL avidity. Eur J Immunol 39:1819–1830
20. Ranasinghe C, Eyers F, Stambas J et al (2011) A comparative analysis of HIV-specific mucosal/systemic T cell immunity and avidity following rDNA/rFPV and poxvirus-poxvirus prime boost immunisations. Vaccine 29: 3008–3020
21. Ranasinghe C, Trivedi S, Stambas J et al (2013) Unique IL-13Ralpha2-based HIV-1 vaccine strategy to enhance mucosal immunity, CD8(+) T-cell avidity and protective immunity. Mucosal Immunol 6(6):1068–1080. doi:10.1038/mi.2013.1
22. Green MR, Sambrook J (2012) Molecular cloning: a laboratory manual, 4th edn. Cold Spring Harbor Laboratory Press, Spring Harbor, NY
23. Coupar BE, Andrew ME, Boyle DB (1988) A general method for the construction of recombinant vaccinia viruses expressing multiple foreign genes. Gene 68:1–10
24. Heine HG, Boyle DB (1993) Infectious bursal disease virus structural protein VP2 expressed by a fowlpox virus recombinant confers protection against disease in chickens. Arch Virol 131:277–292
25. Coupar BE, Purcell DF, Thomson SA et al (2006) Fowlpox virus vaccines for HIV and SHIV clinical and pre-clinical trials. Vaccine 24:1378–1388
26. Dale CJ, Thomson S, De Rose R et al (2006) Prime-boost strategies in DNA vaccines. Methods Mol Med 127:171–197
27. Boyle DB, Anderson MA, Amos R et al (2004) Construction of recombinant fowlpox viruses carrying multiple vaccine antigens and immunomodulatory molecules. Biotechniques 37(104–106):108–111
28. Ranasinghe C, Trivedi S, Stambas J et al (2013) Unique IL-13Rα2 based HIV-1 vaccine strategy to enhance mucosal immunity, CD8+ T cell avidity and protective immunity. Mucosal Immunol 6(6):1068–1080
29. Kozak M (1987) An analysis of 5′-noncoding sequences from 699 vertebrate messenger RNAs. Nucleic Acids Res 15:8125–8148
30. Coupar BE, Andrew ME, Both GW et al (1986) Temporal regulation of influenza hemagglutinin expression in vaccinia virus recombinants

and effects on the immune response. Eur J Immunol 16:1479–1487

31. Townsend A, Bastin J, Gould K et al (1988) Defective presentation to class I-restricted cytotoxic T lymphocytes in vaccinia-infected cells is overcome by enhanced degradation of antigen. J Exp Med 168:1211–1224
32. Earl PL, Moss B (2001) Characterization of recombinant vaccinia viruses and their products. Curr Protoc Mol Biol 43:16.18.11–16.18.11
33. Earl PL, Moss B, Wyatt LS et al (2001) Generation of recombinant vaccinia viruses. Curr Protoc Mol Biol 43:16.17.11–16.17.19
34. Earl PL, Cooper N, Wyatt LS et al (2001) Preparation of cell cultures and vaccinia virus stocks. Curr Protoc Mol Biol 43:16.16.11–16.16.13
35. Buller RM, Smith GL, Cremer K et al (1985) Decreased virulence of recombinant vaccinia virus expression vectors is associated with a thymidine kinase-negative phenotype. Nature 317:813–815
36. Andrew ME, Coupar BE, Boyle DB (1989) Humoral and cell-mediated immune responses to recombinant vaccinia viruses in mice. Immunol Cell Biol 67:331–337
37. Listvanova S, Temmerman S, Stordeur P et al (2003) Optimal kinetics for quantification of antigen-induced cytokines in human peripheral blood mononuclear cells by real-time PCR and by ELISA. J Immunol Methods 281:27–35
38. Day SL (2008) Evaluation of interferon responses and activity for effective vaccine therapy. Ph.D., Australian National University
39. Trivedi S, Jackson RJ, Ranasinghe C (2012) Interleukin-3 and granulocyte-macrophage colony-stimulating factor expression, a biomarker of memory CD8+ T cell immunity and vaccine efficacy. J Vaccines Vaccination 3:166–173

Chapter 6

Development of Antibiotic-Free Selection System for Safer DNA Vaccination

Jeremy M. Luke, Aaron E. Carnes, and James A. Williams

Abstract

The use of antibiotic-resistance markers in DNA vaccines is discouraged by regulatory agencies due to various theoretical safety concerns. This chapter presents methodologies for the design and cloning of synthetic antigen genes into RNA-OUT encoding antibiotic-free DNA vaccine vectors that are additionally optimized to improve protein expression, and immunogenicity, compared to alternative kanamycin-resistant vectors.

First, antigen targeting considerations are discussed in the context of immune response customization through MHC class I or class II directed antigen presentation; the example NTC868 series RNA-OUT vector system allows simultaneous cloning into multiple vectors that feature various transgene intracellular targeting destinations. Then a detailed flowchart for codon optimization and synthetic transgene design is presented. Finally in-depth methodologies for cloning transgenes into the NTC868 series RNA-OUT vector system are presented. The resultant antibiotic-free DNA vaccine vectors are a more potent, safer alternative to existing kanamycin resistance marker encoding vectors.

Key words DNA vaccination, Plasmid, Antibiotic-free, Nonviral, Fermentation, RNA-OUT, Vaccine, Ubiquitin, LAMP1, TPA

1 Introduction

The recent licensure of four DNA vaccine products for animal health demonstrates the utility of DNA vaccination in large animals. DNA vaccines are inherently safe since the vectors encode and express only the target antigen and are non-replicating and therefore cannot revert to virulence as with viral vectors. Unlike viral vector particles, DNA vaccines do not induce anti-vector immunity, and therefore may be utilized with multiple products intended for the same patient. A DNA product is highly stable and DNA vaccine manufacture much easier and faster than alternative vaccine platforms. DNA vaccines are well tolerated and have an excellent safety profile in human clinical studies with no reported safety concerns such as autoimmunity, antigen tolerance, or plasmid integration into the host genome [1–3].

Monica Rinaldi et al. (eds.), *DNA Vaccines: Methods and Protocols*, Methods in Molecular Biology, vol. 1143,
DOI 10.1007/978-1-4939-0410-5_6, © Springer Science+Business Media New York 2014

DNA vaccine vectors consist of (1) a eukaryotic region that directs expression of the transgene in the target organism and (2) a bacterial region that provides selection and propagation in the *Escherichia coli* (*E. coli*) host.

The eukaryotic region contains a promoter upstream, and a polyadenylation (polyA) signal downstream, of the antigen transgene. Upon transfection into the cell nucleus, the promoter directs transcription of an mRNA that includes the transgene. The polyA signal mediates mRNA cleavage and polyadenylation, which leads to efficient mRNA export to the cytoplasm. Inclusion of a Kozak sequence (gccgccRcc**ATG**G consensus, transgene ATG start codon within the Kozak sequence is bolded, critical residues in caps, R = A or G) ensures ribosome recruitment in the cytoplasm to effect efficient transgene translation. The expressed antigen is presented to the immune system by either major histocompatibility complex (MHC) class I or II; this may be influenced by antigen targeting [1].

Within the eukaryotic region, the constitutive human Cytomegalovirus (CMV) promoter is the most common DNA vaccine promoter since it is highly active in most mammalian cells. PolyA signals derived from the rabbit β-globin or bovine growth hormone genes are typically used. These signals contain accessory sequences upstream and downstream of the polyA site (AATAAA or ATTAAA) that increase polyadenylation efficiency resulting in increased mRNA levels, and improved transgene expression. Transgene expression is further increased by inclusion of an intron within the eukaryotic region 5′ untranslated region (UTR) [4, 5].

The bacterial region combines a high copy replication origin, most usually the pUC origin, with a selectable marker. First-generation DNA vaccine vectors typically used the kanamycin resistance (kanR) gene as a selectable marker. However, antibiotic-resistance markers in DNA vaccines have potential safety concerns [6, 7]. These include production mediated environmental contamination with the antibiotics used in fermentation culture or with the plasmid encoded markers, transfer of antibiotic resistance to a patient's endogenous microbial flora (e.g., topically applied plasmid DNA may transfect skin resident microorganisms), or marker expression from host cell promoters after spurious incorporation into the patient's genome. Ampicillin is generally not acceptable due to potential hyperreactivity to residual trace β lactam antibiotics in the product. The European Pharmacopeia states "Unless otherwise justified and authorized, antibiotic-resistance genes used as selectable genetic markers, particularly for clinically useful antibiotics, are not included in the vector construct. Other selection techniques for the recombinant plasmid are preferred" [8]. The European Medicines Agency (EMA) concluded that kanamycin and neomycin cannot be classified as having minor therapeutic relevance for veterinary and human use [9]. To address these

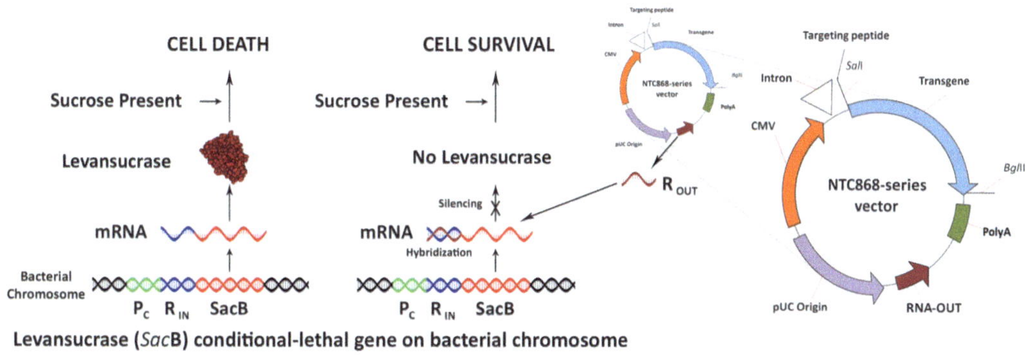

Fig. 1 Sucrose selection with RNA-OUT selection marker encoding NTC868 series DNA vaccine plasmids. *Left*: Sucrose selection. NTC868 series plasmid borne 70 bp RNA-OUT antisense RNA binds a chromosomally encoded constitutively expressed mRNA that contains the RNA-IN target sequence in the leader. This prevents translation of the downstream levansucrase (sacB), allowing growth on sucrose media. *Right*: NTC868 series vectors. *Purple arrow* in bacterial region is pUC replication origin, *brown arrow* is the RNA-OUT selection marker. Eukaryotic region CMV promoter, transgene, and rabbit β globin polyA are depicted with *orange arrow*, *blue arrow*, and *green box*, respectively. The *Sal*I–*Bgl*II cloning cassette for NTC8682, NTC8684, and NTC8685-EGFP vectors are shown, with the insertion point for the targeting peptides in NTC8682 and NTC8684 indicated with an *arrow*

regulatory concerns, alternative non-antibiotic selection methods are needed.

The use of any protein-based selection marker raises the concern that it may be unintentionally expressed and translated in the vaccinated organism. While a number of antibiotic-free (AF) plasmid retention systems have been developed in which the vector-encoded selection marker is not protein based [4, 10], superior expression and manufacture have been observed with DNA vaccine vectors that incorporate RNA based antibiotic-free selection markers. For example, the NTC868 series sucrose selection vectors encode RNA-OUT, a small 70 bp antisense RNA (Fig. 1) [11]; the pMINI vector utilizes the ColE1 origin-encoded RNAI antisense RNA [12]; while pFAR4 and pCOR vectors encode a nonsense suppressor tRNA marker [13, 14]. These plasmid borne RNAs regulate the translation of a host chromosome encoded selectable marker allowing plasmid selection (Fig. 1). Of these, high yield fermentation processes (>500 mg/L) have been developed for RNA-OUT vectors (1,800 mg/L) [15] and pMINI (900 mg/L) [12]. In all these vectors, replacement of the kanR antibiotic selection marker resulted in dramatically increased transgene expression in the target organism compared to kanR comparator vectors such as pVAX1, demonstrating elimination of antibiotic selection to meet regulatory criteria may unexpectedly also improve product performance [13, 16]. Antibiotic-free RNA-OUT vectors also improve DNA vaccine induced immune responses compared to kanR vectors [17], have raised no safety issues in completed biodistribution and nonclinical toxicology evaluations, and are utilized in ongoing human clinical trials.

Table 1
NTC868 series targeting vectors

Protein destination	Targeting tag	Vector	Synthetic gene design	Targeting–synthetic gene
Endosome	N-and C-terminal LAMP1	NTC8681	*XhoI-transgene-EcoRI*	Kozak-**ATG**-N- LAMP1-*Xho*I-***Transgene***-*Eco*RI-C-LAMP1- **TAA**
Secreted	N-terminal TPA Signal peptide	NTC8682 (Fig. 3)	*SalI-transgene*-STOP-*Bgl*II	Kozak- **ATG** -TPA-*Sal*I-***Transgene- TAA*** *Bgl*II
	Proteasome N-terminal Ubiquitin A76	NTC8684	*SalI-transgene*-STOP-*Bgl*II	Kozak- **ATG** -UbA76-*Sal*I-***Transgene- TAA*** *Bgl*II
Cytoplasmic	ATG	NTC8685-EGFP	*Sal*I-ATG transgene-STOP-*Bgl*II	Kozak- *Sal*I ***ATG*** **-** ***Transgene-TAA*** *Bgl*II

All vectors are also available in a RIG-I activating immunostimulatory dsRNA (eRNA41H) version

In this chapter, optimized methods for designing and cloning transgenes into Nature Technology Corporation (NTC) RNA-OUT sucrose selection vectors, using the NTC868 series antigen targeting vector family, are described. A flowchart for synthetic transgene design to generate a codon-optimized antigen gene, and methodologies for cloning transgenes into RNA-OUT vectors, and characterization of recombinant RNA-OUT plasmids, are presented.

1.1 Targeting Vector Selection

DNA vaccines, due to in vivo antigen expression, have the advantage that vaccinologists may easily customize encoded antigens through rational transgene design. This allows antigens to be modified to remove potential oncogenic sequences, attenuate virulence, and inactivate nuclear localization, enzymatic, or DNA binding activities [4], or, for multiple serotype pathogens, to design novel cross-neutralizing antigens such as consensus immunogens [18].

After an antigen protein sequence is designed, the optimal targeting vector is selected from the NTC868 series family of antigen targeting RNA-OUT selection marker encoding DNA vaccine plasmids. These vectors feature compatible cloning into plasmids encoding either N-terminal Tissue Plasminogen Activator (TPA) signal peptide (secretion targeting; NTC8682), N-terminal and C-terminal Lysosomal-associated membrane protein 1 (LAMP1) (endosomal targeting; NTC8681), N-terminal destabilizing Ubiquitin A76 (proteasome targeting; NTC8684), or no targeting (NTC8685) (Table 1).

1.2 Synthetic Transgene Design and Bioinformatics Analysis

Once the antigen protein sequence is finalized and the targeting vector (s) selected a synthetic gene sequence encoding the antigen of interest is designed (Fig. 2), synthesized and cloned into the targeting vector. Commercial gene synthesis is now rapid and inexpensive, thus enabling synthetic codon-optimized antigen

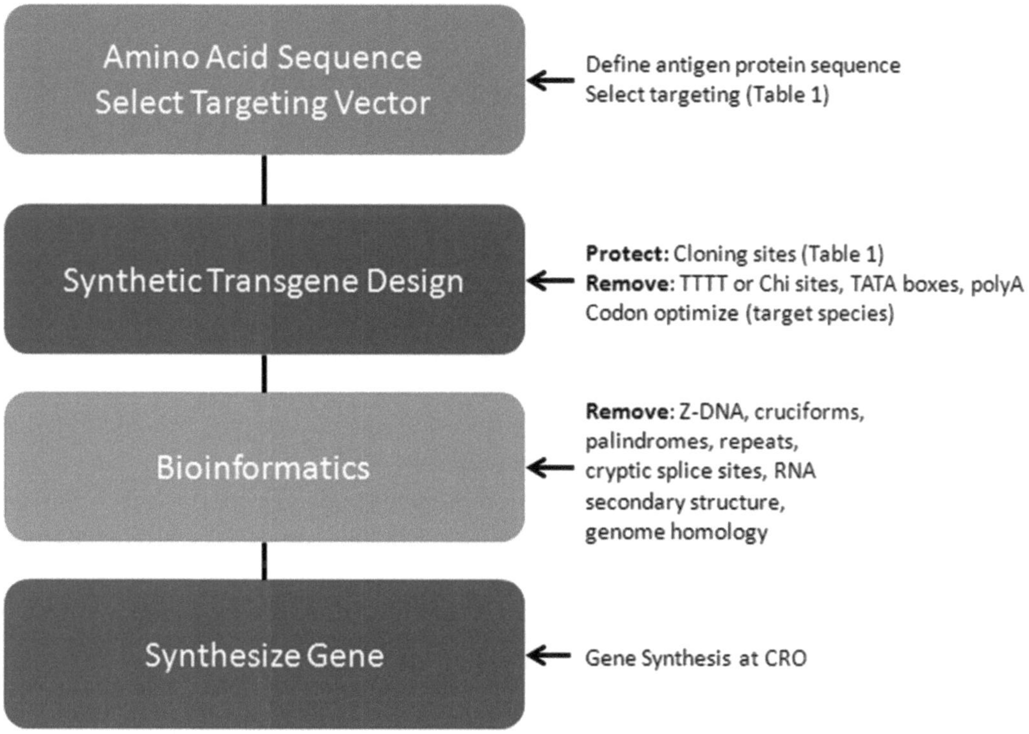

Fig. 2 Synthetic gene design flowchart

gene synthesis, vector construction, and vaccine manufacture on a highly compressed timeline.

The antigen protein sequence is reverse translated into a gene sequence, selecting optimal codon usage for the target species. Codon optimization to match high use codons for the target species has been shown to dramatically increase transgene expression and immunogenicity [19, 20].

1.3 Cloning into RNA-OUT Vectors

Plasmid selection is performed in the presence of sucrose using engineered *SacB* expression strains from NTC (Fig. 1). Vector borne RNA-OUT represses expression of a sucrose counter-selectable marker (*SacB*) from the host chromosome. *SacB* encodes a levansucrase, which is toxic in the presence of sucrose.

1.4 Recombinant Clone Verification

Seed stocks of recombinant RNA-OUT plasmids are created and validated by restriction mapping and DNA sequencing.

The presented methodologies are generic and, by modification of the transgene flanking restriction sites, can be used to design and clone transgenes into alternative RNA-OUT selection marker containing vectors. Additionally, the cloning and characterization methods can be adapted to replace an antibiotic-resistance marker in an existing plasmid with an RNA-OUT antibiotic-free selection marker as described [16].

2 Materials

2.1 Targeting Vector Selection

1. NTC8681 RNA-OUT endosome targeting vector (NTC, Lincoln, NE, USA).
2. NTC8682 RNA-OUT secretion targeting vector (NTC, Lincoln, NE, USA).
3. NTC8684 RNA-OUT proteasome targeting vector (NTC, Lincoln, NE, USA).
4. NTC8685-EGFP RNA-OUT cytoplasm targeting vector (NTC, Lincoln, NE, USA).

2.2 Synthetic Transgene Design and Bioinformatics Analysis

1. Sequence analysis and design software (e.g., VectorNTI, Life Technologies, Grand Island, NY, USA).
2. Codon optimization program (e.g., GeneArt® GeneOptimizer® [21], Life Technologies, Grand Island, NY, USA; Optimum Gene™ algorithm, GenScript, Piscataway, NJ, USA).
3. Molecular biology bioinformatics mRNA secondary structure analysis online server (e.g., Mfold program [22] at Mobyle [23]).
4. Molecular biology bioinformatics mRNA splicing analysis online server (e.g., NetGene2 or GeneSplicer [24]).
5. Molecular biology bioinformatics *E. coli* promoter analysis online server (e.g., BProm program, Softberry Mount Kisco, NY, USA).
6. Molecular biology bioinformatics human promoter analysis online server (e.g., TSSG program, Softberry Mount Kisco, NY, USA).
7. Molecular biology bioinformatics DNA repeat analysis online server (e.g., palindrome, Repeats 1.1, equicktandem, and etandem programs at Mobyle [23]).
8. Molecular biology bioinformatics non B DNA structure analysis online server (e.g., nBMST programs [25]).
9. Molecular biology bioinformatics DNA helical stability online server (e.g., WebSIDD [26] or WEB-Thermodyn [27]).
10. National Center for Biotechnology (NCBI) BlastN online server [28].
11. Gene synthesis Contract Research Organization (CRO) (e.g., Life Technologies, Grand Island, NY, USA; GenScript, Piscataway, NJ, USA).

2.3 Cloning into RNA-OUT Vectors

1. Synthetic gene and NTC868 series vector (from Subheading 2.1).
2. Restriction enzymes: *Sal*I, *Bgl*II, *Xho*I, *Eco*RI and 10× buffers.
3. T4 DNA ligase and 10× buffer.

4. Agarose, loading dye, and nucleic acid stain suitable for gel electrophoresis.
5. Agarose gel electrophoresis system: for 150 mL of 1.0 % agarose gel, use 1.5 g of ultrapure agarose (electrophoresis grade) with 150 mL of 1× TAE. Prepare 1 L of 10× TAE stock solution in ultrapure water with 48.4 g of Tris base, 3.72 g disodium EDTA, and adjust to pH 8.5 with glacial acetic acid.
6. Gel documentation system.
7. QIAquick Gel Extraction Kit (Qiagen).
8. Electrocompetent cells: *E. coli* NTC4862 DH5α attλ::$P_{5/6\ 6/6}$-RNA-IN-*SacB*, *catR* or NTC5402 XL1-Blue attλ::$P_{5/6\,6/6}$-RNA-IN-*SacB*, *catR* (NTC, Lincoln, NE, USA) for selection and propagation of NTC RNA-OUT vectors.
9. Electroporator: Bio-Rad GenePulser®II, BTX®ECM®630 or equivalent.
10. 0.1 cm electroporation cuvette.
11. SOC medium: To 900 mL of distilled H_2O add 20 g Bacto Tryptone, 5 g Bacto Yeast Extract, 2 mL of 5 M NaCl, 2.5 mL of 1 M KCl, 10 mL of 1 M $MgCl_2$, 10 mL of 1 M $MgSO_4$, 20 mL of 1 M glucose. Adjust to 1 L with distilled H_2O. Sterilize by autoclaving and store up to 3 months at room temperature.
12. 50 % Sucrose (wt/vol): Sterilize by filtering through a sterile 0.2-μm membrane and store at room temperature for several months.
13. Sucrose medium plates: To 700 mL of distilled H_2O add 10 g tryptone, 5 g yeast extract, 15 g agar. Adjust to 880 mL with distilled H_2O. Sterilize by autoclaving. When cooled to ~50 °C, add 120 mL of 50 % Sucrose (wt./vol.), mix and pour into sterile 100 mM × 15 mM petri plates. Store refrigerated for up to 3 months.
14. Incubator for growth of plates.
15. 14 mL round-bottom snap-cap tubes, sterile.
16. Shaking incubator for growth of liquid cultures.

2.4 Recombinant Clone Verification

1. LB/sucrose medium: To 700 mL of distilled H_2O add 10 g tryptone, 5 g yeast extract, Adjust to 880 mL with distilled H_2O. Sterilize by autoclaving. When cooled to ~50 °C, add 120 mL 50 % Sucrose (wt./vol.), cool to room temperature before use. Store at room temperature for up to 3 months.
2. 50 % glycerol (wt./vol.): Sterilize by autoclaving and store at room temperature for several months.
3. Sterile inoculation loops.
4. 14 mL round-bottom snap-cap tubes, sterile.

5. Sterile cryovials.
6. Shaking incubator for growth of cultures.
7. Plasmid miniprep kit: Qiagen Qiaprep miniprep kit including optional PB wash or equivalent.
8. UV spectrophotometer.
9. Restriction enzymes: *Sal*I, *Bgl*II, *Xho*I, *Eco*RI and 10× buffers.
10. Agarose, loading dye, DNA molecular weight ladder, and nucleic acid stain suitable for gel electrophoresis.
11. Agarose gel electrophoresis system: for 150 mL of 1.0 % agarose gel, use 1.5 g of ultrapure agarose (electrophoresis grade) with 150 mL of 1× TAE. Prepare 1 L of 10× TAE stock solution in ultrapure water with 48.4 g of Tris base, 3.72 g disodium EDTA, and adjust to pH 8.5 with glacial acetic acid.
12. Gel documentation system.
13. pVAC5′ sequencing primer: GCTTTTCTGCCAGGTGCTGA.
14. pVAC3′ sequencing primer: GCCAGAAGTCAGATGCTCAA.
15. LAMPFseq sequencing primer: TGCTCTTCCAGTTCGGGATG.

3 Methods

The antigen protein sequence is finalized and the targeting vector (s) selected (Subheading 3.1), a synthetic gene sequence encoding the antigen of interest is designed (Fig. 2) and synthesized (Subheading 3.2), cloned into the targeting vector backbone (Subheading 3.3) and recombinant clones isolated and characterized (Subheading 3.4).

3.1 Targeting Vector Selection

1. Select antigen protein sequence.
2. If nuclear antigen, inactivate nuclear localization sequences in the antigen protein sequence (*see* **Note 1**).
3. Determine optimal antigen targeting (Table 1) (*see* **Note 2**).
4. Identify optimal NTC868 series family targeting vector (Table 1) (*see* **Notes 2–4**).
5. Remove or inactivate signal peptides from the antigen protein sequence if using NTC8681, NTC8682, or NTC8684 vectors (*see* **Note 5**).
6. If endosomal targeting (NTC8681), additionally remove or inactivate internal transmembrane domains from the antigen protein sequence (*see* **Note 6**).

3.2 Synthetic Transgene Design and Bioinformatics Analysis

1. Open codon optimization program, name project and enter antigen protein sequence.

2. Select target species for codon optimization.
3. Enter two stop codons (TAA or TGA, not TAG) if cloning into NTC8682, NTC8684, or NTC8685 (*see* **Note 7**).
4. Enter flanking restriction sites for cloning into the selected vector (Table 1) and protect these sites to block creation of internal sites during codon optimization (*see* **Note 8**).
5. Enter the following sequences to be excluded: GCTGGTGG (Chi site) GTTGTAAC (ter site core), AATAAA, ATTAAA (polyA sites), TATA (consensus eukaryotic promoter core), TTAGGG (immunosuppressive telomeric motif), GCCGTCTGAA AAGTGCGGT ACAAGCGGTC (DNA uptake sequences), AGGT (consensus splice donor), CAGG (consensus splice acceptor), AAAAA (polyA binding proteins consensus), and (TTATCCACA (DnaA binding site) (*see* **Note 9**).
6. Codon-optimize gene and download resultant synthetic gene sequence (*see* **Note 10**).
7. Download NTC868 series vector sequence (s) (*see* **Note 11**).
8. Use sequence analysis and design software to make a vector sequence file of the synthetic gene sequence cloned into the relevant NTC868 series vector (s).
9. Make a DNA sequence file of the predicted spliced mRNA (NTC868 series vector encoded exon 1, exon 2 linked to synthetic gene). Bioinformatics screen the predicted spliced mRNA, using a mRNA secondary structure analysis program, for stable mRNA snapback secondary structures that mask the Kozak sequence. Redesign as necessary to incorporate silent codon changes that eliminate Kozak masking secondary structures (*see* **Note 12**).
10. Make a DNA sequence file of the predicted unspliced mRNA (NTC868 series vector encoded exon 1, intron, exon 2 linked to synthetic gene). Bioinformatics screen this unspliced mRNA for cryptic splice donor or splice acceptor sites, using a mRNA splicing analysis program. Redesign as necessary to eliminate splice sites (*see* **Note 13**).
11. Bioinformatics screen both orientations of the synthetic gene insert (as redesigned in steps 9 and 10) for *E. coli* and human promoters and redesign as necessary (*see* **Note 14**).
12. Bioinformatics screen the synthetic gene insert (as redesigned in **steps 9–11**) for direct or inverted DNA repeats and palindromes (*see* **Note 15**).
13. Bioinformatics screen the synthetic gene insert (as redesigned in **steps 9–11**) within the NTC868 series vectors for non B DNA structures, and DNA helical stability (e.g., cruciforms) (*see* **Note 16**).

14. BlastN screen the finalized synthetic gene insert for human genome identity (*see* **Note 17**).
15. Send synthetic gene insert sequence to CRO for synthesis.

3.3 Cloning into RNA-OUT Vectors

1. Digest 1 μg of NTC868 series vector and 1 μg of synthetic gene vector from CRO with appropriate enzymes for cloning the insert (Table 1) (*see* **Notes 18** and **19**). Inactivate the enzymes by heating to 65 °C for 20 min (min).
2. Load restriction digests on agarose gel, resolve and gel purify vector and insert fragments using the QIAQuick kit, eluting fragments with 30 μL of elution buffer.
3. Set ligation reaction. Typically, for a 10 μL ligation, 3.6 μL of the gel purified vector fragment and 5 μL of the gel purified insert fragment are mixed with 1 μL of 10× ligation buffer and 0.4 μL of T4 DNA ligase. The ligation is incubated overnight at 10–22 °C and then heat-killed at 65 °C for 20 min (*see* **Note 20**).
4. Dilute 3 μL of the ligation with 7 μL of sterile water and add the entire 10 μL to a chilled microcentrifuge tube. Prechill on ice one 0.1 cm electroporation cuvette for each sample to be electroporated (*see* **Notes 21** and **22**).
5. Thaw NTC4862 or NTC5402 electrocompetent cells on wet ice (*see* **Notes 23** and **24**).
6. When cells are thawed, mix cells by tapping gently. Add 25 μL of cells to each chilled microcentrifuge tube containing the DNA to be transfected.
7. Refreeze any unused cells. Although the cells are refreezable, subsequent freeze–thaw cycles will decrease transformation efficiency.
8. Pipette the cell–DNA mixture into a prechilled 0.1 cm cuvette and wipe off water condensate from the outside of the cuvette with a paper towel.
9. Insert the cuvette into the electroporator and electroporate. For the BTX®ECM®630 or Bio-Rad GenePulser®II electroporator, use the following electroporator conditions: one pulse at 1.8 kV, 200 Ω, 25 μF (Ec1 setting on Bio-Rad MicroPulser).
10. To the cells in the cuvette, quickly add 0.5–1.0 mL of room temperature SOC medium and transfer the solution to a 14 mL snap-cap tube.
11. Shake vigorously at 225–300 rpm in a shaking incubator for 1–2 h at 30 °C.
12. Aseptically spread 50–100 μL of transformation outgrowth on sucrose medium plates. Hold remaining transformation outgrowth at room temperature.

13. Incubate plates overnight at 30 °C.
14. Check plates for single colonies. If no colonies are observed, replate 200 μL of room temperature held transformation outgrowth on sucrose medium plates and incubate overnight at 30 °C. If confluent growth is observed, dilute transformation outgrowth 1:10 and spread 50–100 μL of dilution on sucrose medium plates and incubate overnight at 30 °C (*see* **Notes 25–27**).

3.4 Recombinant Clone Verification

1. Identify 4 well isolated single colonies from the transformation plates. Using sterile inoculation loops, pick each colony and aseptically inoculate each into a 6 mL culture of LB/Sucrose medium in a 14 mL snap-cap tube.
2. Grow the four cultures by shaking vigorously at 225–300 rpm in a shaking incubator at 30 °C for 16–20 h.
3. Visually verify cultures are saturated. Make glycerol stocks of each culture by aseptically mixing 200 μL of 50 % glycerol with 800 μL of culture. Aseptically transfer to a cryovial and freeze at −80 °C (*see* **Note 27**).
4. Isolate plasmid DNA from the remaining culture.
5. Determine plasmid DNA concentration by spectrophometry.
6. Identify clones with correct insert size by restriction digestion. Include a lane with undigested DNA to allow monomer: dimer ratio to be estimated (*see* **Notes 27–29**).
7. Validate monomer plasmid clone with correct insert size by sequencing the 5′ and 3′ termini of the transgene insert using the pVAC5′ (5′ termini) and pVAC3′ (3′ termini) sequencing primers, or, for NTC8681, LAMPFseq sequencing primer rather than pVAC5′ for 5′ termini sequence verification (*see* **Note 30**).
8. Grow culture of validated clone to isolate plasmid for in vitro and in vivo testing (*see* **Notes 30–33**).

4 Notes

1. For nuclear localized antigens, nuclear localization sequences may interfere with TPA, LAMP1 or Ubiquitin targeting and may need to be removed. Inactivation of nuclear localization sequences in influenza nucleoprotein increased cytoplasmic protein levels for proteasome processing resulting in increased MHC class I presentation, and improved levels of antigen reactive $CD8^+$ T-cells [29].
2. Adaptive immune responses may be improved by enhancing antigen processing and MHC class I and/or class II presentation

[1, 2]. This can be accomplished by the addition of a targeting peptide that routes antigens to various intracellular destinations. DNA vaccine antigens are most commonly targeted to the secretion pathway using a signal peptide. This may use a heterologous secretion signal, or, in the case of a secreted protein, the native secretion signal. Use of an optimized signal sequence may dramatically improve expression over the native sequence. Improvement has been observed using an optimized tissue plasminogen activator (TPA) signal peptide [19] or IgE [2] gene leader. An optimized TPA secretion tag is included in the NTC8682 vector and the transgene is cloned downstream and in frame with the signal peptide. Alternatively a different signal peptide may be included when designing the synthetic gene and cloned into NTC8685. To promote MHC class I presentation, and antigen reactive $CD8^+$ T-cell response, antigens may be targeted to the proteasome using an N-terminal ubiquitin tag (terminal ubiquitin G76 residue altered to A76 to destabilize the fusion protein; NTC8684) [30]. To increase MHC class II antigen presentation endosomal targeting by transgene insertion within the LAMP1 protein (NTC8681) is used [31].

3. All NTC868 series vectors are also available in a Retinoic Acid Inducible Gene 1 (RIG-I) activating immunostimulatory dsRNA (eRNA41H) version, which encodes an RNA based adjuvant in the vector backbone (Table 1). eRNA41H vectors have been shown to improve DNA vaccination induced antigen-specific humoral and cellular responses [32].
4. For gene therapy applications using a human therapeutic protein to avoid potential immune response the protein and secretion signal (if applicable) amino acid sequences are typically not extensively modified. Synthetic codon-optimized transgenes are therefore cloned into NTC8685 without eRNA41H.
5. Many antigens contain signal peptides. Signal peptides may be identified by bioinformatics for example using the SignalP predictor [33]. It is critical that the antigen encoded signal peptide is removed for cloning into the TPA signal peptide vector NTC8682 since a dual signal peptide transgene will not be properly exported. Likewise, antigen encoded signal peptides should also be removed for cloning into NTC8681 or NTC8684 since the signal peptide may interfere with LAMP1 trafficking, or ubiquitin function, respectively.
6. Internal hydrophobic transmembrane domains should be removed for cloning into NTC8681, since these may interfere with LAMP1 mediated endosomal trafficking. Internal transmembrane may be identified by bioinformatics, for example using the Phobius Web server [34], the TopPred program at Mobyle [23], or TMpred on the EMBnet server [35].

7. The transgene is inserted internally within the LAMP1 luminal domain within NTC8681. No stop codons should be included within the transgene.
8. A single insert may be cloned into NTC8682, NTC8684 and NTC8685 vectors if an ATG start codon is included immediately after the *Sal*I site. This ATG start codon is optional for NTC8682 and NTC8684 but is required for NTC8685 cloning since in this leaderless vector the *Sal*I–ATG (gtcgac**ATG**) site serves as an effective Kozak sequence (gccRcc**ATGG**). Including an ATG therefore allows evaluation of secretion, cytoplasmic or proteasome targeting of a single antigen synthetic gene to determine the optimal targeting. For antigens cloned into NTC8685, to ensure maximal expression from the synthetic gene, where necessary consider modifying the second amino acid residue to a Gxx codon (Val, Ala, Asp, Glu, Gly) to match the Kozak **ATG**G consensus. This is not necessary with the other NTC868 series vectors where the antigen is cloned downstream of a leader peptide or protein that encodes the Kozak sequence.
9. The removed sequences are potentially deleterious to bacterial production, plasmid quality, or eukaryotic function. Additionally, high levels of TT or TA dinucleotide motifs reduce mRNA stability and protein expression [36]. The frequency of TT or TA motifs is naturally reduced by codon optimization for human or other mammalian species *see* ref. [4] for a detailed review
10. Elimination of extensive RNA secondary structure is important. Some codon optimization programs such as the GeneArt GeneOptimizer® process combine RNA and codon optimization [20, 21].
11. Vector sequences and fully annotated VectorNTI vector maps are available from NTC upon request. The annotated VectorNTI map of NTC8682 is shown in Fig. 3.
12. An important consideration is that RNA secondary structure between the synthetic gene and the vector 5′ UTR is not screened by gene synthesis companies. Secondary structure between the synthetic gene and the 5′ UTR encoded Kozak sequence may interfere with ribosome recruitment and reduce transgene expression [21]. Such hybrids may be detected using a program such as mfold [22].
13. The authentic NTC868 series vector encoded intron splice sites will be identified so the relevant strength of cryptic transgene splice sites is evaluated versus the known intron splice sites. For weak predicted splice sites, check an additional splicing analysis program to determine if authentic (i.e., predicted by both programs). To remove cryptic splice

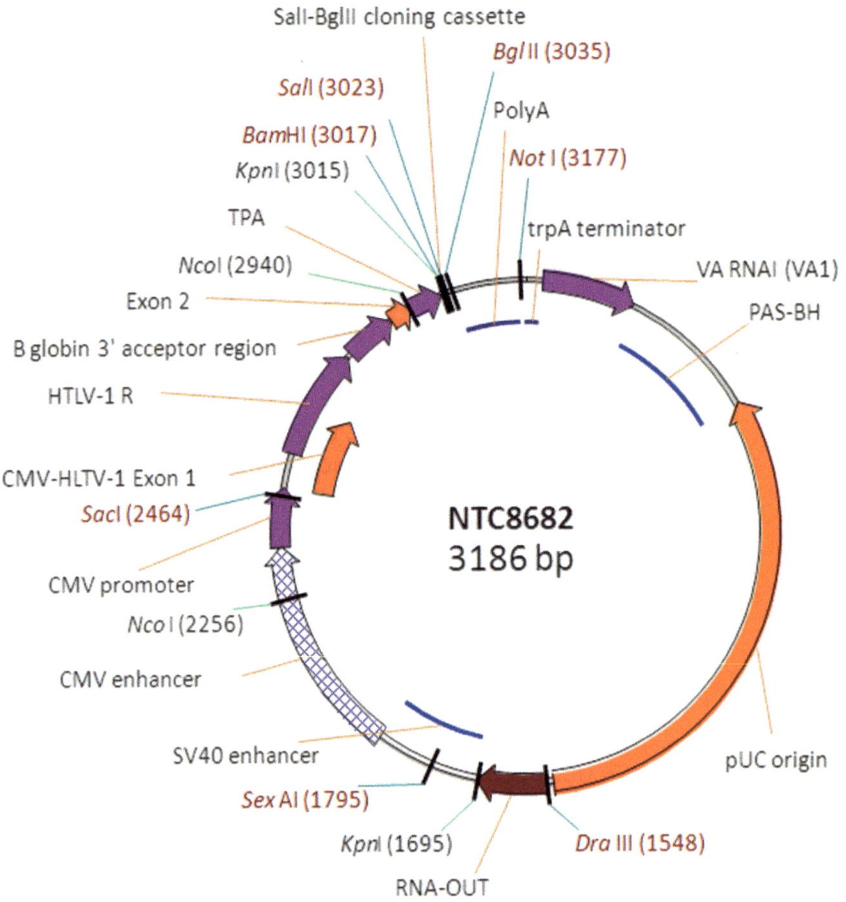

Fig. 3 Annotated NTC8682 VectorNTI plasmid map

sites predicted by two or more programs, translate the gene and make silent codon changes that decrease the identity to the consensus sequences (splice donor consensus = AGGT; splice acceptor consensus = poly pyrimidine tract upstream of CAGG).

14. Both orientations are screened since complementary strand *E. coli* promoters could express toxic peptides in the *E. coli* host that reduce plasmid copy number or stability [37] while complementary eukaryotic promoters would generate antisense RNA that could form dsRNA with the transgene mRNA to reduce transgene expression through translation inhibition or RNA interference. Translate the gene and make silent codon changes that decrease the identity to the consensus sequences (e.g., *E. coli* promoter consensus for -35 is **TTG**ACA, for -10 is **TA**AT**AT**, most critical residues are bolded).

15. Codon-optimized genes rarely contain significant repeats unless the target antigen is repetitive. For repetitive antigens, codon optimization should reduce identity between repeats sufficiently to enable stable propagation in the *E. coli* host.
16. Codon-optimized genes rarely contain significant non B DNA structures such as Z DNA, G quadruplexes, or cruciforms. The entire vector is used for bioinformatics so the strength of the insert encoded structures can be evaluated versus known vector destabilizing sequences within the pUC origin (and weaker sites within RNA-OUT and the poly A site) and Z DNA (within exon 1).
17. BLASTN searching is a precaution to reduce potential regulatory agency concern regarding the theoretical risk of insertional mutagenesis of the host genome. A codon-optimized synthetic gene typically contains regions with only short tract homology of less than 30 bps of perfect identity to the target genome. These short tracts of homology should not be an issue since characterization of plasmid DNA integration into the genome using repeat-anchored integration capture (RAIC) PCR has demonstrated short homology driven integration events are extremely rare [38].
18. With synthetic inserts the vector and synthetic genes will be digested with *Sal*I and *Bgl*II (NTC8682, NTC8684, NTC8685-EGFP) or *Xho*I and *Eco*RI (NTC8681). In the case of NTC8685-EGFP, double digestion is verified by release of the 730 bp EGFP "stuffer." Double digestion cannot be verified with the other vectors that contain a short polylinker between the cloning sites. To ensure vector double digestion, it is recommended to digest the vector 1 h at 37 °C, add a second aliquot of each enzyme and digest a further 30 min at 37 °C.
19. Alternatively, rather than using a synthetic gene, an existing gene for insertion into the NTC868 series vector can be cloned using standard PCR amplification using a proofreading thermostable polymerase such as Pfu DNA polymerase. Primers are designed to amplify the gene and incorporate a 5′ *Sal*I site and a 3′ *Bgl*II site (NTC8682, NTC8684, NTC8685-EGFP) or 5′ *Xho*I and 3′ *Eco*RI (NTC8681). The sequence to be amplified is included immediately after the enzyme sites exactly as in the case of a synthetic gene. To ensure restriction enzyme digestion after PCR amplification an additional 6 bp should be included at the beginning of the primers before the enzyme site. Example primers to amplify an existing gene from ATG start to TAA stop codon for cloning into *Sal*I/*Bgl*II digested NTC868 series vectors are: Forward *Sal*I (GTCGAC) containing primer GCGCAT**GTCGAC**ATG—Gene of interest 5′ end; and Reverse

*Bgl*II (AGATCT) containing primer GCAGAA**AGATCT**TTA—Gene of interest (reverse complement) 3′ end. Since cleavage of the short PCR product ends cannot be verified by agarose gel analysis, to ensure vector double digestion it is recommended to digest the vector 1 h at 37 °C, add a second aliquot of each enzyme and digest a further 30 min at 37 °C. If the existing gene contains one or more internal sites for the cloning enzymes, the 4 bp overhangs for cloning can be generated using certain type IIS enzyme sites in the primers. For example, *Aar*I cleaves after +4 and +8 bp downstream of the AarI recognition sequence (CACCTGC) to create any designed 4 bp 3′-recessed sticky end. Example *Aar*I type IIS containing primers to PCR amplify a gene from ATG start to TAA stop codon to generate 4 bp complementary ends for cloning into *Sal*I/*Bgl*II digested NTC868 series vectors are as follows: *Aar*I forward primer 5′ CTCCAG**CACCTGC**CTAT*TCGA*CATG—Gene of interest 5′ end; and *Aar*I reverse primer 5′ CGTGAG **CACCTGC**AACG*GATC*TTA—Gene of interest (reverse complement) 3′ end. Alternatively, if the gene of interest encodes internal *Bgl*II sites, *Bam*HI or *Bcl*I restriction enzymes can be utilized in the PCR primers since they create *Bgl*II compatible sticky ends. Note that both enzyme sites will be lost in the resulting construct. *Xho*I is not recommended as an alternative to *Sal*I for cloning into NTC8685-EGFP, since the ligated hybrid (GTCGAGATG) has not been established as an effective Kozak sequence.

20. Ligations may be performed using a variety of conditions. Successful results have been obtained using a thermocycler programmed for a gradient ligation with steps from 10 °C (90 min), 12 °C (90 min), 14 °C (90 min), 16 °C (180 min), 18 °C (90 min), 20 °C (90 min), and 22 °C (90 min), heat kill 65 °C (20 min), hold 4 °C. Alternatively, incubate ligation in an ice bucket starting with chilled ice water which slowly warms to room temperature overnight.
21. Transformation efficiencies will be 10- to 100-fold lower for ligation mixtures than for an intact control plasmid. Salts and buffers severely inhibit electroporation by increasing the risk of arcing.
22. Transform using more plasmid DNA than is typically used with antibiotic selection. For electrocompetent cells transform at least 30–50 ng from a cloning ligation, or 5 ng purified RNA-OUT plasmid, into 25 μL electrocompetent cells.
23. Alternatively, prepare electrocompetent cells using standard methodology using single colonies isolated on LB agar plates from cell line glycerol stocks available from NTC. Single colony isolation is critical to maintain low levels of cells that contain

mutated genomic RNA-IN-*SacB* that give rise to sucrose-resistant colonies without plasmid (empty colonies; *see* **Note 28**). The cell lines without plasmid are NOT sucrose resistant, so it is critical that competent cells are prepared in standard LB medium rather than mutation selecting sucrose containing medium. For a 50 mL LB medium culture, inoculate with a well isolated single colony and grow at 30 °C until the OD_{600} of the culture is 0.4–0.6. Cells are chilled on ice for 10 min, and centrifuged 10 min at 3,000 × *g*. The cell pellet is resuspended in 50 mL of ice-cold 10 % glycerol and centrifuged 10 min at 3,000 × *g*. The cell pellet is resuspended in 25 mL of ice-cold 10 % glycerol and centrifuged 10 min at 3,000 × *g*. After washing cell pellets, resuspend electrocompetent cells to 30 OD_{600}/mL in ice-cold 10 % glycerol. For example, if the OD_{600} of the 50 mL culture is 0.5, resuspend the washed pellet in 0.83 mL. Chemical competent cells prepared using the Z-comp kit (Zymo Research, Irvine, CA, USA) can also be used. However, electrocompetent cells are preferred with ligation reactions since higher efficiency transformation electrocompetent cells require less cells to be plated than with chemical transformation, thus reducing the frequency of empty colony clones.

24. The DH5α derived cell line NTC4862 is the default cell line for use. XL1Blue derived NTC5402 is an alternative host that may improve fermentation yields with vectors encoding small (<1 kb) transgene inserts [16]. For gene therapy applications use higher expressing dcm- cell lines NTC48165 DH5α dcm attλ::$P_{5/6\,6/6}$-RNA-IN-*Sac*B, *cat*R or NTC54208 XL1-Blue dcm attλ::$P_{5/6\ 6/6}$-RNA-IN-SacB, *catR* (NTC, Lincoln, NE, USA).

25. If confluent growth is observed, check that the sucrose medium and sucrose medium plates did not contain NaCl since salt inhibits *Sac*B mediated toxicity. As well, ensure that the 50 % sucrose solution was sterilized by filtration rather than autoclaving which may partially degrade the sucrose into glucose and fructose. Finally, ensure that the competent cells were not contaminated during use—only the engineered RNA-IN-*Sac*B strain is sucrose sensitive so environmental bacteria and other *E. coli* strains can grow on the sucrose plates. Since the RNA-IN-*Sac*B cell lines are chloramphenicol resistant (*cat*R), if desired supplement plates and/or media with 6 μg/mL chloramphenicol to prevent contaminant growth.

26. Plate transformations using fewer cells than is typically used with antibiotic selection to avoid satellite colony formation (i.e., Sucrose-resistant colonies that do not grow in liquid media). When transforming with an existing supercoiled plasmid preparation instead of a ligation, plate 10 μL of outgrowth (or 100 μL of 1:10 diluted outgrowth) rather than 100 μL.

If confluent growth is observed, replate fewer cells from the transformation outgrowth to get individual colonies.

27. 30 °C for outgrowth and selective plating is recommended rather than 37 °C to reduce metabolic burden and prevent stationary phase induced plasmid multimerization during transformation. Metabolic burden during transformation can lead to toxicity with certain transgene inserts in a variety of vector backbones. This toxicity can be overcome by lowering the temperature during transformation conditions to 30 °C [39].
28. Sucrose-resistant colonies without plasmid (empty colonies) are due to either (1) competent cell contamination (*see* **Note 25**) or (2) mutations to the host strain genomic RNA-IN-*Sac*B cassette that prevent SacB expression. Such colonies should be rare if the competent cells are properly prepared from single colony isolates (*see* **Note 23**) and if sufficient ligation is transformed. Frequent empty colonies are often indicative that the ligation is poor or that too little ligation reaction was transformed. Ethanol-precipitate the ligation to concentrate and retransform. Rarely, high frequency empty colonies are due to a toxic insert.
29. If an insert cannot be recovered the encoded antigen may encode a toxic protein that is expressed in the *E. coli* host from cryptic bacterial promoters within the insert. Alternatively, the insert may contain unusual DNA structures such as large palindromes that inhibit cell growth or deletion prone inverted repeats. Such sequences should be detected by bioinformatics (*see* Subheading 3.2) but, if the insert was generated from an existing clone by PCR amplification, bioinformatics as described in Subheading 3.2 can be used to investigate the basis for toxicity.
30. The genomically integrated *att*λ::$P_{5/6\ 6/6}$-RNA-IN-*Sac*B, *cat*R genome cassette in the *E. coli* cell lines can be identified by testing for chloramphenicol resistance (*cat*R) using plates and/or media supplemented with 6 μg/mL chloramphenicol. If desired, 6 μg/mL chloramphenicol can be included in shake flask or fermentation media used for plasmid production for in vitro or in vivo testing. The single copy insert in the cell lines does not confer resistant to the high levels of chloramphenicol (34 μg/mL) used to select multicopy *cat*R plasmids.
31. It is recommended that the entire transgene insert be sequence verified if the transgene was obtained by PCR amplification rather than a synthetic gene insert. PCR amplified DNA may contain mutations, even when using a proofreading thermostable polymerase in the PCR reaction. For transgenes larger than 500 bp, this will require design of transgene-internal sequencing primers.

32. To activate the temperature sensitive pUC origin to increase copy number, which aids plasmid purification, plasmid for in vitro or in vivo testing should be isolated from cultures grown at 37 °C rather than 30 °C. Growth at 37 °C also ensures physiological levels of negative supercoiling (the underwinding of a DNA helix). Plasmid grown at 30 °C may have lower negative supercoiling [40] which can affect promoter activity and transgene expression [41].
33. Antibiotic-free NTC868 series RNA-OUT vectors can be produced to high yield (1.8 g/L [15]) using the NTC HyperGRO™ inducible fed-batch fermentation process [42, 43] which has been utilized to manufacture clinical grade DNA for various plasmids and is generally available for commercial production of research grade (NTC) or clinical grade plasmid DNA through licenses to several current good manufacturing practice (cGMP) plasmid manufacturers, including Aldevron, Eurogentec, and VGXI.

Acknowledgements

We thank Justin Vincent for preparation of the figures. This chapter described work supported by NIH grants R44 GM072141-03 and R43 GM080768-01 to JAW.

References

1. Liu MA (2011) DNA vaccines: an historical perspective and view to the future. Immunol Rev 239:62–84
2. Kutzler MA, Weiner DB (2008) DNA vaccines: ready for prime time? Nat Rev Genet 9:776–788
3. Lu S, Wang S, Grimes-Serrano JM (2008) Current progress of DNA vaccine studies in humans. Expert Rev Vaccines 7:175–191
4. Williams JA, Carnes AE, Hodgson CP (2009) Plasmid DNA vaccine vector design: impact on efficacy, safety and upstream production. Biotechnol Adv 27:353–370
5. Mairhofer J, Grabherr R (2008) Rational vector design for efficient non-viral gene delivery: challenges facing the use of plasmid DNA. Mol Biotechnol 39:97–104
6. FDA (2007) Guidance for industry: considerations for plasmid DNA vaccines for infectious disease indications. (USA Food and Drug Administration), FDA
7. EMA (2001) Note for guidance on the quality, preclinical and clinical aspects of gene transfer medicinal products. (European Medicines Agency)
8. EMA (2011) Gene transfer medical products for human use (European Pharmacopeia 7.0), 5.14
9. EMA (2007) Presence of the antibiotic resistance marker gene nptII in GM plants and food and feed uses. (European Medicines Agency)
10. Vandermeulen G, Marie C, Scherman D, Preat V (2011) New generation of plasmid backbones devoid of antibiotic resistance marker for gene therapy trials. Mol Ther 19:1942–1949
11. Luke J, Carnes AE, Hodgson CP, Williams JA (2009) Improved antibiotic-free DNA vaccine vectors utilizing a novel RNA based plasmid selection system. Vaccine 27:6454–6459
12. Mairhofer J et al (2010) Marker-free plasmids for gene therapeutic applications–lack of antibiotic resistance gene substantially improves the manufacturing process. J Biotechnol 146: 130–137
13. Marie C et al (2010) pFARs, plasmids free of antibiotic resistance markers, display high-level transgene expression in muscle, skin and tumour cells. J Gene Med 12:323–332

14. Soubrier F et al (1999) pCOR: a new design of plasmid vectors for nonviral gene therapy. Gene Ther 6:1482–1488
15. Nelson CA et al (2013) Antibiotic-free production of a Herpes Simplex Virus 2 DNA vaccine in a high yield cGMP process. Hum Vaccine Immunother 9:10,1–5
16. Carnes AE et al (2010) Critical design criteria for minimal antibiotic-free plasmid vectors necessary to combine robust RNA Pol II and Pol III-mediated eukaryotic expression with high bacterial production yields. J Gene Med 12: 818–831
17. Luke JM et al (2011) Improved antibiotic-free plasmid vector design by incorporation of transient expression enhancers. Gene Ther 18:334–343
18. Laddy DJ et al (2009) Electroporation of synthetic DNA antigens offers protection in nonhuman primates challenged with highly pathogenic avian influenza virus. J Virol 83: 4624–4630
19. Wang S et al (2006) Relative contributions of codon usage, promoter efficiency and leader sequence to the antigen expression and immunogenicity of HIV-1 Env DNA vaccine. Vaccine 24:4531–4540
20. Fath S et al (2011) Multiparameter RNA and codon optimization: a standardized tool to assess and enhance autologous mammalian gene expression. PLoS One 6:e17596
21. Raab D, Graf M, Notka F, Schodl T, Wagner R (2010) The GeneOptimizer Algorithm: using a sliding window approach to cope with the vast sequence space in multiparameter DNA sequence optimization. Syst Synth Biol 4: 215–225
22. Zuker M (2003) Mfold web server for nucleic acid folding and hybridization prediction. Nucleic Acids Res 31:3406–3415
23. Neron B et al (2009) Mobyle: a new full web bioinformatics framework. Bioinformatics 25:3005–3011
24. Pertea M, Lin X, Salzberg SL (2001) GeneSplicer: a new computational method for splice site prediction. Nucleic Acids Res 29:1185–1190
25. Cer RZ et al. (2012) Searching for non-B DNA-forming motifs using nBMST (non-B DNA motif search tool). *Curr Protoc Hum Genet* Chapter 18, Unit 18 17 11–22
26. Bi C, Benham CJ (2004) WebSIDD: server for predicting stress-induced duplex destabilized (SIDD) sites in superhelical DNA. Bioinformatics 20:1477–1479
27. Huang Y, Kowalski D (2003) WEB-THERMODYN: sequence analysis software for profiling DNA helical stability. Nucleic Acids Res 31:3819–3821
28. Boratyn GM et al (2013) BLAST: a more efficient report with usability improvements. Nucleic Acids Res 41:W29–W33
29. Ohba K et al (2007) Mutant influenza A virus nucleoprotein is preferentially localized in the cytoplasm and its immunization in mice shows higher immunogenicity and cross-reactivity. Vaccine 25:4291–4300
30. Rodriguez F et al (1998) DNA immunization with minigenes: low frequency of memory cytotoxic T lymphocytes and inefficient antiviral protection are rectified by ubiquitination. J Virol 72:5174–5181
31. Wu TC et al (1995) Engineering an intracellular pathway for major histocompatibility complex class II presentation of antigens. Proc Natl Acad Sci U S A 92:11671–11675
32. Luke JM et al (2011) Coexpressed RIG-I agonist enhances humoral immune response to influenza virus DNA vaccine. J Virol 85:1370–1383
33. Petersen TN, Brunak S, von Heijne G, Nielsen H (2011) SignalP 4.0: discriminating signal peptides from transmembrane regions. Nat Methods 8:785–786
34. Kall L, Krogh A, Sonnhammer EL (2007) Advantages of combined transmembrane topology and signal peptide prediction–the Phobius web server. Nucleic Acids Res 35: W429–W432
35. Falquet L, Bordoli L, Ioannidis V, Pagni M, Jongeneel CV (2003) Swiss EMBnet node web server. Nucleic Acids Res 31:3782–3783
36. Al-Saif M, Khabar KS (2012) UU/UA dinucleotide frequency reduction in coding regions results in increased mRNA stability and protein expression. Mol Ther 20:954–959
37. Luke JM, Carnes AE, Hodgson CP, Williams JA (2011) Vector insert-targeted integrative antisense expression system for plasmid stabilization. Mol Biotechnol 47:43–49
38. Wang Z et al (2004) Detection of integration of plasmid DNA into host genomic DNA following intramuscular injection and electroporation. Gene Ther 11:711–721
39. Williams JA et al (2009) Generic plasmid DNA production platform incorporating low metabolic burden seed-stock and fed-batch fermentation processes. Biotechnol Bioeng 103: 1129–1143
40. Goldstein E, Drlica K (1984) Regulation of bacterial DNA supercoiling: plasmid linking

numbers vary with growth temperature. Proc Natl Acad Sci U S A 81:4046–4050

41. Lim HM, Lewis DE, Lee HJ, Liu M, Adhya S (2003) Effect of varying the supercoiling of DNA on transcription and its regulation. Biochemistry 42:10718–10725

42. Carnes AE et al (2011) Plasmid DNA fermentation strain and process-specific effects on vector yield, quality, and transgene expression. Biotechnol Bioeng 108:354–363

43. Carnes AE, Williams JA (2011) US Patent 7943377.

Part II

Delivery System

Chapter 7

Electroporation-Based DNA Delivery Technology: Methods for Gene Electrotransfer to Skin

Anita Gothelf and Julie Gehl

Abstract

DNA delivery to for example skin and muscle can easily be performed with electroporation. The method is efficient, feasible, and inexpensive and the future possibilities are numerous. Here we present our protocol for gene transfection to mouse skin using naked plasmid DNA and electric pulses.

Key words Electroporation, Gene electrotransfer, Skin, Luciferase, In vivo, Bioimaging, Plasmid, Naked DNA

1 Introduction

Gene electrotransfer, or electroporation-based DNA transfer, is an efficient and simple means for transfecting DNA into tissues [1]. With the combination of injection of naked plasmid DNA and subsequent delivery of short electric pulses, it is possible to obtain a protein expression which can be evaluated both locally at the treated area as well as systemically in for example serum [2, 3]. The technique is widely accepted in the preclinical setting and a large number of clinical trials have been conducted or are underway [4, 5]. Gene electrotransfer is a highly interesting method with many future possibilities since the outcome of the transfection depends on the plasmid. If the plasmid encodes a therapeutic protein, e.g., an anticancer molecule, the result of the transfection will be production of the transgene protein, which subsequently can act as an anticancer drug. If the plasmid instead is encoding known antigens from infectious agents, the transfection will act as a DNA vaccine and the effect will be an immunological response.

Muscle and skin are the preferred organs for gene electrotransfer [6, 7] and a wide number of techniques have been used, but common to them all is the injection of plasmid and subsequent delivery of electric pulses. The injection technique varies from conventional

Monica Rinaldi et al. (eds.), *DNA Vaccines: Methods and Protocols*, Methods in Molecular Biology, vol. 1143, DOI 10.1007/978-1-4939-0410-5_7, © Springer Science+Business Media New York 2014

intradermal or intramuscular injections to needle-free jet injection devices. However much more variability exists in terms of electrodes used and electric pulses delivered [8]. Up to now there is no consensus regarding which treatment protocol is the most efficient, and many different research groups have received excellent results with different injection volumes and electrical pulse parameters.

We have extensive experience regarding particularly transfection of naked DNA into skin and have mainly worked with plasmids encoding reporter proteins or hormones. In this chapter we will present our treatment protocols regarding transfection of luciferase plasmid to mouse skin evaluated in vitro on tissue homogenates and in vivo bioimaging.

2 Materials

2.1 Plasmid Purification

1. Plasmid: Firefly luciferase plasmid (pCMV-luc, Promega).
2. Plasmid purification kit: Qiafilter Plasmid Maxiprep kit (Qiagen).
3. Dissolving agent: PBS for a plasmid concentration of 1 μg/μL (*see* **Note 1**).

2.2 Animal Preparation

1. Mice: 9–15 week old NMRI.
2. Anesthesia: Hypnorm (Janssen Saunderton), Dormicum (Roche).
3. 27 G syringe.
4. Shaving machine.
5. Depilatory cream.

2.3 Equipment for Electroporation

1. 29 G insulin syringe.
2. Electroporator: Cliniporator (IGEA, Carpi, Italy).
3. Custom-made plate electrodes (Fig. 1).
4. Ultrasound gel: EKO-GEL (Ekkomarine Medico, Holstebro, Denmark).

2.4 In Vitro Analysis

1. Lysis buffer: Luciferase Reporter Gene Assay (Roche).
2. Tissue homogenization: Tissuelyzer (Qiagen).
3. 2 mL tubes (Eppendorf).
4. Microfuge.
5. Stainless steel beads, 5 mm diameter.
6. Reporter gene assay: Luciferase Reporter Gene Assay (Roche).
7. Luminometer: Lumistar Galaxy (BMG Labtech).

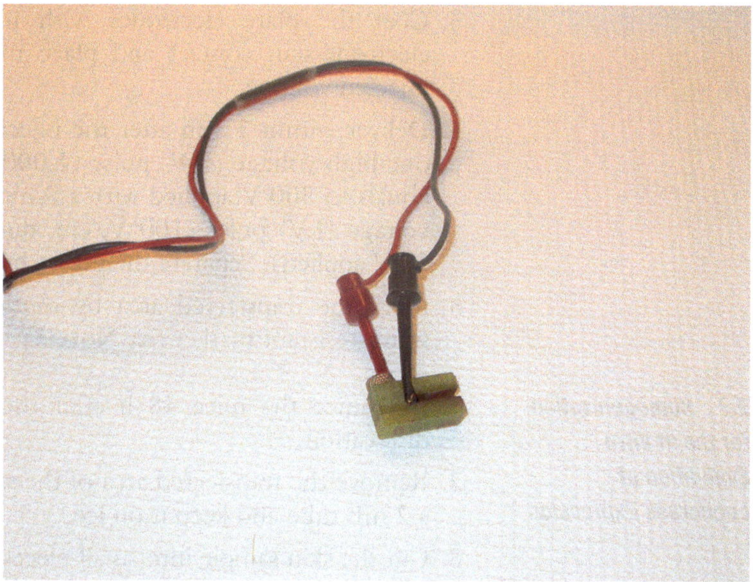

Fig. 1 Custom-made electrode for gene electrotransfer to skin. The distance between the electrode blades can be adjusted from 2 to 4 mm. For gene electrotransfer to mouse skin a distance of 3 mm between the blades is suitable when the volume injected is 100 μL

8. White 96-well plate (Nunc).
9. Luciferase standard concentrations: Recombinant luciferase (Promega Corporation), Luciferase Reporter Gene Assay (Roche), bovine serum albumin (BSA) 2.5 mg/mL.

2.5 In Vivo Analysis

1. Luciferin 10 mg/mL: Beetle luciferin potassium salt (Promega Corporation), 1× PBS.
2. 27 G syringe.
3. Bioimaging: Optix MX-2 Time Domain Optical Imaging (ART Advanced Research Technologies, Montreal, Canada).

3 Methods

3.1 Gene Electrotransfer Procedure

1. Anesthetize mice with a combination of Hypnorm and Dormicum, 0.4 mL/kg and 2 mg/kg respectively via intraperitoneal injection using a 27 G syringe. Shave patches of skin of approximately 3 × 2 cm and remove remaining hair with depilatory cream.
2. Inject 100 μL of plasmid solution intradermally using a 29 G insulin syringe (*see* **Note 2**).

3. Coat the plate electrodes with ultrasound gel to secure electrode-skin contact and place the electrodes around the injected volume.
4. Deliver within 2 min after the injection the electric pulses as one high voltage (HV) pulse (1,000 V/cm, duration 100 μs; equals to 300 V applied with a 3 mm electrode) and one low voltage (LV) pulse (100 V/cm, duration 400 ms; equals to 30 V applied). Separate the pulses by a lag of 1 s.
5. Mark the transfected area by outlining the borders with a water-resistant marker (*see* **Note 3**).

3.2 Homogenization for the In Vitro Evaluation of Luciferase Expression

1. Euthanize the mice 48 h after the transfection by cervical dislocation.
2. Remove the transfected area of the skin (*see* **Note 4**), put it in a 2 mL tube and keep it on ice.
3. Cut the skin sample into small pieces and note the weight.
4. Cover the skin sample with 0.5 mL of lysis buffer and leave it at room temperature for 10–20 min.
5. Put the tube on ice until further processing.
6. Add the stainless steel beads to the 2 mL tube and homogenize the samples 3 × 3 min at 30 Hz. (*see* **Note 5**).
7. Add additional 0.5 mL of lysis buffer and spin the samples for 10 min at 3,000 rpm.
8. Remove the supernatant for evaluation of the activity.

3.3 In Vitro Evaluation of Expression in a Luminometer

1. Place 50 μL of each sample in wells in a white 96-well plate.
2. Measure the luciferase activity in counts in relative light units (RLU).
3. Convert the RLUs to microgram of luciferase per sample by a standard curve of known luciferase concentrations (*see* **Note 6** and 7).

3.4 In Vivo Bioluminescence Evaluation of Luciferase Expression

1. Anesthetize the mice and remove the hair as described above (*see* **Note 8**).
2. Inject 300 μL of luciferin intraperitoneally using a 27 G syringe.
3. Place the mice in the bioimager after 15–20 min and scan for bioluminescence.

4 Notes

1. In the experiments described above we have transfected a plasmid encoding luciferase into skin. For reasons explained below luciferase may not always be the optimal choice.

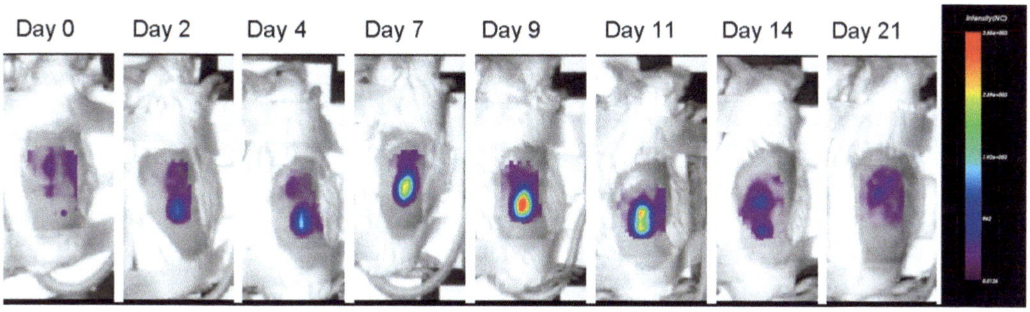

Fig. 2 In vivo bioimaging of gene electrotransfer with Katushka to skin. It is possible with bioimaging to perform repeated scans and hence monitor the same animal over time. The present figure depicts an animal scanned at different time points and a peak in expression observed at day 9 [9]

We have worked extensively with a plasmid encoding the erythropoietin hormone (EPO), which has many advantages: (a) you can measure the transgene protein directly in the blood by an ELISA and (b) you can evaluate the effect of the transgene protein by measuring the end product of the erythropoietin hormone, namely, hemoglobin [3]. Another plasmid, which is a relevant alternative, is the pTurboFP635-c, which encodes Katushka, a protein in the far-red area [9]. Katushka can also be evaluated by in vivo bioimaging, which allows for sequential in vivo scans of the same animal (Figs. 2 and 3). An advantage compared to luciferase is that it is not necessary to inject a substrate (like luciferin), and hence, there are no issues concerning distribution and degradation.

2. Mice skin is very thin and one has to be very careful when making intradermal injections; the injected volume can easily pass to the subcutaneous layer. We found that holding the skin and lift it with a set of tweezers and then inject the volume decreased the fraction of subcutaneous injection remarkably. The injected volume is visible as a bleb in the skin if injected correctly.
3. We have in our studies with gene transfection to mice skin used noninvasive plate electrodes for electroporation (Fig. 1). It can be an advantage to use invasive needle electrodes in thicker skin such as pig skin (Fig. 4) since a higher level of expression can be achieved [10]. Furthermore, it is an advantage to dip the needles in a well of ink before inserting them into then skin; that creates a perfect landmark for the transfection.
4. The protocol described can, if including minor changes, be applicable in other tissues such as muscle. Particular the pulsing parameters and the amount of plasmid needed to obtain an expression are different. The HV pulse is delivered with 800 V/cm instead of 1,000 V/cm and the plasmid is in the dose range of 1–10 μg compared to 75–100 μg in skin.

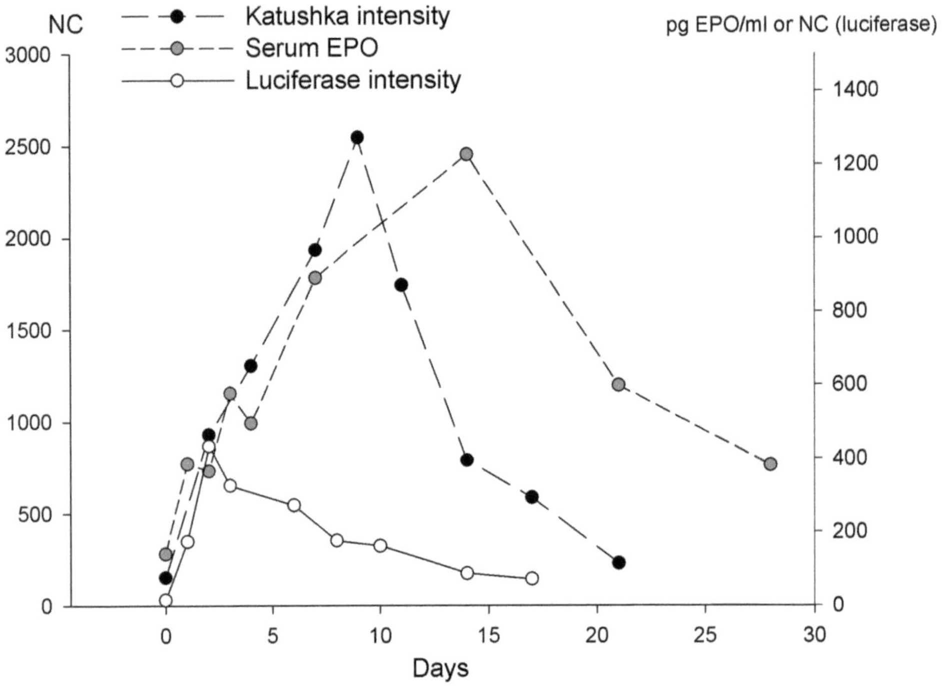

Fig. 3 Comparison of gene electrotransfer to skin with plasmids encoding luciferase, Katushka, and EPO. Katushka is evaluated with bioimaging and the EPO expression is evaluated by measuring the serum EPO level by ELISA. With Katushka a peak in expression (in NC) is reached after 9 days, whereas serum EPO (in pg EPO/mL) seems to peak after 14 days, modified from ref. 9. In contrast, it is observed that the luciferase expression peaks after few days

5. Instead of using a Tissuelyzer, or similar equipment, the skin samples can be grinded manually. First, they are placed in a mortar and covered with liquid nitrogen. When snap-frozen, the skin is then grinded quickly in the mortar. It has to be very quick, since the small skin samples thaw easily. Afterwards, the grinded skin is placed in a 2 mL tube and covered with 1 mL of lysis buffer. Finally the mixture is homogenized with a homogenizer (e.g., S8N-5G homogenizer, IKA-Werke, Germany), spun for 10 min at 3,000 rpm. The luciferase expression can then be measured in the supernatant as described above.
6. Luciferase expression after gene transfection displays, particular when evaluated in vitro, a large variability, often of several logs [10, 11]. Furthermore, there is variability between studies, even when conducted with similar conditions, which must be taken into consideration when reporting the results.
7. Luciferase is evaluated by bioluminescence and in our studies we found that at least two factors can quench the signal: (a) initially, we used C57Bl/6 mice for gene transfection to skin, but observed that they have black spots in their skin.

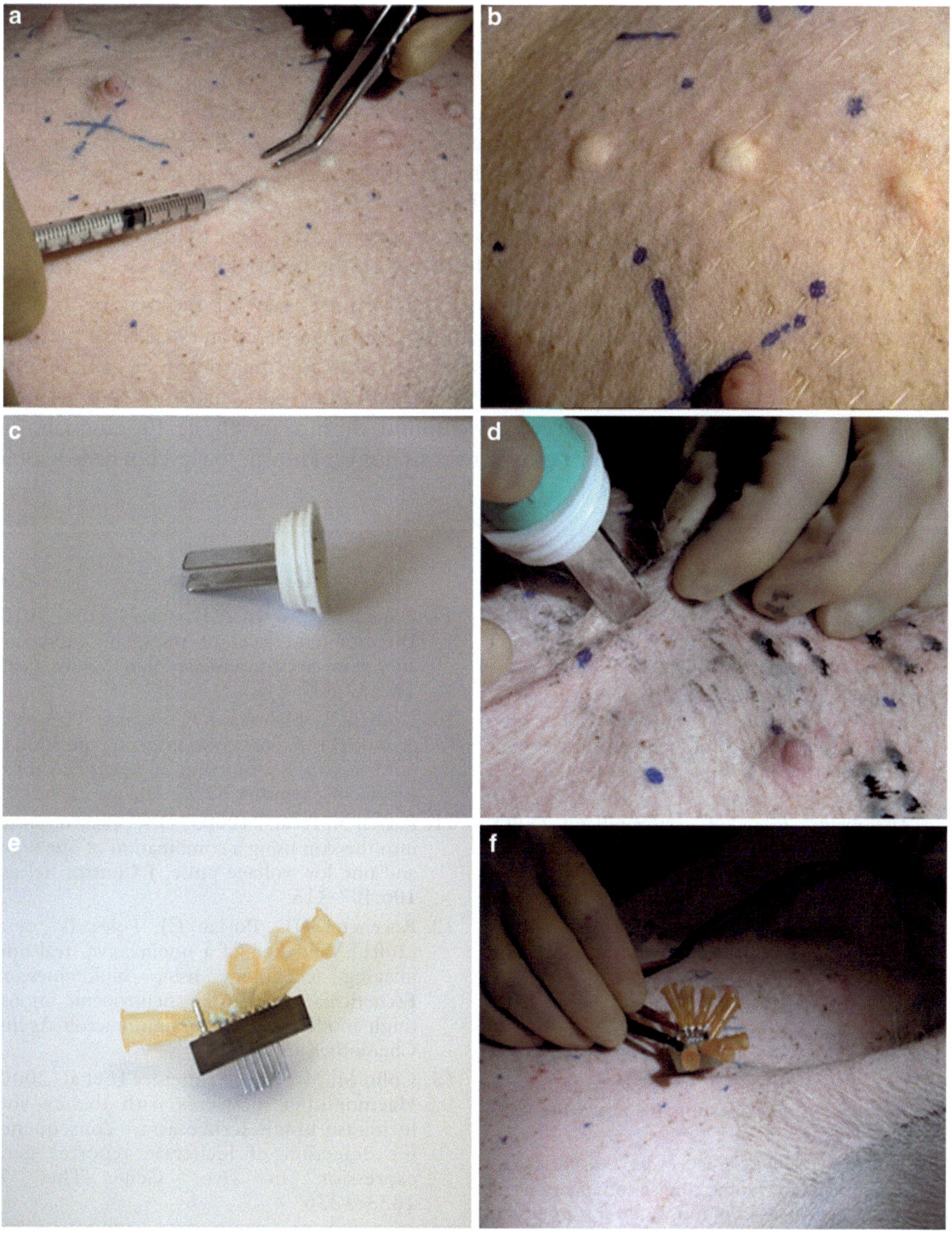

Fig. 4 The gene electrotransfer procedure depicted in a porcine model: (**a**) injection of the plasmid; (**b**) the visible bleb after plasmid injection; (**c**) commercially available plate electrode (IGEA, Carpi, Italy); (**d**) delivery of pulses with the plate electrode; (**e**) custom-made needle electrode with 2 × 4 needles, 6 mm between the rows; (**f**) delivery of pulses with the needle electrode [10]

We found that the expression of luciferase was lower or absent in skin samples with dark pigmentation and we thus changed the animal model to NMRI [12]. (b) We have conducted gene transfection to pig skin in vivo, and at the time for evaluation of the transfection efficacy, the transfected area was removed with an 8 mm punch biopsy. If the biopsy was too deep, and hence it contained superfluous amount of subcutaneous fat tissue and blood, the luciferase expression was lower than expected, probably due to interference with the hemoglobin [13].

8. If the animals are anesthetized, not for bioimaging but for example blood sampling after transfection with EPO plasmid, caution should be made regarding measurement of hemoglobin levels. In our work with gene transfection to skin with the EPO plasmid, we found during anesthesia the animals get hypoxic, and for reasons not yet known, the level of hemoglobin drops [14].

References

1. Mir LM, Moller PH, Andre F et al (2005) Electric pulse-mediated gene delivery to various animal tissues. Adv Genet 54:83–114
2. Hojman P, Gissel H, Gehl J (2007) Sensitive and precise regulation of haemoglobin after gene transfer of erythropoietin to muscle tissue using electroporation. Gene Ther 14:950–959
3. Gothelf A, Hojman P, Gehl J (2010) Therapeutic levels of erythropoietin (EPO) achieved after gene electrotransfer to skin in mice. Gene Ther 17:1077–1084
4. Daud AI, DeConti RC, Andrews S et al (2008) Phase I trial of interleukin-12 plasmid electroporation in patients with metastatic melanoma. J Clin Oncol 26:5896–5903
5. Gothelf A, Gehl J (2012) What you always needed to know about electroporation based DNA vaccines. Hum Vaccin Immunother 8:1694–1702
6. Cemazar M, Sersa G (2007) Electrotransfer of therapeutic molecules into tissues. Curr Opin Mol Ther 9:554–562
7. Andre FM, Gehl J, Sersa G et al (2008) Efficiency of high- and low-voltage pulse combinations for gene electrotransfer in muscle, liver, tumor, and skin. Hum Gene Ther 19:1261–1271
8. Gothelf A, Gehl J (2010) Gene electrotransfer to skin; review of existing literature and clinical perspectives. Curr Gene Ther 10:287–299
9. Gothelf A, Eriksen J, Hojman P et al (2010) Duration and level of transgene expression after gene electrotransfer to skin in mice. Gene Ther 17:839–845
10. Gothelf A, Mahmood F, Dagnaes-Hansen F et al (2011) Efficacy of transgene expression in porcine skin as a function of electrode choice. Bioelectrochemistry 82:95–102
11. Pavselj N, Preat V (2005) DNA electrotransfer into the skin using a combination of one high- and one low-voltage pulse. J Control Release 106:407–415
12. Rocchetta HL, Boylan CJ, Foley JW et al (2001) Validation of a noninvasive, real-time imaging technology using bioluminescent Escherichia coli in the neutropenic mouse thigh model of infection. Antimicrob Agents Chemother 45:129–137
13. Colin M, Moritz S, Schneider H et al (2000) Haemoglobin interferes with the ex vivo luciferase luminescence assay: consequence for detection of luciferase reporter gene expression in vivo. Gene Ther 7: 1333–1336
14. Gothelf A, Hojman P, Gehl J (2009) Change in hemoglobin levels due to anesthesia in mice: an important confounder in studies on hematopoietic drugs. Biol Proced Online 11: 325–330

Chapter 8

DNA Vaccination in Skin Enhanced by Electroporation

Kate E. Broderick, Amir S. Khan, and Niranjan Y. Sardesai

Abstract

DNA vaccines are a next generation branch of vaccines which offer major benefits over their conventional counterparts. However, to be effective in large mammals and humans, an enhancing delivery technology is required. Electroporation is a physical technique which results in improved delivery of large molecules through the cell membrane. In the case of plasmid DNA, electroporation enhances both the uptake and expression of the delivered DNA. The skin is an attractive tissue for DNA vaccination in a clinical setting due to the accessibility of the target, the ease of monitoring, and most importantly the immunocompetent nature of the dermis. Electroporation in the skin has the benefit of being minimally invasive and generally well tolerated. Previous studies have determined that optimized electroporation parameters (such as electrical field intensity, pulse length, pulse width, and plasmid formulation) majorly impact the efficiency of DNA delivery to the skin. We provide an overview of DNA vaccination in skin and muscle. In addition, we detail a protocol for the successful intradermal electroporation of plasmid DNA to guinea pig skin, an excellent dermatological animal model. The work detailed here suggests that the technique is safe and effective and could be highly applicable to a clinical setting.

Key words Electroporation, Skin, Plasmid DNA, Guinea Pig, DNA vaccine

1 Introduction

Electroporation (EP) involves the application of brief electrical pulses that result in the creation of aqueous pathways within the lipid bilayer membranes of mammalian cells. This allows the passage of large molecules, including DNA and other macromolecules, through the cell membrane which would otherwise unable to cross. As such, EP increases both the uptake and the extent to which drugs and DNA are delivered to the target tissue of interest [1–4]. Historically, EP has been primarily targeted to muscle tissue and currently multiple clinical trials are being conducted using this route of delivery [5, 6].

EP as a mode of enhanced delivery provides a platform for the in vivo manufacture of the desired gene product which, in the case of DNA vaccines, can lead to both antibody and cellular immune responses. Optimized DNA plasmid vectors through

Monica Rinaldi et al. (eds.), *DNA Vaccines: Methods and Protocols*, Methods in Molecular Biology, vol. 1143,
DOI 10.1007/978-1-4939-0410-5_8, © Springer Science+Business Media New York 2014

codon optimization, RNA optimization, the addition of leader sequences and optimized consensus sequences, are vital to elicit robust responses [7–11]. However, the resulting immune responses from such plasmids are significantly enhanced with EP when compared to injection alone [12, 13].

Intradermal (ID) EP provides a convenient and accessible site for the delivery of antigenic proteins [14, 15]. Since the target tissue of skin is considerably shallower than skeletal muscle from a depth perspective (*see* **Note 1**), dermal EP devices can be designed to be much less invasive and indeed even completely noninvasive. This has the important implication from a patient tolerability standpoint of not activating deep nerves and muscles.

While EP clearly improves the delivery of DNA in vivo, the electrical parameters must be adjusted to ensure optimal delivery. Previously, the optimal input voltage for the surface EP device (SEP) was investigated [16]. Here we described the optimal EP parameters for efficient delivery of DNA vaccines with a specific emphasis on eliciting cellular immunity in addition to robust humoral responses. Low voltage (5–10 V) EP parameters elicited higher and more sustained cellular immune response as well as robust humoral responses when compared to the higher voltage (above 20 V) EP parameters after two immunizations confirming that low voltage EP using the SEP device is capable of efficient delivery of DNA vaccines into the skin without tissue damage. The SEP targets the epidermal region of skin. Depending on the depth of penetration, other ID EP devices protrude further into the skin and so target cells in the dermis or sub-dermal region.

Influenza vaccine delivered via the SEP device was also able to induce potent cellular responses in a mouse model and most importantly was able to demonstrate 100 % protection in a lethal challenge [17]. Protective HAI titers were also induced in guinea pigs and macaques demonstrating proof of concept efficacy for DNA delivery with this device [17].

In this protocol, we describe administration of a plasmid encoding green fluorescent protein (GFP) at different concentrations to the skin of guinea pigs. The DNA is delivered to the skin through an intradermal injection and uptake of the GFP plasmid is enhanced by surface electroporation. GFP expression is determined over a 2 day period since this was previously determined to be the peak of expression in the skin. Fluorescence microscopy is used to visualize the resulting gene expression in excised skin. We provide strategies to optimize the intradermal injections in guinea pig as well as improve the reproducibility of the EP procedure. In addition, we detail different plasmid doses and the impact on expression.

2 Materials

2.1 Plasmid Preparation

1. Plasmid DNA, pgWIZ-GFP (Aldevron, ND). Plasmid DNA is formulated into a volume of between 50 and 100 μL per injection site.
2. Sterile saline.

2.2 Animals

1. Hartley guinea pigs (Charles River) between 4–6 weeks of age and 400–600 g in weight.
2. Approved animal experimental protocol (in accordance with the NIH, Animal Welfare Act and USDA).
3. Electric clippers (for hair removal).
4. Oxygen and isoflurane delivered through rodent face mask (Stoelting, IL).

2.3 Intradermal Injection

1. 29 G Insulin needle and syringe.
2. Forceps, to assist with the intradermal injection.
3. Lab marker, to indicated treatment area.
4. Sharps container.

2.4 Electroporation Procedure

1. ELGEN pulse generator (Inovio Pharmaceuticals, Blue Bell, PA, USA) (Fig. 1b).
2. Surface EP device (Inovio Pharmaceuticals) (Fig. 1a).

2.5 Tissue Harvest and Analysis

1. Pentobarbital (86 mg/kg).
2. 21 G 1/2 in. needle with 1 mL syringe.
3. Scalpel.

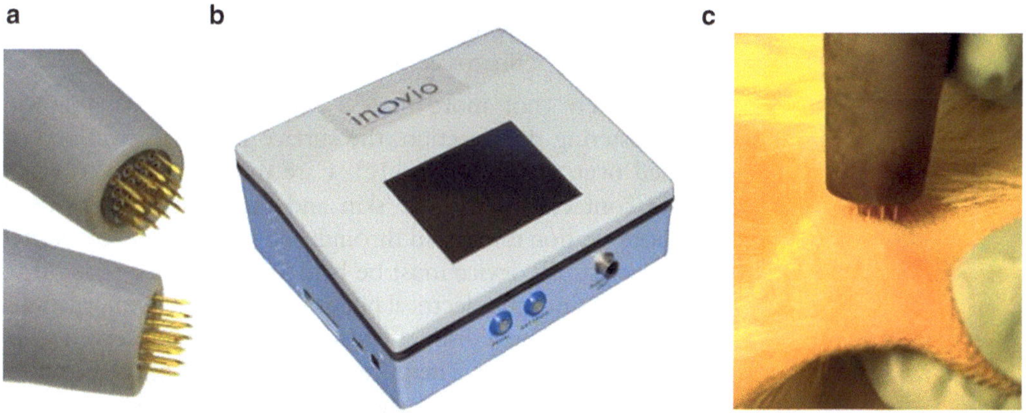

Fig. 1 Skin electroporation procedure. (**a**) Surface EP device (Inovio Pharmaceuticals). Photo shows electrode configuration. (**b**) ELGEN pulse generator (Inovio Pharmaceuticals) which is tethered to the surface EP device and delivers the electrical pulses. (**c**) Positioning of the surface EP device on guinea pig skin

4. Sharps container.
5. Ziplock bag.
6. Fluorescent microscope and camera (MagnaFire software).
7. Adobe Photoshop and Microsoft Excel.

3 Methods

3.1 Plasmid Preparation

1. Dilute plasmid preparations (prepare dose curve of GFP plasmid) in 1 mL injectable saline. Dilutions were 6, 4.5, 3, 2, 1.5, 1, and 0.5 mg/mL.
2. Prepare syringes with the volume of plasmid to be injected.

3.2 Intradermal Injection and Electroporation Procedure

1. House animals in groups or individually based on intuitional cage size with ad lib access to food and water. Randomly assign guinea pigs to treatment groups. Allow acclimatization for 5 days.
2. Anesthetize the animals with inhaled isoflurane (5 %) and maintain light anesthesia with isoflurane (3 %) through appropriately sized face mask.
3. Carefully choose the treatment site, preferably opting for a central flank area (*see* **Note 2**) with no scarring or existing skin irritation.
4. Shave the flank of the animals to be treated. Clean the area with ethanol to ensure full removal of oil, dust, and dander.
5. Initiation of electroporation and DNA injection (Fig. 1). Administer intradermal injection using standard Mantoux technique (*see* **Note 3**). A pair of forceps can used to assist with the positioning of the needle (*see* **Note 4**). A successful intradermal injection will result in a small pea sized (5 mm^2 diameter), glassy bleb. Following the injection, the needle and syringe are disposed of in a Sharps container.
6. Immediately (not more than 2 min; *see* **Note 5**), following administering the injection, the surface EP device will be positioned over the bleb (Fig. 1c). Care must be taken to ensure good contact between the skin and the device (*see* **Note 6**). Electroporation is initiated through activation of the foot switch. The surface EP device must be held firmly in place thought the procedure since the electrical pulses will cause involuntary muscle contractions. Three distinct contractions will be observed with a successful treatment. The pulse generator will sound an audible series of beeps to mark a successful treatment. The device can now be removed from the skin surface.
7. If multiple treatments are planned, the device must be cleaned between treatments with a toothbrush to remove skin debris which builds up on the electrodes (*see* **Note 7**).

8. The animals must be monitored for 2 h to ensure a full recovery from anesthesia.
9. Peak GFP reporter gene expression is detected between 24 and 72 h (*see* **Note 8**).
10. The animals must be euthanized using standard institutional procedures at the desired time point (*see* **Note 9**). Here, anesthetized animals are sacrificed following a cardiac puncture with 0.5 mL pentobarbital.
11. Excise the treated skin from the animal using a scalpel blade. Remove with care excess skin around the treatment area. The excised skin can then be kept flat or rolled (depending on sample size) and stored in a Ziplock bag on ice. If the skin will not be processed immediately, it can be frozen at −20 °C. Mark the skin with standard lab marker pens but only black ink should be used (*see* **Note 10**).
12. To view GFP expression, mount the excised skin on a flat surface (*see* **Note 11**) and view it under low magnification fluorescence microscopy. The entire treatment site can be captured in its entirety or each individual site addressed. Images can be saved as TIFF or BITMAP files (Fig. 2a).
13. Using commercially available pixel counting software (Photoshop), each treatment site can be analyzed for expression (Fig. 2b).

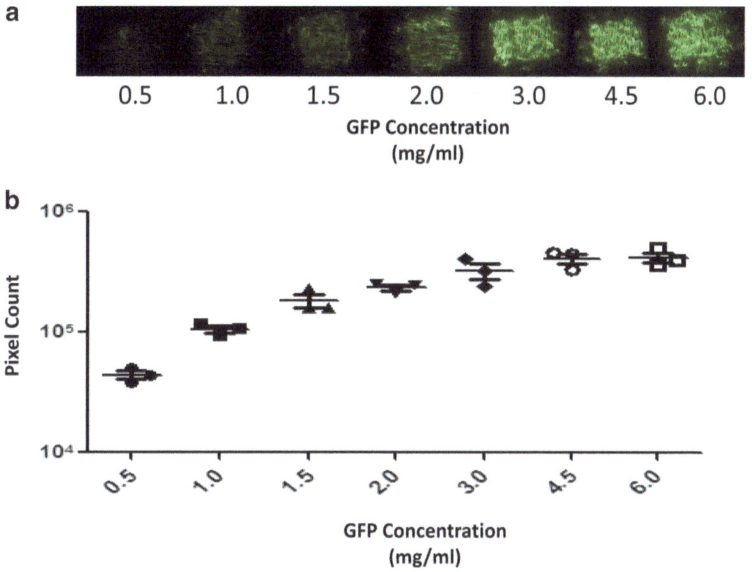

Fig. 2 GFP expression visualization and quantification. (**a**) Dose varied GFP plasmid expression in guinea pig skin visualized under fluorescent microscope. The treatments sites shown here are representative examples. (**b**) Graphing of the pixel count numbers relating to expression of GFP at different doses. Adobe Photoshop was used to determine the pixel number over background

14. Statistical analysis between treatments can be performed using the Student's *t*-test program in the Microsoft Excel statistics package.

4 Notes

1. Aspects of DNA vaccination related to delivery in either skin or muscle are summarized in Table 1.
2. The thickness of the skin can directly affect the level of reporter gene expression. As such, ensuring consistent skin thickness between treatment sites will result in more reproducible results. Using a central location on the flank of the animal, keeping away from the spine and belly regions, is optimal. A 10 $cm^2 \times 5$ cm^2 region on either side of the animal can usually be used allowing for approximately 40 treatment sites.
3. The Mantoux injection method [18] involves the insertion of a thin gauge (28 G or less) needle, bevel up in a parallel orientation to the skin. This results in an ultra-shallow delivery of drug to the epidermal region of skin. This technique requires practice and skill and is frequently performed incorrectly [19].
4. To assist with the needle positioning when injecting intradermally, a pair of forceps can be used to pinch up the skin into a U-shaped formation. The injection needle can then be inserted

Table 1
DNA vaccination in skin and muscle

	Intramuscular	Intradermal
Method of injection	Needle and syringe Jet injection	Needle and syringe (Mantoux method) Jet injection Gene gun Microneedles
Injection depth	1,000–5,000 μm	10–500 μm
Injection volume	1,000–2,000 μL	Typically less than 100 μL
Targeted tissue region	Skeletal Muscle	Epidermis Dermis Sub-dermis
Targeted cell type	Myocytes	Dendritic cells Keratinocytes
Max. dose (estimated on 10 mg/mL formulation)	10–20 mg	Less than 1 mg

into the skin between the arms of the forceps, allowing for a larger target surface.

5. The intradermally delivered drug will begin to dissipate at the treatment site approximately 2 min following injection. It is imperative to EP before this occurs. It is best practice to inject and immediately EP. In addition, since the skin is particularly rich in endonucleases, EP should occur as fast as possible to avoid degradation of the plasmid DNA.
6. Good contact is required for the EP procedure to work effectively. The skin can be pinched from behind to allow the application of back pressure behind the device. The patterning of the device electrodes should be imprinted in the skin following a successful usage of this device.
7. Although the device can be used multiple times between cleanings, it is good practice to remove debris from the electrodes on a regular basis. The oil and skin cell build up can become "baked" onto the electrodes and reduce the effectiveness of the device.
8. GFP expression can be detected in the skin as early as 1 h post treatment (microscopically) and persist out to 12 days.
9. Here we chose an intercardiac injection of pentobarbital. The animals are monitored for 10 min to ensure no life signs.
10. A 15 cm × 15 cm square of Perspex plastic is ideal.
11. Lab markers come in a variety of colors. However, the ink in the red and green pens auto-fluoresce under blue light illumination and can cause confusion when visualizing GFP positive skin. Care should be taken only to use black ink to ring treatment sites or label animals.

Acknowledgments

The authors would like to thank Janess Mendoza for help with the animal procedures. This work was supported by Inovio Pharmaceuticals (Blue Bell, PA).

References

1. Widera G, Austin M, Rabussay D et al (2000) Increased DNA vaccine delivery and immunogenicity by electroporation in vivo. J Immunol 164:4635–4640
2. Prud'homme GJ, Draghia-Akli R, Wang Q (2007) Plasmid-based gene therapy of diabetes mellitus. Gene Ther 14:553–564
3. Otten G, Schaefer M, Doe B et al (2004) Enhancement of DNA vaccine potency in rhesus macaques by electroporation. Vaccine 22:2489–2493
4. Mathiesen I (1999) Electropermeabilization of skeletal muscle enhances gene transfer in vivo. Gene Ther 6:508–514
5. Bagarazzi ML, Yan J, Morrow MP et al (2012) Immunotherapy against HPV16/18 generates potent TH1 and cytotoxic cellular immune responses. Sci Transl Med 4:155ra138

6. El-Kamary SS, Billington M, Deitz S et al (2012) Safety and tolerability of the Easy Vax clinical epidermal electroporation system in healthy adults. Mol Ther 20:214–220
7. Andre S, Seed B, Eberle J et al (1998) Increased immune response elicited by DNA vaccination with a synthetic gp120 sequence with optimized codon usage. J Virol 72:1497–1503
8. Deml L, Bojak A, Steck S et al (2001) Multiple effects of codon usage optimization on expression and immunogenicity of DNA candidate vaccines encoding the human immunodeficiency virus type 1 Gag protein. J Virol 75: 10991–11001
9. Muthumani K, Zhang D, Dayes NS et al (2003) Novel engineered HIV-1 East African Clade-A gp160 plasmid construct induces strong humoral and cell-mediated immune responses in vivo. Virology 314:134–146
10. Schneider R, Campbell M, Nasioulas G et al (1997) Inactivation of the human immunodeficiency virus type 1 inhibitory elements allows Rev-independent expression of Gag and Gag/protease and particle formation. J Virol 71: 4892–4903
11. Yang JS, Kim JJ, Hwang D et al (2001) Induction of potent Th1-type immune responses from a novel DNA vaccine for West Nile virus New York isolate (WNV-NY1999). J Infect Dis 184:809–816
12. Miyazaki S, Miyazaki J (2008) In vivo DNA electrotransfer into muscle. Dev Growth Differ 50:479–483
13. Draghia-Akli R, Khan AS, Cummings KK et al (2002) Electrical enhancement of formulated plasmid delivery in animals. Technol Cancer Res Treat 1:365–372
14. Gothelf A, Gehl J (2010) Gene electrotransfer to skin; review of existing literature and clinical perspectives. Curr Gene Ther 10:287–299
15. Charoo NA, Rahman Z, Repka MA et al (2010) Electroporation: an avenue for transdermal drug delivery. Curr Drug Deliv 7:125–136
16. Lin F, Shen X, Kichaev G et al (2012) Optimization of electroporation-enhanced intradermal delivery of DNA vaccine using a minimally invasive surface device. Hum Gene Ther Methods 23:157–168
17. Broderick KE, Shen X, Soderholm J et al (2011) Prototype development and preclinical immunogenicity analysis of a novel minimally invasive electroporation device. Gene Ther 18: 258–265
18. Andersen P, Doherty TM (2005) The success and failure of BCG — implications for a novel tuberculosis vaccine. Nat Rev Microbiol 3:656–662
19. Flynn PM, Shenep JL, Mao L et al (1994) Influence of needle gauge in Mantoux skin testing. Chest 106:1463–1465

Chapter 9

Intradermal Vaccination by DNA Tattooing

Joost H. van den Berg, Koen Oosterhuis, Ton N.M. Schumacher, John B.A.G. Haanen, and Adriaan D. Bins

Abstract

DNA vaccination is an attractive vaccination method. First, the production of plasmid DNA as a vaccine is considerably more cheap and simple than the production of recombinant protein. Second, the expression cassette of DNA vaccines can readily be modified, making DNA vaccines highly flexible. Finally, in animal models, DNA vaccination is able to induce potent cellular immune responses.

Over the past decade, the focus in the DNA vaccination field has in large part moved from intramuscular immunization towards dermal administration. As a natural "porte d'entrée" for pathogens, the skin is rich in antigen-presenting cells, which are required for generating an efficient antigen-specific immune response.

This chapter describes a DNA vaccination protocol that utilizes a simple tattooing device for the dermal delivery of plasmid DNA. This technique, called DNA tattooing, is capable of generating high frequencies of antigen-reactive T cells in mice and macaques.

Key words DNA vaccination, Tattooing, Intradermal delivery, Immune response, Plasmid DNA

1 Introduction

DNA vaccines have an exceedingly simple design, solely consisting of plasmid DNA that encodes an antigen of interest under the control of a eukaryotic viral promoter. Upon introduction of the DNA vaccine, both professional antigen-presenting cells (pAPCs) and non-APCs can take up the plasmid and produce the encoded antigen. Similar to conventional vaccines, the production of DNA vaccine-encoded antigens can lead to the induction of an antigen-specific immune response, and DNA vaccination is considered a particularly efficient method for the activation of the cellular arm of the adaptive immune system [1].

The fact that DNA vaccines can induce strong cellular immune responses—at least in preclinical models—makes them attractive candidates for vaccination against intracellular pathogens and cancer. Because the expression cassette of DNA vaccine plasmids

Monica Rinaldi et al. (eds.), *DNA Vaccines: Methods and Protocols*, Methods in Molecular Biology, vol. 1143,
DOI 10.1007/978-1-4939-0410-5_9, © Springer Science+Business Media New York 2014

can readily be modified by conventional molecular cloning techniques, it is straightforward to tailor/optimize the characteristics of the produced protein. As such, DNA vaccines could potentially be utilized to induce immune activity in personalized vaccine approaches, for instance to target patient-specific mutant antigens in human cancer [2, 3].

The successful induction of an immune response against the vaccine-encoded antigens requires antigen processing and presentation by professional APCs. Since the skin, like other epithelia, is a natural entry site for pathogens, a high number of APCs is present in the upper layers of the skin to guard against invading pathogens. The most abundant APCs in the skin are the epidermal Langerhans cells (LCs) and the dermal dendritic cells (dermal DCs) [4]. In addition, macrophages can infiltrate into infected/inflamed skin and can also function as APC [5]. The exact contribution of these different APC types is currently under investigation [6–9]. Nevertheless, the observation that expression of antigen in nonmigratory skin cells can lead to strong immune responses is consistent with antigen cross presentation by migratory APCs [10].

Over the past years, several devices for the (epi)dermal administration of plasmids have been developed, such as the gene gun, jet injectors, micro-needle systems, and electroporation-based devices [11].

The technique that is described in this chapter, called DNA tattooing [12], relies on the use of a permanent makeup device. This device, which is commonly applied in cosmetic tattooing, delivers the plasmid solution into the skin by thousands of small punctures. Following DNA tattooing, expression of the vaccine-encoded antigen is observed in skin cells, primarily keratinocytes in the epidermis. Importantly, even though the number of antigen-expressing cells is low and expression is transient [12, 13], DNA tattoo results in a very strong antigen-specific cellular immune response in animal models. Moreover short interval DNA tattooing yields high levels of antigen-specific T cells within 2 weeks after the start of vaccination, considerably faster than what is achieved by conventional intramuscular vaccination regimens [12]. Besides a detailed description of DNA tattooing in mice this chapter describes techniques to determine antigen expression and antigen-specific T cell responses.

2 Materials

2.1 Intradermal DNA Tattoo Vaccination

1. Animal model: C57BL/6 mice, normally 6–10 weeks of age.
2. Vaccine-encoding plasmid: Gene of interest in a eukaryotic expression vector (*see* **Note 1**) amplified in *E. coli* DH5α and purified (*see* **Note 2**). Resuspend the isolated DNA to a final concentration of 2 mg/mL in dH_2O.

3. Depilatory cream or hair trimmer.
4. Tattoo device and needle cartridge (*see* **Note 3**).
5. Anesthesia: Inhalation of mixture of 3 % isoflurane and 97 % oxygen or another accepted form of (intraperitoneal) anesthetic.

2.2 Intravital Imaging Luciferase Expression

1. D-Luciferin dissolved in phosphate-buffered saline (PBS) at 15 mg/mL, filtered through 0.2 μm filter and aliquoted. Sterile aliquots are stored at −20 °C.
2. Sterile syringe and needle (29 G).
3. Anesthesia: Inhalation of mixture of 3 % isoflurane and 97 % oxygen or another accepted (intraperitoneal) anesthetic.
4. CCD camera bioluminescence camera with software.

2.3 Flow Cytometry-Based Immunomonitoring

1. Scissors.
2. Mouse restrainer.
3. FACS buffer: 1× PBS, 0.5 % bovine serum albumin (BSA), 0.02 % sodium azide.
4. 1.5 mL tubes (Eppendorf) or a similar container.
5. Heparin dissolved in PBS, 5 I.U./mL.
6. Erylysis buffer: 156 mM NH_4Cl, 10 mM $KHCO_3$, 0.1 mM EDTA, pH 7.4. In case sterile buffer is required, filtering through 0.2 μm is preferred over heat sterilization.
7. Sterile 96-well plate, round bottom.
8. Microfuge and centrifuge for well plates.
9. Vortex.
10. BD Cytofix/Cytoperm Kit™ (BD Bioscience, San Jose, CA) or a similar fixation and permeabilization protocol. BD Cytofix/Cytoperm Kit™ contains Cytofix/Cytoperm solution, Perm/wash buffer, and Golgistop.
11. PBMC culture medium: Iscove's modified Dulbecco's medium (IMDM), 5 % fetal calf serum (FCS), 100 I.U./mL penicillin, 100 μg/mL streptomycin.
12. Recombinant human interleukin-2 (IL-2) (*see* **Note 4**).
13. MHC class I- and/or MHC class II-binding peptides of interest.
14. Fluorescently labeled antibodies for flow cytometry (*see* **Note 5**).
15. Fluorescently labeled recombinant MHC class I tetramers (H2-K^b or H2-D^b allele for C57BL/6 mice), loaded with epitopes of interest for MHC tetramer-based monitoring of antigen-specific $CD8^+$ T cell responses (*see* **Note 6**).
16. Propidium iodide solution 20×: 0.2 mg/mL, 1× PBS.
17. Multicolor flow cytometer.

3 Methods

3.1 Intradermal DNA Tattoo Vaccination

1. 1 day before DNA tattoo vaccination remove hair at the planned administration site by shaving or use of depilatory cream (*see* **Note 7**).
2. Anesthetize animals to allow proper fixation for tattoo vaccination.
3. Fixate hind foot between middle finger and thumb, and support the hind leg with index finger (*see* Fig. 1).
4. Apply a droplet of 10–20 μL of DNA solution on the skin of the calf muscle (*see* **Note 8**).
5. Administer the DNA solution using the tattoo device while applying slight pressure. Standard settings for in vivo DNA tattoo vaccination are oscillation frequency of 100 Hz and needle depth of 1 mm with an application time of 30 s. The tattooed area should cover approximately 1 cm^2 (*see* **Note 9**).

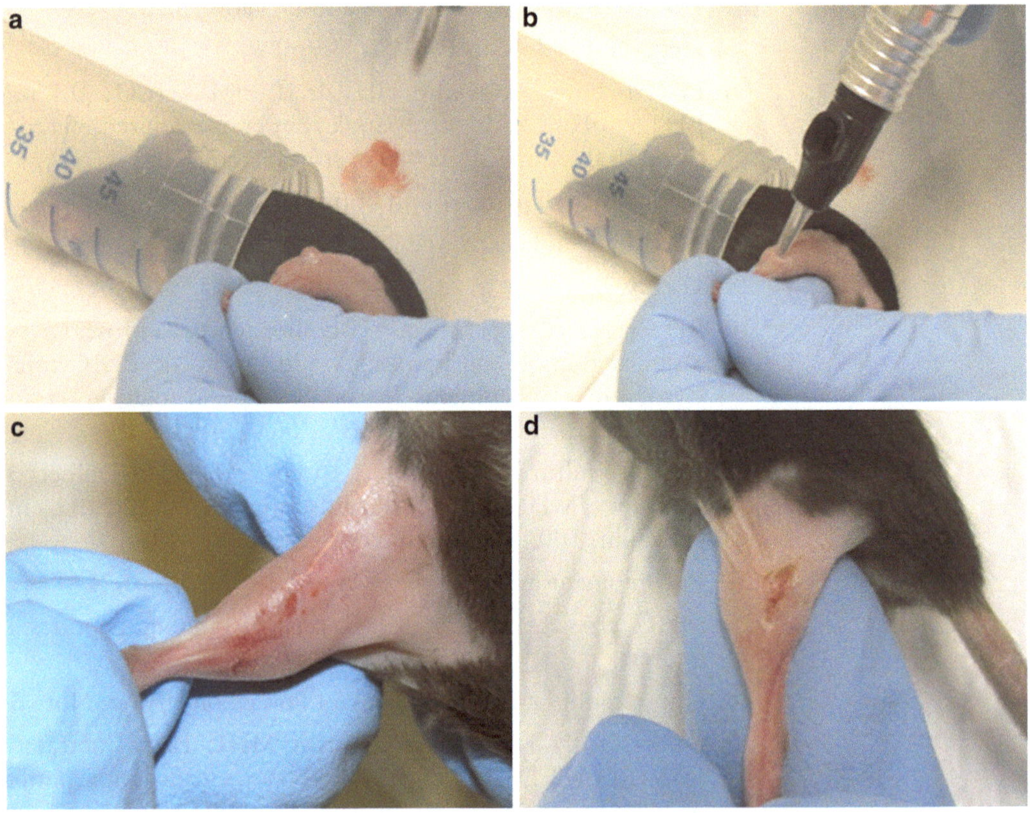

Fig. 1 General DNA tattoo procedure for mice. (**a**) Fixation and application of plasmid solution under anesthesia, (**b**) DNA tattooing of fixated leg with tattoo device, (**c**) superficial damage directly after DNA tattoo administration, (**d**) crust formation, 1 day after DNA tattoo administration

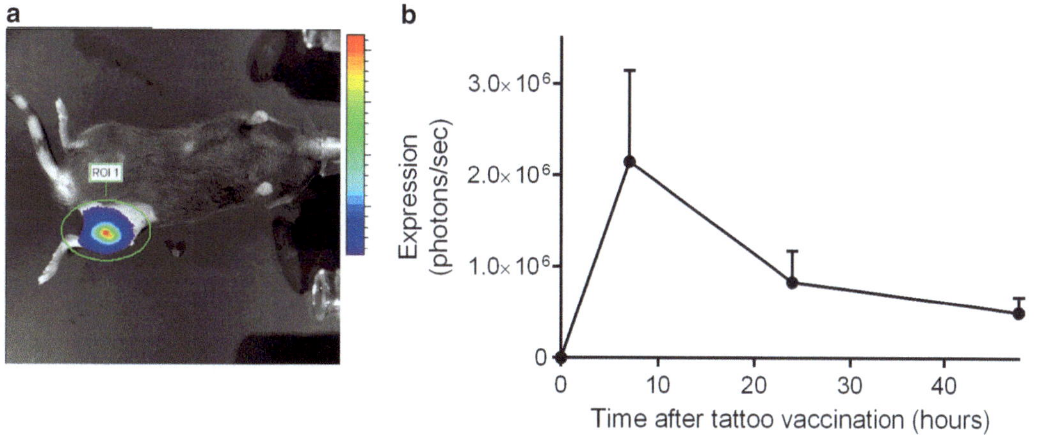

Fig. 2 Typical intravital image (**a**) and kinetics (**b**) of luciferase expression upon DNA tattoo vaccination with a luciferase-encoding plasmid (pVAX-Luc). Plot (**b**) shows the mean of eight animals plus standard deviation

6. On days 3 and 6, tattoo the same hind leg as on day 0, but at a slightly different location (d0: ventral/d3: medial/d6: dorsal), in order to prevent repeated tattooing of the same skin area (*see* **Notes 10** and **11**).

3.2 Intravital Imaging of Luciferase Expression

1. 1 day after tattooing, measure the expression of luciferase at the tattooed area (*see* **Note 12** and Fig. 2).
2. Inject pre-warmed luciferin substrate solution intraperitoneally at 150 mg/kg. Anesthetize the animals using isoflurane or another suitable anesthetic. Quantify luciferase expression 18 min after luciferin administration for a period of 1 min (*see* **Note 13**). Signal intensity is quantified as the sum of all detected light within the region of interest, after subtraction of background luminescence.

3.3 Flow Cytometry-Based Immunomonitoring

3.3.1 Detection of Antigen-Specific $CD8^+$ T Cell Response Using MHC Tetramers

1. Withdraw a small venous blood sample of approximately 50 μL (corresponding to 5–10 small drops) using a tail clip. Removal of a small piece of skin suffices to take this amount of blood; do not clip the bone. Drops of blood can be collected by stroking the sides of the tail from base to tip. Alternatively, blood collection can be performed based on local preferences/protocols. Collect the blood in a 1.5 mL tube, or any other suitable blood collection container, prefilled with 1 mL PBS, supplemented with 5 I.U./mL heparin. Keep samples on ice after collection.
2. Spin down the samples for 5 min at 800×*g*, and remove supernatant. Vortex the cell pellet for a few seconds.
3. Add 500 μL of pre-chilled erylysis buffer and incubate for 2–3 min on ice.

4. Spin down the samples for 5 min at 800 × *g*, and remove supernatant.
5. Resuspend samples in 200 μL of erylysis buffer and transfer to 96-well plate (round bottom).
6. Centrifuge plate for 2 min at 200 × *g*, and remove supernatant. Dab plate upside down onto paper towel to remove residual supernatant from the wells.
7. Add 200 μL of FACS buffer, resuspend, and centrifuge for 2 min at 200 × *g*. Remove supernatant, and vortex cell pellets carefully. Make sure that the cell pellet is no longer reddish (if still red, a second erylysis step followed by a washing step with FACS buffer should be performed).
8. Add mixture of fluorescently labeled antibody and MHC tetramer, using 0.1 μL anti-mouse CD8 antibody (0.2 mg/mL stock) and 2 μL MHC tetramer (25 μg/mL stock) in 20 μL of FACS buffer per sample. Make sure that the cells and staining mixture are properly mixed.
9. Stain for 20 min at room temperature in the dark.
10. Wash off the staining solution by adding 200 μL of FACS buffer and centrifuge for 2 min at 200 × *g*. Remove supernatant. Repeat this step.
11. Resuspend cell pellets in 50–100 μL FACS buffer for analysis.
12. Add propidium iodide (final concentration 10 μg/mL) or another suitable dead/live marker a few minutes before sample acquisition with flow cytometer to exclude dead cells.
13. Antigen-specific $CD8^+$ T cell responses are quantified by calculating the percentage of MHC tetramer-positive cells of total $CD8^+$ cells (*see* Fig. 3 and **Note 14**).

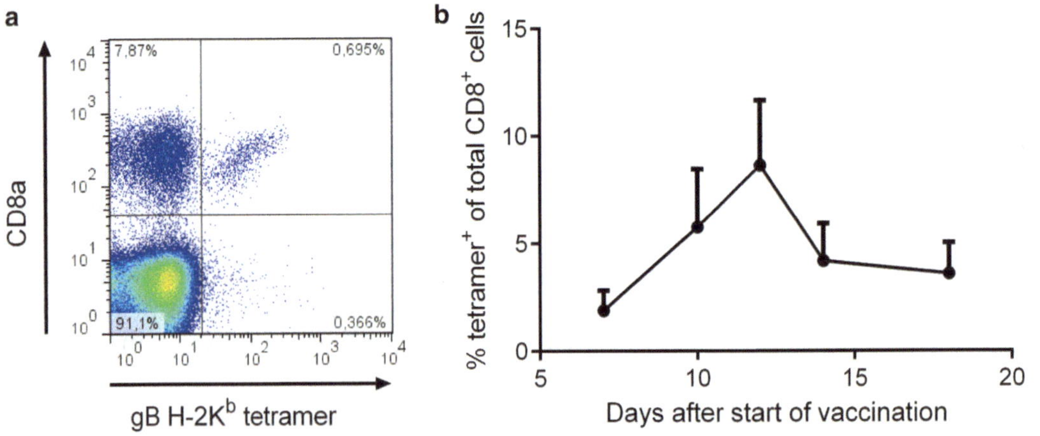

Fig. 3 Typical $CD8^+$ T cell response (**a**) and kinetics of $CD8^+$ T cell response (**b**) upon DNA tattoo vaccination of C57BL/6 mice with a fusion construct of TTFC with an HSV-1 glycoprotein B epitope. Plot (**b**) shows the mean of six animals plus standard deviation

*3.3.2 Detection of Antigen-Specific T Cell Responses by Intracellular Cytokine Staining (See **Note 15**)*

1. Prepare suspensions of peripheral blood mononuclear cell (PBMC), *see* **steps 1–8** in Subheading 3.3.1, and transfer cells to a round-bottom 96-well plate. Keep cells sterile.
2. Resuspend PBMCs in 200 μL medium containing 1 μL/mL Golgiplug™, 1 μg/mL of the T cell epitope of interest, and 20 I.U./mL recombinant IL-2. Take along controls without peptide or an irrelevant peptide.
3. Incubate the samples for 3–16 h at 37 °C, 5 % CO_2.
4. Centrifuge plate for 3 min at 200 × *g*, and remove supernatant. Dab plate upside down onto paper towel to remove residual media from the wells. Vortex cells gently.
5. Add 100 μL of Cytofix/Cytoperm™ and incubate on ice for 20 min.
6. Add 100 μL of Perm/Wash™ solution, spin cells for 3 min at 200 × *g*, and remove supernatant. Wash two more times with Perm/Wash™.
7. Add antibodies of interest diluted in 20 μL of Perm/Wash™ (*see* **Note 16**).
8. Stain for 30 min on ice in the dark.
9. Wash cells once with Perm/Wash™ and once with FACS buffer.
10. Resuspend cell pellets in 50–100 μL FACS buffer for analysis.
11. Analyze cells by flow cytometry without the addition of propidium iodide (*see* **Note 17**).

4 Notes

1. In general, the gene of interest is cloned into a standard expression vector with a CMV promoter (such as pVAX1 or pcDNA3.1). For certain experimental designs, a cell type-specific promoter may be considered. In order to increase immunogenicity, carboxy-terminal fusion of the epitope/antigen to a carrier protein of foreign origin may be considered (e.g., tetanus toxin fragment C). The increased immunogenicity of the resulting vaccines is thought to result from both increased antigen stability and increased $CD4^+$ T cell help [14]. In addition, our lab recently developed minimal carrier protein-encoding gene cassettes that provide $CD4^+$ help and ER localization/retention that together greatly enhance immunogenicity of HPV antigens as compared to these antigens in isolation [15].
2. Vaccine-encoding plasmids are expressed and amplified in *E. coli* DH5α and are purified using a Qiagen Endotoxin-free Maxi/Mega kit to rule out any possible effects of contaminating

bacterial LPS. This concern might however be exaggerated as we demonstrated that the presence of residual LPS in a model DNA vaccine applied by DNA tattooing did not significantly influenced antigen expression or CTL response in a mouse model [16].

3. Most standard available tattoo devices, used for cosmetic or decorative tattooing of humans, are suitable. Our laboratory standard is the PMU device (MT.DERM, Berlin, Germany), with a 9-needle cartridge.
4. Human and mouse IL-2 are mutually cross-reactive and are both suitable.
5. For intracellular cytokine staining upon peptide stimulation we routinely use anti-mouse CD8a and anti-mouse IFN-γ, but other effector cytokines such as IL-2 and tumor necrosis factor-α (TNF-α) can be measured in parallel by multicolor flow cytometry. The standard end concentration of antibody during staining is 1 μg/mL, but this should be optimized for each assay.
6. MHC tetramers are routinely produced using the UV exchange technique. *See* Toebes et al. for detailed protocol on MHC tetramer generation using this technology [17]. The standard end concentration of MHC tetramers during staining is 2.5 μg/mL.
7. We have observed that antigen expression is the highest when the hair on the vaccination site is removed using depilatory cream rather than by shaving. The skin becomes hypersensitized after hair removal, so leave the animals for at least 1 day to allow the skin to recover. In case depilatory cream is used, carefully remove the cream after a few minutes of incubation and wash with water to prevent extensive skin irritation.
8. The majority of tattoo devices adsorb a volume of plasmid solution in the needle cartridge by capillary forces. Because of this, 10–20 μL of plasmid solution is applied onto the needle before the first animal is tattooed to compensate for this loss.
9. For inducing an optimal immune response, it is important to induce superficial damage to the tattooed skin area (*see* Fig. 1c), leading to some crust formation the day after tattooing (*see* Fig. 1d). It may be speculated that this superficial damage increases local immune activation as a result of "danger signals." To support this speculation, we have shown that DNA tattooing by itself is able to induce high levels of serum IL-6 [16].
10. In a standard vaccination protocol, the same hind leg is tattooed three times (on days 0, 3, and 6). A tattoo regimen of days 0, 3, and 6 leads to the induction of very strong $CD8^+$ T cell responses within a short time span. In case different limbs,

draining to separate lymph nodes, are tattooed, the magnitude of the induced immune response will be reduced [12].

11. It is possible to simultaneously vaccinate one animal with two different DNA vaccines (with the two vaccines applied to a different hind leg). Note that immune competition between the vaccines cannot be excluded in this experimental setup.

12. When DNA vaccines are used that encode fusion proteins/peptides that include luciferase, levels of luciferase can be quantified intravitally as a measure of antigen expression. Generally, luciferase expression levels are the highest during the first day after DNA tattoo vaccination and stay detectable above background for 3–5 days (*see* Fig. 2). In case luciferase-based monitoring is not feasible, other methods for quantification of antigen expression levels may be utilized:
    - In vitro luciferase assays or western blot analysis on skin digests.
    - Detection of fluorescent reporter proteins in skin-derived single-cell suspensions [13].
    - Intravital imaging of fluorescent reporter proteins by confocal or two-photon microscopy [18].

13. Luminescence normally reaches a plateau between 15 and 20 min after luciferin injection.

14. When the initial immune response has reached the memory phase (in general after 2–3 months, when antigen-specific $CD8^+$ T cell responses are stable at around 10–20 % of the peak response), animals can be re-vaccinated with a single homologous DNA tattoo vaccination. The secondary T cell response upon this single tattoo will generally develop faster than the initial T cell response, with a comparable or even higher frequency of antigen-specific T cells at the peak of the response [15, 19].

15. In case a functional readout is preferred, intracellular cytokine production may be measured upon peptide stimulation. Normally, production of interferon-γ (IFN-γ) is measured, but other effector cytokines such as IL-2 and TNF-α can be measured in parallel by multicolor flow cytometry.

    The protocol described is based on the BD Cytoperm/Cytofix™ protocol, but a similar fixation/permeabilization protocol can be used as well.

16. We normally stain for cell surface markers and intracellular cytokines at the same time, but staining with antibodies against cell surface markers before the permeabilization step can be considered. Standard staining conditions are 0.1 μL anti-mouse CD8a antibody (0.2 mg/mL stock) and 1 μL anti-IFN-γ (0.2 mg/mL stock) antibody in a volume of 20 μL.

17. A fixable dead/cell marker can be used during intracellular cytokine staining if preferred. Such a marker should be added before permeabilization.

References

1. Nagata T, Aoshi T, Uchijima M et al (2004) Cytotoxic T-lymphocyte-, and helper T-lymphocyte-oriented DNA vaccination. DNA Cell Biol 23:93–106
2. van Rooij N, van Buuren MM, Philips D et al (2013) Tumor exome analysis reveals neoantigen-specific T-cell reactivity in an ipilimumab-responsive melanoma. J Clin Oncol 31:e439
3. Lu YC, Yao X, Li YF et al (2013) Mutated PPP1R3B is recognized by T cells used to treat a melanoma patient who experienced a durable complete tumor regression. J Immunol 190: 6034–6042
4. Kanitakis J (2002) Anatomy, histology and immunohistochemistry of normal human skin. Eur J Dermatol 12:390–399
5. Kataru RP, Jung K, Jang C et al (2009) Critical role of CD11b+ macrophages and VEGF in inflammatory lymphangiogenesis, antigen clearance, and inflammation resolution. Blood 113:5650–5659
6. Shklovskaya E, Roediger B, Fazekas de St GB (2008) Epidermal and dermal dendritic cells display differential activation and migratory behavior while sharing the ability to stimulate CD4+ T cell proliferation in vivo. J Immunol 181:418–430
7. Stoecklinger A, Grieshuber I, Scheiblhofer S et al (2007) Epidermal langerhans cells are dispensable for humoral and cell-mediated immunity elicited by gene gun immunization. J Immunol 179:886–893
8. Stoitzner P, Tripp CH, Eberhart A et al (2006) Langerhans cells cross-present antigen derived from skin. Proc Natl Acad Sci U S A 103: 7783–7788
9. Liard C, Munier S, Joulin-Giet A et al (2012) Intradermal immunization triggers epidermal Langerhans cell mobilization required for CD8 T-cell immune responses. J Invest Dermatol 132:615–625
10. Bins AD, Wolkers MC, van den Boom MD et al (2007) In vivo antigen stability affects DNA vaccine immunogenicity. J Immunol 179:2126–2133
11. Kis EE, Winter G, Myschik J (2012) Devices for intradermal vaccination. Vaccine 30: 523–538
12. Bins AD, Jorritsma A, Wolkers MC et al (2005) A rapid and potent DNA vaccination strategy defined by in vivo monitoring of antigen expression. Nat Med 11:899–904
13. van den Berg JH, Nujien B, Beijnen JH et al (2009) Optimization of intradermal vaccination by DNA tattooing in human skin. Hum Gene Ther 20:181–189
14. Wolkers MC, Toebes M, Okabe M et al (2002) Optimizing the efficacy of epitope-directed DNA vaccination. J Immunol 168: 4998–5004
15. Oosterhuis K, Aleyd E, Vrijland K et al (2012) Rational design of DNA vaccines for the induction of human papillomavirus type 16 E6- and E7-specific cytotoxic T-cell responses. Hum Gene Ther 23:1301–1312
16. van den Berg JH, Quaak SG, Beijnen JH et al (2010) Lipopolysaccharide contamination in intradermal DNA vaccination: toxic impurity or adjuvant? Int J Pharm 390:32–36
17. Toebes M, Rodenko B, Ovaa H et al (2009) Generation of peptide MHC class I monomers and multimers through ligand exchange. Curr Protoc Immunol 87:18.16.1–18.16.20
18. Bins AD, van Rheenen J, Jalink K et al (2007) Intravital imaging of fluorescent markers and FRET probes by DNA tattooing. BMC Biotechnol 7:2
19. Oosterhuis K, Ohlschlager P, van den Berg JH et al (2011) Preclinical development of highly effective and safe DNA vaccines directed against HPV 16 E6 and E7. Int J Cancer 129: 397–406

Chapter 10

Microneedle Applications for DNA Vaccine Delivery to the Skin

Hae-yong Seok, Hyemee Suh, Sunghyun Baek, and Yeu-Chun Kim

Abstract

Microneedles were initially developed as pretreatment tools for the delivery of therapeutic drugs to intradermal locales in the human skin. Over time, variations in microneedle forms and functions burgeoned through the works of many researchers worldwide. The four major types of microneedles in use today are solid, dissolving, coating, and hollow microneedles. The emergence of different types of microneedles also paved the way for a flourishing diversification of microneedle applications, one of the most remarkable of which deals with the transcutaneous delivery of prophylactic vaccines. Here, we describe fabrication methods of microneedles and DNA vaccine loading methods on the microneedle surface. Furthermore, in the latter part of this chapter, in vivo test protocols for assessing the efficacy of gene delivery using microneedles are described.

Key words Microneedle, DNA vaccine, Gene delivery, Electroporation, Nanoparticle

1 Introduction

Today, intramuscular (IM) injection by hypodermic needles is one of the most commonly practiced methods of vaccine delivery against disease-causing pathogens. Although this method possesses numerous benefits, including rapid kinetics and low cost, the IM method also has many disadvantages, such as the necessity of professional healthcare personnel, pain, and fear of needles, which leads to patient compliance issues. Furthermore, the problem of needle reuse in developing countries can aggravate the spread of disease, rather than preventing it, in impoverished populations where the supply of vaccines may not meet immediate requirements [1].

Vaccination using microneedles has been proposed as a solution to the abovementioned issues. The human skin is composed of the *stratum corneum*, epidermis, and dermis. Of these, the epidermis and dermis are enriched in Langerhans cells and dermal dendritic cells, which play an important role in the initiation and modulation of immune responses [2]. The micron-scale dimensional parameters

Monica Rinaldi et al. (eds.), *DNA Vaccines: Methods and Protocols*, Methods in Molecular Biology, vol. 1143, DOI 10.1007/978-1-4939-0410-5_10, © Springer Science+Business Media New York 2014

of microneedles are appropriate for specifically targeting the stimulation of these immune cells. Therefore, in some cases, microneedles may be a more potent stimulator of immune responses than conventional intramuscular injection techniques. Furthermore, this method offers the possibility of self-vaccination without professional training and also resolves the problem of vaccine shortage, owing to its dose-sparing advantages. Another advantage of microneedle vaccination is that there is nearly no perceived pain when the needles puncture the skin [3].

The efficacy and potency of DNA-coated microneedles have been widely investigated by many research groups in terms of their ability to induce gene expression in mammalian systems in vivo. More specifically, the use of microneedles for gene delivery has been acclaimed by scientists for its powerful stimulation of immune responses and remarkable dose-sparing advantages compared to intramuscular injection [4]. In addition, dissolving-type microneedles, which are distinct from coated microneedles were used as carriers for influenza vaccines. In the case of dissolving microneedles for influenza vaccination, similar levels of immune response could be induced compared to coated microneedles [5].

1.1 Preparation of DNA Vaccine

Preparation of DNA vaccines for microneedle coating depends on the particular needs of the experiment and should vary according to the specification parameters outlined by different laboratories. Primarily, two factors should be considered in preparing DNA vaccines for use in microneedles: the design of plasmid vaccine vectors and the formulation of the vaccine solution. Typically, recombinant DNA vaccines contain an insert gene that encodes one or more immunogens and transcription regulatory elements such as viral promoters, enhancers, and termination signals that drive high-level expression in eukaryotic cells [6].

1.2 Solid Microneedles for Pretreatment

In the pretreatment approach of microneedle vaccination, solid microneedles are inserted into the skin to make micron-scale punctures in the *stratum corneum*, which acts as the passageway for delivery of therapeutic macromolecules. The formation of pores in the skin facilitates the transport of a variety of molecules through the crevices of minutely damaged skin and into the epidermis and dermis. In general, the formulation of interest (e.g., drug, DNA, or RNA) is topically applied onto the skin over the pretreated region, following removal of the microneedles [7].

1.3 Coated Microneedles

Dry-coating the surface of microneedles with DNA vaccine and injecting into skin is one of the simplest and most verified methods of microneedle-mediated DNA immunization. The process of coating microneedles involves dip-coating using an in-plane microneedle row-coating device. Coating formulations can include a range of different viscosity enhancers to enhance the thickness of coatings, surfactants to facilitate wetting and deposit a uniform coating on the surface of microneedles, and, in some cases, stabilizers to prevent damage of biomolecules during the coating and drying process. The exact compositions of such DNA formulations should vary depending on the needs and specifications of the experiment [8].

Polyelectrolyte multilayers (PEMs) can be deposited onto the microneedle surface to fabricate nanostructured films, composed of a variety of natural and synthetic polyelectrolytes, which allows for controlled and localized release of DNA, proteins, and other macromolecules. Ideally, an optimal microneedle coating method should deposit a controlled dosage of drug, employ gentle processes to protect sensitive chemicals, provide fine-tunable drug release kinetics, and possess high drug loading capacity while minimizing the necessity of excipients. The "layer-by-layer" (LbL) method of coating DNA has been devised to meet such needs. This approach has been optimized through many studies that investigated the physicochemical parameters of DNA vaccines using the PEM method. In the LbL method, it is standard practice to include a biodegradable PEM coating that endows controlled-release DNA properties in vivo. Many research groups have investigated and confirmed the benefits of using the hydrolytically degradable cationic polymer, PBAE (polymer-1), to this end. Polymer-1 is used as the release layer as well as an effective DNA transfection agent [9].

Another novel DNA coating method, the "gas jet" approach, employs a gas jet to control the dispersion of coating solution on the microneedle patch, so that the coating solution specifically wets the surfaces of microneedle projections only, without clogging the space between the projections, and thereby creating a uniformly dispersed layer of coated DNA on the projections of the patch. In addition, the gas flow can be utilized to dry away any excess remaining coating solution in the crevices of the microneedle patch in a relatively short amount of time. The gas can be nitrogen, argon, air, or other inert gases depending on the reactivity of the coating formulations [10, 11].

1.4 Dissolving Microneedles

Dissolving microneedles can be produced by water-soluble or biodegradable polymers which can load drugs or other molecules into the microneedle matrix. After insertion of dissolving microneedles into the skin, the loaded materials are released from microneedles by dissolving or degradation of needle tips.

1.5 Microneedle Array for Electroporation

To produce a novel microneedle array that can deliver genes effectively to cells by electroporation, a specific micromolding method is applied. In this technique, a double-layer micromold approach is combined with metal-transfer micromolding to produce a microneedle array suitable for electroporation [12]. Generally, the process of making electroporation-applicable microneedles is as follows: (1) The rigid mold is fabricated by using photolithography; (2) the master structure is fabricated from the rigid mold; (3) the metal-patterned microneedle structure is replicated by using metal-transfer micromolding; (4) metal wires are attached to the electrically linked structure through laser-drilled micro-channel [13].

1.6 Particle Approach for Microneedle Application

Because nucleic acids are negatively charged owing to its phosphate group backbones, they need to be conjugated to polycationic nanocarriers in order to be transfected effectively into cells. Furthermore, conjugation with nanocarriers enhances the stability of nucleic acids in vivo, which are naturally prone to degradation by serum nucleases. Therefore, application of nanocarriers is vital for the effective transfer of genes into target cells. Poly(lactic-co-glycolic acid) (PLGA) is a representative biodegradable, biocompatible polymer and known to induce high gene transfection efficiency and stability of nanoparticles when polymer nanoparticles are produced by blending with polycations [14]. In this chapter, a gene delivery method of using PLGA nanoparticles for microneedle-based vaccination is introduced.

1.7 In Vivo Tests

In order to monitor the efficacy of DNA immunization administered via microneedles, mice are injected with the DNA-coated microneedles until the DNA is completely dissolved off the microneedles and to evaluate the parameters of the induced immune response. In order to check the transfection efficiency of coated DNA, microneedles can be coated with a plasmid DNA expressing a reporter luciferase and imaged under an in vivo imaging system. As a more direct quantification of humoral immune responses, serum antibody levels are determined at least 3 weeks after immunization by enzyme-linked immunosorbent assay (ELISA). In some cases, serum antibody quantification is generally conducted at weeks 3, 5, and 7 post-immunization and day 4 after virus challenge.

2 Materials

2.1 DNA Vaccine Preparation

1. Vaccine plasmid DNA.
2. Selective antibiotic.
3. Competent *E. coli* DH5α strain.
4. 50 mL centrifuge tubes.
5. Anion-exchange affinity column.
6. Resuspension buffer: RNase A, glucose, EDTA, Tris–HCl.
7. Alkaline lysis buffer: SDS, NaOH.

8. Neutralization buffer: $KC_2H_3O_2$.
9. Wash buffer: 70 % ethanol.
10. Elution buffer (low salt).
11. Isopropanol.
12. 70 % ethanol.
13. Autoclaved double-distilled water or TE buffer.

2.2 Solid Microneedles for Pretreatment

1. Abdominal or breast skin samples, formulation of interest.
2. Surgical scalpel.
3. Normal saline.
4. Single-row strip of silicon microneedles with a height-to-base ratio of 3:2.
5. A microneedle applicator [15, 16].

2.3 Coated Microneedles

2.3.1 CMC Coating

1. Microneedles.
2. Coating solution: Plasmid DNA expressing antigenic gene, 1 % (w/v) carboxymethyl cellulose (CMC) sodium salt, 15 % (w/v) trehalose, 0.5 % (w/v) Lutrol F-68 NF.
3. Dip-coating apparatus [8].

2.3.2 DNA Only

1. Microneedles.
2. Plasmid DNA expressing antigenic gene: 8 mg/mL.
3. Dip-coating apparatus.

2.3.3 DNA as Coating Formulation

1. Microneedles.
2. Coating solution: Plasmid DNA expressing antigenic gene, inactivated influenza virus, trehalose, deionized (DI) water.
3. Fluorescein isothiocyanate–bovine serum albumin (BSA) conjugated.
4. Placebo (inert) DNA.
5. Dip-coating apparatus.

2.3.4 Layer by Layer

1. Microneedles.
2. Sodium poly(4-styrenesulfonate) (SPS) solution: 5 mM SPS, 20 mM NaCl, 1× PBS, pH 5.0.
3. Protamine sulfate (PS) solution: 2 mg/mL PS, 100 M sodium acetate (NaOAC), 1× PBS, pH 5.0.
4. Release-layer polymer-1 solution: 2 mg/mL poly(ß-amino ester) (PBAE), 100 mM NaOAc, 1× PBS, pH 5.0.
5. Poly(dimethylsilocane) (PDMS) molds with arrays of conical cavities.

6. Cy3-labeled fluorescent pLuc.
7. Vaccine-encoding plasmid DNA: 1 mg/mL DNA, 100 mM NaOAc.
8. PBS wash buffer: 1× PBS.
9. NaOAc wash bath buffer: 100 mM NaOAc, 1× PBS, pH 5.0.

2.3.5 Gas Jet Coating for Microneedle Patches

1. Microneedle patches made of solid silicon.
2. A mixture of glycerol and water: 1:1 ratio.
3. Coating solutions of methylcellulose (MC).
4. A detergent (poloxamer 188 or Quil-A).
5. Active agents (vaccines or fluorescent dyes).
6. A gas jet applicator.

2.4 Dissolving Microneedles

1. Pin headers.
2. Glass substrate.
3. 20 % polyvinyl alcohol (PVA) solution.
4. Razor blade.
5. Fine scissors.
6. Nucleic acid solution.

2.5 Microneedle Array for Electroporation

1. SU-8.
2. Glass substrate.
3. Photomask.
4. Propylene glycol monomethyl ether acetate (PGMEA).
5. Ti/Cu.
6. Ni.
7. Copper etching solution (saturated $CuSO_4$ in NH_4OH).
8. Au.
9. PS container.
10. PDMS.
11. Poly-lactic acid (PLA).
12. Polyimide tape.
13. Laser drill.
14. Silver paste.
15. Plasmid DNA-encoding GFP.
16. 1× PBS.

2.6 Particle Approach for Microneedle Application

1. β-galactosidase gene-encoding pCMV-β [17].
2. Anthrax protective antigen (PA63)-encoding pGPA plasmid [18].
3. PLGA.

4. Acetone.
5. 1,2-dioleoyl-3-trimethylammonium-propane, chloride (DOTAP).
6. Pluronic F68.
7. 1 % methyl orange.
8. Chloroform.
9. Citric acid and disodium hydrogen orthophosphate buffer solution.
10. Ultracentrifuge.

2.7 In Vivo Tests

2.7.1 Immunization

1. 6–8-week-old BALB/c mice.
2. Depilatory cream.
3. 70 % ethanol.
4. Vaccine-coated microneedles.
5. Hypodermic needles.

2.7.2 Bioluminescence Imaging

1. 6–8-week-old BALB/c mice.
2. GFP-expressing plasmid DNA (phMGFP/CBL).
3. D-Luciferin substrate: 30 mg/mL.
4. Anesthesia: Inhalation of mixture of 3 % isoflurane and 97 % oxygen.
5. In vivo imaging system.
6. Vaccine-coated microneedles.

2.7.3 Determination of Serum Antibody Responses by Enzyme-Linked Immunosorbent Assay

1. 96-well microtiter plates.
2. Coating buffer: 4 μg/mL inactivated virus, 0.1 M $NaHCO_2$, pH 9.5.
3. Blocking buffer: 3 % BSA, 1× PBS, pH 7.0.
4. Dilution buffer: 0.05 % Tween-20, 1× PBS, pH 7.0.
5. Washing buffer: 0.05 % Tween-20, 1× PBS, pH 7.0.
6. Horseradish peroxidase-labeled goat anti-mouse antibodies (IgG, IgG1, IgG2a).
7. Substrate *o*-phenylenediamine solution.
8. H_2SO_4 (0.5 M).

3 Methods

3.1 DNA Vaccine Preparation

1. Integrate the gene of interest into the vaccine plasmid construct. The gene of interest should be composed of the DNA sequence of one or more antigenic subunits that encode for naturally expressed proteins of the pathogenic virus (*see* **Note 1**).

2. For best results, append vital transcriptional regulatory elements in the vaccine plasmid that maximize gene expression levels in eukaryotic cells, such as the potent cytomegalovirus enhancer/chicken ß-actin (CAG) promoter and woodchuck hepatitis virus posttranscriptional regulatory element (WPRE) to increase transgene expression from viral vectors.
3. Depending on the experiment design, load appropriate excipients into the DNA vaccine formulation to comprise the bulk of the vaccine solution. Typically, CMC is used to elevate the viscous quality of the DNA sample without the bulky volume, for easy coating onto microneedles.
4. If necessary, introduce an inactivated virus component into the DNA formulation to modulate immune responses. In such a case, trehalose is necessary to prevent the destabilization of inactivated viruses during the coating and drying process [19].
5. Include an antibiotic-resistance gene for plasmid selection (*see* **Note 2**).
6. Feed the vaccine plasmid DNA into transformation competent *E. coli* DH5α strain.
7. Seed the bacteria that have taken up plasmid DNA into 2 mL LB broth with antibiotics. Incubate the culture at 37 °C for 9 h. Transfer the bacterial culture into 200 mL LB broth with antibiotics and incubate for 16 h.
8. Harvest the bacterial culture by centrifuging at 6,000 × *g* for 15 min. Discard the supernatant.
9. Secure an anion-exchange resin column into a 50 mL centrifuge tube.
10. Add 20 mL of resuspension buffer, and resuspend the pellet by vortexing. The RNase A will remove any unwanted RNA.
11. Add 20 mL alkaline lysis buffer, and mix the suspension by inverting the tube 10–15 times. Do not vortex, as this may cause shearing of genomic DNA. Incubate for 5 min at room temperature until the solution forms a clear suspension. The detergent in the lysis buffer will denature the phospholipids of the cell membrane and any membrane proteins, whereas NaOH will denature plasmid and genomic DNA.
12. Add 20 mL of neutralization buffer, and immediately mix the suspension by inverting the tube 10–15 times. The acidic potassium acetate will neutralize the pH, and a solid precipitate of potassium dodecyl sulfate will form. The covalently bonded circular plasmid DNA will reassociate correctly, whereas the linear genomic DNA will precipitate out due to random association of the strands.
13. Centrifuge at 20,000 × *g* for 30 min at 4 °C.

14. Transfer the supernatant to the anion-exchange resin. Allow the column to empty by gravity flow. Discard the flow-through.
15. Add 30 mL of wash buffer to the anion-exchange resin. Discard the flow-through.
16. Add 15 mL of a low-salt elution buffer to the anion-exchange resin. Collect the flow-through that contains the eluted plasmid DNA.
17. Precipitate DNA by adding 11 mL of isopropanol to the eluted solution. Isopropanol lowers the solubility of DNA in solution. Mix well by vortexing.
18. Centrifuge at 15,000 × *g* for 30 min at 4 °C.
19. Without disrupting the DNA pellet, carefully discard the supernatant and wash away any residual salt with 5 mL of 70 % ethanol.
20. Remove the supernatant, and air-dry the DNA pellet for 5–10 min under sterile conditions.
21. Dissolve the DNA pellet in 300 μL autoclaved water or TE buffer.

3.2 Solid Microneedles for Pretreatment

1. In order to insert the solid microneedles into skin samples, solid-type microneedles and full-thickness skin samples were prepared.
2. Remove residual subcutaneous adipose tissue using a surgical scalpel.
3. Wash the samples with normal saline, and store the samples at below −20 °C.
4. In order to deliver DNA into skin using microneedle pretreatment, topically apply a small amount of the formulation of interest onto the skin samples using a pipette.
5. Insert the microneedle arrays into the prepared skin samples with an applicator or manually.
6. Hold the microneedle arrays in the position for 1 min to form micron-sized holes in the skin surface, and remove the arrays from the skin smoothly [3, 7, 15, 16].

3.3 Coated Microneedles

3.3.1 CMC Coating

1. Adjust the position of the micropositioners to enable in-plane alignment of microneedles with the dip-hole reservoir in the microneedle row-coating device.
2. Using a pipette, fill the reservoir with the prepared coating solution (*see* **Note 3**).
3. While observing through a microscope, dip the microneedles horizontally into the reservoir, and rapidly withdraw to adsorb a thin film of coating solution onto the microneedle surface. Dry for 30 s.

4. Repeat the dipping and coating procedure until the microneedles have acquired a desirable amount of coating thickness.
5. Dry the coated microneedles in air for 24 h at room temperature [8].

3.3.2 DNA Only

1. Formulate DNA-only vaccine solutions that do not contain excipients such as CMC as a viscosity enhancer, as this may compromise gene expression of vaccine plasmid as well as lead to some activity loss of viral hemagglutinin. Rather than depending on the presence of excipients to enhance the thickness of the solution, the concentration of the DNA itself should be boosted to achieve the desired viscosity for easy coating onto microneedles.
2. Secure the microneedles onto a dip-coating device, and deposit a base layer onto the microneedles by rapidly dipping and withdrawing, in succession, the microneedles into an adjacently situated wedge filled with a coating solution consisting purely of plasmid DNA vaccine. Dry for 30 s between each dipping-and-coating cycle. Repeat at least three times (*see* **Note 4**).
3. After drying the base layer in air for 5 min, dip-coat the microneedles nine times into the DNA solution.
4. For quantifying the amount of DNA vaccine loaded on the microneedles, incubate the DNA-coated microneedles in distilled water for 12 h at 4 °C to dissolve off the DNA.
5. Determine the amount of DNA by measuring UV spectrophotometric absorption at 260/280 nm [20].

3.3.3 DNA as Coating Formulation

1. Prepare a coating formulation containing both the plasmid DNA expressing the antigenic gene of interest and the inactivated virus vaccine. The standard coating solution contains inactivated influenza virus, DNA vaccine, and stabilizer such as trehalose in DI water (*see* **Notes 5** and **6**).
2. Conduct dip-coating process of microneedles by horizontally dipping the microneedles into the coating solution nine times. Dry for 30 s between each dipping-and-coating cycle.
3. To quantify the coated amount of inactivated virus vaccine on the microneedles, incubate coated microneedles in deionized water for 12 h at 4 °C.
4. Measure the amount of released proteins by a BCA protein assay kit and spectrofluorimetry [21].

3.3.4 Layer by Layer

1. Pre-coat the microneedles with ten bilayers of SPS solution, a synthetic polyanion, and PS solution, the cationic counterpart. Alternatively submerge the microneedles in SPS solution and PS solution for 10 min, separated by a 1-min PBS wash step. This process will coat the microneedles with a base film to provide a uniform initial surface charge density before depositing

DNA or protein films. All solutions are in PBS and adjusted to pH 5.0 before dipping.

2. PEMs are constructed on top of the base layer via the alternating adsorption of polymer-1 (the release layer) and plasmid DNA. Deposit polymer-1/DNA layers as follows. First, submerge the substrates in a solution of polymer-1 for 5 min.
3. Remove substrates and immerse in the NaOAc wash bath buffer for 1 min.
4. Transfer the substrates to a DNA solution for 5 min.
5. Rinse again. Repeat the cycle until the desired number of layers is deposited. Typically, 16 bilayers are deposited.
6. Following the final rinse step, air-dry the substrates under filtered air and store at room temperature [9, 22, 23].

3.3.5 Gas Jet Coating for Microneedle Patches

1. Typically, a gas jet (~10 m/s) is used to adjust the position of coating solution on the patch so that the coating solution wets the projections without being held between the projections and thereby to disperse the coating solutions uniformly on the projections of the patch. In addition, the gas flow can be utilized to dry the coating solution on the patch in a short time and get rid of excess coating solution. The gas can be nitrogen, argon, air, or other inert gases depending on the reactivity of the coating formulations.
2. Clean the patch in the mixture of glycerol and water for 10 min, and wash the patch with an excess of water.
3. Apply the prepared coating solution onto the patch, and use a gas jet (6–8 m/s) to evenly spread and dry the coating solution on the patch. Additionally, the gas jet can be used to eliminate excess coating solution between microneedle projections.

3.4 Dissolving Microneedles: Protrusion Array Device Preparation for Delivery and Application

3.4.1 PAD Production

1. Prepare protrusion arrays by using an assembly of "pin headers" as fabrication templates. Fabricate microneedles by touching the tips of the pin header projections with a 1 mm thick film of 20 % polyvinyl alcohol solution on glass substrate.
2. Once the pin header projections have contacted with PVA film, pull the projections from the glass substrate. The standard experimental conditions are 1 cm of distance over 13 s under 3.0 m/s of airflow at 45 °C. From these fabrication conditions, microneedle structures can be produced on film surfaces with hollow or grooved structures that take advantage of capillary action. Dry the strands of viscous polymer solution to form an array of dried microneedle structure.
3. Separate the pre-protrusions of needle structure from the template array by a razor blade. Mechanically trim each pre-needle with fine scissors to form sharp tips with 45° angles and 750 μm length above the film surface [24].

3.4.2 PAD Loading

1. The fabricated microneedles contain hollow channels which can be utilized to deliver macromolecules by capillary action. Therefore, therapeutic nucleic acid solutions which have good wettability properties can be loaded into protrusion array device (PAD) needles by capillary action.
2. Once the nucleic acid solution is loaded into microneedles, keep PADs in a 50 °C vacuum oven that is evacuated to −18 in. mercury pressure for 4 h. PVA needles will harden to facilitate skin penetration while being sufficiently flexible for skin applications. Although PVA is a flexible polymer, it has sufficient mechanical strength to penetrate stratum corneum layer.

3.4.3 PAD Application

1. Manually insert the PAD microneedles to the outermost part of skin surface to allow the needle tips to pierce the *stratum corneum*. Make sure that the needles have fully penetrated the skin by flicking the microneedles a few times with fingers. Leave the microneedles injected in the skin for 20 min. During this time, the needle tips are hydrated by the interstitial fluid beneath the skin surface and softened to form a viscous gel structure.
2. After hydration period of microneedles, remove the microneedles by pulling on the dried outer portion. The gel plugs are left in the injected site with loaded materials. The loaded materials are released from the hydrated gel plug to intradermal sites.

3.5 Microneedle Array for Electroporation

3.5.1 Fabrication of the Rigid Mold

1. The rigid mold for the microneedle array is composed of two layers and produced by using a negative photoresist of SU-8. Conduct inclined lithography [25] to define the inverse shape of microneedles, and perform conventional UV lithography to form protrusion structures on the second layer.
2. Prepare the first layer of SU-8 as 800 μm thickness on glass substrate by baking at 100 °C for 24 h. Expose the sample to UV light through a photomask that has a 10-by-10 needle array of 200 μm square pattern while the sample is rotating at a 20° inclined angle.
3. Conduct spin coating of second SU-8 layer on top of the first layer. Expose the second layer to UV light with a photomask which is aligned with the first layer to make the inverse form of the desirable metal pattern.
4. Perform etching process to simultaneously develop the two-layer SU-8 mold by using PGMEA as etching solution [13].

3.5.2 Fabrication of the Master Structure

1. Deposit a Ti/Cu seed layer of 500 Å/5,000 Å ratio on the fabricated SU-8 mold to make a master structure by using a CVC DC sputterer.
2. Conduct electroplating of nickel on top of the Ti/Cu seed layer at room temperature with 15 μm/h of deposition rate for 20 h.

3. Soak the sample by submerging in the copper etching solution to remove the copper seed layer for 24 h. Detach the fabricated Ni master structure from the SU-8 mold upon complete dissolution of the Cu seed layer.

3.5.3 Fabrication of the Metal-Patterned Microneedle Array

1. Conduct casting process of PDMS by placing the Ni master structure into a PS container and filling the container with PDMS. Store at 37 °C for 24 h. Separate PDMS from Ni master by using tweezers.
2. Upon separation of PDMS, metalize PDMS mold through electron beam deposition with Au/Ti (2,000 Å/500 Å). Subsequently, use adhesive tape to remove the metal layer on top of the protrusive parts. It is important to perform this process repeatedly to completely remove the metal layer on the protrusion sites.
3. Cast PLA by PDMS mold for biocompatibility. For this process, stack PLA pellets in the mold, and incubate this mold in a vacuum oven heated to 195 °C for 1 h.
4. Conduct degassing process again to remove any remaining bubbles on the sample surface. After this step, cool down until the sample reaches room temperature for 2 h, and separate the fabricated PLA needle structure from the PDMS mold.
5. Perform electroplating process of Ni to enhance structural rigidity. For this step, attach the sample onto a glass slide which has properties that make it endurable against the pre-deposition of Ti/Cu seed layer (300 Å/3,000 Å).
6. Cure silver paste onto the edge of the sample, and cover glass slide with polyimide tape, except for the sample area for Ni deposition. Conduct Ni electroplating for 30 min with a 10 mA/cm^2 current density.
7. Cover the backside of the device with polyimide tape of 25 μm thickness, and perform laser drilling on the backside of the device to create an electrical connection between the pulse generator and the microneedle device.
8. Fill the prepared hollow micro-channels with conductive paste under a vacuum at 25 Torr for 1 min. Subsequently, remove any excess conductive paste by attaching polyimide tape.
9. Attach copper wire to the electrical channel by using silver paste, and fix the device onto a glass slide for easy handling for cell transfection experiments.

3.5.4 Electroporation Protocols

1. Add pGFP to the cell suspension, and homogeneously mix by using vortexing. The final concentration of pGFP should be 100 μg/mL.
2. For the transfection experiment, first add 6 μL of the cell suspension onto the microneedle surface as a hemispherical shape.

After applying voltage for gene transfection ten times, collect each sample in microcentrifuge tubes to conduct transfection analysis.

3. After keeping the collected sample in a 37 °C water bath for 10 min to help cells recover, wash suspension samples with PBS and centrifuge at 3,500 ×*g* for 5 min, three times.
4. After removing the supernatant, suspend the precipitated cells into complete cell culture medium, and keep this sample in a 5 % CO_2 incubator at 37 °C for 20 h. Resuspend cells in 200 μL of PBS solution for data analysis.

3.6 Particle Approach for Microneedle Application

3.6.1 Preparation of Cationic PLGA Nanoparticles

1. The PLGA-based nanoparticles which have positive charge density on particle surface were prepared by nanoprecipitation–solvent evaporation method.
2. Dissolve the PLGA polymer and DOTAP in acetone, and add this organic phase dropwise into a 10 mL aqueous solution that contains pluronic F68 as stabilizing agent. Gently stir for 3 h at room temperature for evaporation of acetone.
3. After the organic solvent is totally evaporated, use an ultracentrifuge machine for removing excess DOTAP at 16,000 ×*g* at 4 °C for 45 min.
4. Carefully resuspend the precipitated nanoparticles into 200 μL water to avoid particle aggregation. Control PLGA nanoparticles without DOTAP are prepared by a similar method [26].

3.6.2 Determination of the Final Concentration of DOTAP in Cationic Nanoparticles

1. The concentration of DOTAP is determined through colorimetric method.
2. Mix 1 % methyl orange with DOTAP in the presence of chloroform and a buffer solution that contains citric acid and disodium hydrogen orthophosphate for preparing yellow-color complexes.
3. Because the yellow light intensity is proportional to the concentration of methyl orange–DOTAP, a standard curve can be obtained and the concentration of DOTAP measured by spectrophotometer at 415 nm [27].

3.6.3 Coating of Plasmid DNA on the Surface of the Cationic Nanoparticles

1. Conduct coating process for the plasmid DNA of pCMV-β or pGPA on the surface of the cationic nanoparticles by gentle stirring.
2. Mix equal volumes of cationic polymer nanoparticles with pDNA solution to obtain 200 μg/mL of pDNA concentration. In order to find the appropriate ratio of pDNA and coated nanoparticles, mix increasing amounts of nanoparticles to a

fixed amount of DNA until a final concentration is reached in which the zeta potential of pDNA-coated nanoparticles is reversed from negative to positive charge. Keep mixtures at room temperature for 30 min to allow pDNA to adsorb onto the particle surface [28].

3.7 In Vivo Tests

3.7.1 Immunization

1. Using a depilatory cream, remove the back hair of 6-week-old BALB/c mice. Wipe the back skin with 70 % ethanol to remove any contaminants.
2. Manually insert the DNA-coated microneedles into the skin of mice. Leave for at least 20 min to allow the vaccine DNA coating to completely dissolve off the microneedle into the interstitial fluid of mice skin.
3. For comparison, include an intramuscular group of mice that receive hypodermic needle injections of the same dose of DNA vaccines and a naïve group of mice that receive no immunization treatment at all [20, 21, 29].

3.7.2 Bioluminescence Imaging

1. In order to check for efficient DNA transfection and complete dissolution of microneedle DNA coating in vivo, dip-coat microneedles with fluorescent reporter plasmids (phMGFP/CBL), and manually insert the coated microneedles into the back skin of mice. Leave for 20 min for complete dissolution of DNA.
2. After 24 h, inject 100 μL of D-luciferin substrate into the peritoneal cavity of isoflurane-anesthetized mice [20].
3. Ten minutes later, observe for fluorescent GFP expression using an in vivo imaging system.

3.7.3 Determination of Serum Antibody Responses by Enzyme-Linked Immunosorbent Assay

1. Coat 96-well microtiter plates with 100 μL of the coating buffer at 4 °C overnight.
2. Block with the blocking solution to prevent unspecific surface binding of antibodies, at 37 °C for 1 h.
3. Dilute serum samples (1:100) in the diluent buffer. Load serum samples onto wells and incubate for 1 h at 37 °C. Rinse three times with the washing solution.
4. After washing, add horseradish peroxidase-conjugated goat anti-mouse IgG antibodies (+IgG1 and IgG2a for Ab isotype determination), and incubate wells for 1 h at 37 °C.
5. After washing, drop the *o*-phenylenediamine substrate solution into wells. Develop at room temperature for 30 min. Check for the change in color as the substrate develops, and stop the reaction if necessary by adding H_2SO_4.
6. Measure the absorbance at 450 nm using ELISA software [21, 29].

4 Notes

1. Several studies [21, 29] with DNA vaccines used a recombinant that was engineered to express the hemagglutinin (HA) protein of a number of influenza virus strains.
2. The spectrum of viscosity enhancers, surfactants, and stabilizers used by different research groups are listed in Table 1.
3. One study has shown that inclusion of CMC may cause aggregation of DNA, which leads to compromised gene expression and also the loss of activity of influenza hemagglutinin during the coating and drying process [21]. Therefore, we found that coating of DNA vaccine on microneedle by the aid of CMC viscosity enhancer should be avoided due to CMC's negative effect on DNA activity.
4. For DNA-only coating microneedle, it is standard practice to initially dip-coat the microneedles into highly concentrated DNA solution at least three times in order to render the microneedle surface properties favorable for coating additional layers. In this case, the DNA solution concentration generally used in this part of the procedure is 8–10 mg/mL, but variations may exist by practice.
5. When highly concentrated DNA vaccine is used as the coating formulation, the DNA vaccine solution is mixed with an inactivated virus vaccine and trehalose to prevent the destabilization of viral proteins. The standard DNA-only vaccine formulation used is 3 mg/mL inactivated influenza virus, 6 mg/mL DNA, and 3 % trehalose in DI water. The inactivated influenza virus may also be replaced with fluorescent BSA for quantifying the amount of loaded proteins on the microneedle surface.
6. It has been reported that trehalose concentration influences the hemagglutination (HA) activity, as an indirect measure of the inactivated influenza virus vaccine activity, and coating mass. Retention of HA activity was enhanced with increase in trehalose concentration. On the other hand, the extent of virus aggregation was reduced with increases in trehalose concentration.

Table 1
A variety of different viscosity enhancers, surfactants, and stabilizers [3]

Viscosity enhancer	Carboxymethyl cellulose (CMC) sodium salt, methylcellulose, sucrose, hyaluronic acid, sodium alginate, polyvinylpyrrolidone (PVP), glycerol, PLGA, alginic acid, xanthan gum, gum ghatti, karaya gum, and poly[di(carboxylatophenoxy)phosphazene]
Surfactant	Lutrol F-68 NF, Tween 20, Poloxamer 188, and Quil-A
Stabilizer	Trehalose, sucrose, glucose, insulin, and dextran

Furthermore, the amount of the coated vaccine on the microneedles also had a similar dependence on trehalose concentration. Increasing trehalose concentration resulted in the reduction of the mass of the coated antigen. Therefore, exact percentage of trehalose used in coating formulations should be optimized to meet the experimental specifications.

References

1. Drucker E, Alcabes PG, Marx PA (2001) The injection century: massive unsterile injections and the emergence of human pathogens. Lancet 358:1989–1992
2. Banchereau J, Briere F, Caux C et al (2000) Immunobiology of dendritic cells. Annu Rev Immunol 18:767–811
3. Kim Y-C, Park J-H, Prausnitz MR (2012) Microneedles for drug and vaccine delivery. Adv Drug Deliv Rev 64:1547–1568
4. Matriano J, Cormier M, Johnson J et al (2002) Macroflux® microprojection array patch technology: a new and efficient approach for intracutaneous immunization. Pharm Res 19:63–70
5. Sullivan SP, Koutsonanos DG, Del Pilar Martin M et al (2010) Dissolving polymer microneedle patches for influenza vaccination. Nat Med 16:915–920
6. Williams JA, Carnes AE, Hodgson CP (2009) Plasmid DNA vaccine vector design: impact on efficacy, safety and upstream production. Biotechnol Adv 27:353–370
7. Ng KW, Pearton M, Coulman S et al (2009) Development of an ex vivo human skin model for intradermal vaccination: tissue viability and Langerhans cell behaviour. Vaccine 27: 5948–5955
8. Gill HS, Prausnitz MR (2007) Coated microneedles for transdermal delivery. J Control Release 117:227–237
9. Saurer EM, Flessner RM, Sullivan SP et al (2010) Layer-by-layer assembly of DNA- and protein-containing films on microneedles for drug delivery to the skin. Biomacromolecules 11:3136–3143
10. Chen X, Prow TW, Crichton ML et al (2009) Dry-coated microprojection array patches for targeted delivery of immunotherapeutics to the skin. J Control Release 139:212–220
11. Xianfeng C, Prow TW, Crichton ML et al (2008) Novel coating of micro-nanoprojection patches for targeted vaccine delivery to skin. In: International conference on nanoscience and nanotechnology, 2008, ICONN 2008, Melbourne, Australia, 25–29 February, pp 105–108. doi:10.1109/iconn.2008.4639257
12. Rajaraman S, Seong OC, McClain MA et al (2011) Metal-transfer-micromolded three-dimensional microelectrode arrays for in-vitro brain-slice recordings. J Microelectromechan Syst 20:396–409
13. Choi S-O, Kim Y-C, Lee JW et al (2012) Intracellular protein delivery and gene transfection by electroporation using a microneedle electrode array. Small 8:1081–1091
14. Shau MD, Shih MF, Lin CC et al (2012) A one-step process in preparation of cationic nanoparticles with poly(lactide-co-glycolide)-containing polyethylenimine gives efficient gene delivery. Eur J Pharm Sci 46:522–529
15. Zhang W, Gao J, Zhu Q et al (2010) Penetration and distribution of PLGA nanoparticles in the human skin treated with microneedles. Int J Pharm 402:205–212
16. Verbaan FJ, Bal SM, van den Berg DJ et al (2008) Improved piercing of microneedle arrays in dermatomed human skin by an impact insertion method. J Control Release 128:80–88
17. MacGregor GR, Caskey CT (1989) Construction of plasmids that express E. coli β-galactosidase in mammalian cells. Nucleic Acids Res 17:2365
18. Gu M-L, Leppla SH, Klinman DM (1999) Protection against anthrax toxin by vaccination with a DNA plasmid encoding anthrax protective antigen. Vaccine 17:340–344
19. Kim Y-C, Quan F-S, Compans R et al (2011) Stability kinetics of influenza vaccine coated onto microneedles during drying and storage. Pharm Res 28:135–144
20. Song J-M, Kim Y-C, Eunju O, Compans RW, Prausnitz MR, Kang S-M (2012) DNA vaccination in the skin using microneedles improves protection against influenza. Mol Ther 20:1472
21. Kim Y-C, Yoo D-G, Compans RW et al (2013) Cross-protection by co-immunization with influenza hemagglutinin DNA and inactivated virus vaccine using coated microneedles. J Control Release 172:579

22. DeMuth PC, Min Y, Huang B et al (2013) Polymer multilayer tattooing for enhanced DNA vaccination. Nat Mater 12:367–376
23. DeMuth PC, Su X, Samuel RE et al (2010) Nano-layered microneedles for transcutaneous delivery of polymer nanoparticles and plasmid DNA. Adv Mater 22:4851–4856
24. Gonzalez-Gonzalez E, Speaker TJ, Hickerson RP et al (2010) Silencing of reporter gene expression in skin using siRNAs and expression of plasmid DNA delivered by a soluble protrusion array device (PAD). Mol Ther 18: 1667–1674
25. Han M, Lee W, Lee S-K et al (2004) 3D microfabrication with inclined/rotated UV lithography. Sensor Actuator A Phys 111:14–20
26. Barichello JM, Morishita M, Takayama K, Nagai T (1999) Encapsulation of hydrophilic and lipophilic drugs in PLGA nanoparticles by the nanoprecipitation method. Drug Dev Ind Pharm 25:471–476
27. Wang L, Langley D (1977) Identification and determination of ionic surface active agents. Arch Environ Contam Toxicol 5:447–456
28. Cui Z, Mumper RJ (2001) Chitosan-based nanoparticles for topical genetic immunization. J Control Release 75:409–419
29. Kim Y-C, Song J-M, Lipatov AS et al (2012) Increased immunogenicity of avian influenza DNA vaccine delivered to the skin using a microneedle patch. Eur J Pharm Biopharm 81: 239–247

Chapter 11

Multivalent DNA-Based Vectors for DNA Vaccine Delivery

Young Hoon Roh, Kwang Lee, Jessica Jane Ye, and Dan Luo

Abstract

DNA can be utilized as a generic delivery vector as well as a traditional biological material for DNA vaccination. Although the use of DNA as an antigen expression vector or a vaccine adjuvant has been intensively studied for several decades, the use of DNA molecules as a delivery carrier has not been explored until recently. This issue is probably due to the topological limitation of DNA in its natural linear or circular structure form. Multivalent DNA-based vector delivery platforms overcome this structural barrier and are particularly suited for DNA vaccine delivery because of their multifunctionality, monodispersity, anisotropicity, and bioconjugation ability with numerous functional moieties. In this chapter, we mainly describe the construction of multivalent DNA-based delivery vectors using DNA engineering methods. Specifically, the synthesis strategies for highly branched dendrimer-like DNA structures in general and methods for their application to DNA vaccine delivery are introduced.

Key words DNA, Dendrimer-like DNA, Drug delivery, Anisotropicity, Multifunctionality, DNA engineering, DNA vaccines

1 Introduction

Nonviral DNA delivery systems using lipid- and polymer-based materials have great therapeutic potential for the treatment of genetic disorders and the improvement of immune responses [1]. However major challenges in the effective cellular and systemic delivery of DNA vaccines still exist due to the limited functions of delivery platforms. Increasing the multifunctionality of delivery vectors is one of the main strategies to overcome these delivery barriers. For that reason, it is important to synthesize a novel delivery platform with functional components in a precisely controlled manner. DNA as a true polymeric material is extremely useful in the construction of highly branched structures and functionalized materials because of its unique chemical and physical properties [2]. Recently multivalent DNA-based vector delivery systems have been developed [3]. These highly branched DNA structures have both the capacity and ability to bear functional components that

Monica Rinaldi et al. (eds.), *DNA Vaccines: Methods and Protocols*, Methods in Molecular Biology, vol. 1143,
DOI 10.1007/978-1-4939-0410-5_11, © Springer Science+Business Media New York 2014

can be either isotropic or anisotropic on demand. In essence, these multivalent DNA vectors start from simple branched DNA building units such as X-shaped DNA (X-DNA) or Y-shaped DNA (Y-DNA). More complicated structures such as dendrimer-like DNA (DL-DNA) and networked DNA that are easily tunable in terms of structure and size can be generated by additional enzyme ligation [4], chemical conjugation [5], and self-assembly [6]. Due to their built-in modularity and the rich biofunctional chemistry, these multivalent DNA structures also make ideal functional delivery carriers. In particular, highly branched and networked structures can incorporate various types of biological functional moieties including peptides, proteins, lipids, and polymers along with DNA and RNA. These various moieties can contribute desired delivery and biological attributes, including high DNA loading, improved stability, cellular targeting, endosomal escape, and nuclear targeting, which eliminate several obstacles in achieving highly efficient and targeted delivery. This chapter mainly focuses on detailed protocols for constructing multivalent DNA structures based on their structural properties and functionalities. Furthermore the efficient delivery methods of multivalent DNA-based platforms for DNA vaccine delivery are described. In more details, the described methods in Subheading 3 include the procedures needed to select the sequences for DNA engineering as a polymeric material, generate the isotropic DL-DNA, synthesize the anisotropic branched cross-linking DNA monomers, attach functional moieties onto multivalent DNA structures, and verify the delivery efficiency of multivalent DNA-based carriers in mammalian cell lines.

2 Materials

2.1 Dendrimer-Like DNA

1. Oligonucleotides (Integrated DNA Technologies).
2. Annealing buffer: 10 mM Tris, pH 8.0, 1 mM ethylenediaminetetraacetic acid (EDTA) and 50 mM NaCl.
3. T4 DNA ligase and buffers.
4. Nuclease-free, sterile Milli-Q water.
5. Streptavidin-coated polystyrene beads with 5.5 μm diameter (Bangs Laboratories, Fishers, IN).
6. Restriction enzymes (*DdeI*).

2.2 Multifunctional ABC Polymers

1. Hybridization buffer, 1× SSC: 150 mM sodium chloride, 15 mM sodium citrate, and 1 % SDS.
2. Poly(ethylene glycol) acrylate (Sigma-Aldrich).
3. Photoinitiator (Sigma-Aldrich).

2.3 Multivalent DNA Structures with Functional Moieties

1. Plasmid pIVEX1.3WG (Roche Applied Science).
2. Restriction enzymes (*NcoI*, *SmaI*, *ApaI*).
3. Peptides, polymers, and lipids (Sigma-Aldrich).
4. Primers and reagents for qRT-PCR (Invitrogen, Life Technologies).
5. Phosphate-buffered saline (PBS) solution: 10 mM phosphate buffer, pH 7.4, 137 mM NaCl, and 2.7 mM KCl.
6. Culture medium (RPMI 1640) with 10 % fetal bovine serum, 2 % l-glutamine, 1 % sodium pyruvate, 1 % penicillin–streptavidin.
7. Opti-MEM medium.
8. Fluorescent nuclear stains (DAPI) and actin stains (GFP or phalloidin) (Molecular Probes, Life Technologies).
9. DNA easy extraction kit (Qiagen).
10. PicoGreen dsDNA quantification kit (Invitrogen, Life Technologies).
11. Protein specific antibodies for Western blot (Abcam).
12. ELISA kit or other assay kit for cytokine production (Novex, Life Technologies).

2.4 Equipment

1. Gel electrophoresis equipment (Bio-Rad).
2. Thermal cycler (Eppendorf).
3. Flow cytometer (BD Biosciences).
4. Spectrophotometer (Eppendorf).
5. Rotary incubator (Thermo Fisher Scientific).
6. UV crosslinker (Spectronics Corporation).
7. Freeze dry system (Labconco Corporation, Kansas City, MO).
8. High Performance Liquid Chromatography (Waters Corporation).

3 Methods

3.1 Basic Considerations for DNA Engineering as a Polymeric Material

This section describes fundamental DNA sequence design parameters and structural evaluation methods as a polymeric material. The following section examines general strategies for DNA engineering and characterization tools.

3.1.1 DNA Sequence Design and Evaluation

A DNA molecule comprises repeating units called nucleotides. There are four main types of nucleotides, including adenine (A), thymine (T), guanine (G), and cytosine (C), distinguished by different nitrogenous bases. The synthesis of multivalent DNA structures starts from the selection of appropriate nucleotides of single-stranded DNA (ssDNA) and double-stranded DNA

Table 1
Fundamental rules for DNA sequence design [9]

	Empirical rules	Key characteristics
1	Length	It should be long enough to form a stable DNA structure [at least more than 8 nucleotides (nt) long]
2	GC content	In general (varies by design), sequences are routinely chosen that constitute about 50 % GC
3	Non-Watson–Crick base pairing	Sequences containing more than two consecutive Gs should be avoided
4	Free energy (Delta G)	A lower free energy is desired; however, intermediate–low DG is also considered. Many websites provide online tools for the calculations (e.g., http://www.idtdna.com)
5	Secondary structure	In general, the least amount of secondary structure is desired
6	Helix geometry	Half-turns are quantum of the design ($5'n$ bp, where $n = 0, 1, 2 \ldots$ are between junctions)
7	Symmetry	The sequence symmetry (e.g., as those occurred in Holliday junctions) of each arm should be avoided
8	Dimerization Triplexation Z-DNA formation	The molecules should not form a self-dime, a triplex, or a Z-DNA

Reproduced by permission of the Royal Society of Chemistry

(dsDNA). Numerous sets of sequences can be chosen by following Watson–Crick complementary base pairing rules such that A pairs with T and G pairs with C. However the structural properties of these predesigned DNA sequences need to be evaluated to avoid formation of unexpected side products and low product yields. Major considerations for this evaluation include sequence length, self-priming, Gibbs free energy, GC content, and secondary structure formation. Although there are no general principles for evaluating DNA sequence design, we introduce several fundamental rules for DNA sequence selection in Table 1. In addition, several programs are available online that provide useful information regarding preselected DNA sequences. For example, a DNA folding software, UNAFold, can be used to predict the hybridization of the complete sequence and draw representative two-dimensional structure configurations [7]. Another software includes calculations of free energy, melting temperature, and possible secondary structure formation from a user-defined sequence. The above DNA sequence evaluation tools can be obtained from the websites listed in ref. 8.

Design of DNA end sequences is particularly important for building highly branched and networked DNA structures.

Single-stranded overhangs (sticky ends) placed at the terminal branch arms allow more DNA building blocks to be linked on by enzyme ligation. Sticky end sequences can be either palindromic or non-palindromic based on one's final DNA structure design. Non-palindromic sequences on the donor branched DNA in particular can be used to give increased control over formation of highly branched DNA structures, both by avoiding self-ligation (or hybridization) of donor branched DNA and by allowing selective ligation of donor to the acceptor DNA structures.

3.1.2 General Strategies for DNA Engineering

DNA, like general synthetic polymers, can exist in one of the four topologies—linear, branched, dendritic, and networked [9]. To overcome the topological limitations of DNA in its natural forms (linear and circular) and improve its usefulness, additional engineering methods are necessary. To precisely manipulate DNA to form a variety of structures, there are three major classes of DNA engineering strategies consisting of physical, biological, and chemical methods. In physical methods, parameters, chiefly the DNA sequence length and melting temperature (T_m), are controlled in the assembly of DNA structures. The flexibility and rigidity of DNA structures can be manipulated by choosing the DNA sequence length and types (e.g., ssDNA or dsDNA). The dsDNA is rigid when it is shorter than its persistence length (150 bases, approximately 50 nm). However, most ssDNAs are flexible because the persistence length of ssDNA is about 1 nm (1–3 bases). The selection of ssDNA and dsDNA combination during DNA construct design is an important strategy to achieve a desired flexibility and rigidity of final DNA structure. In addition the T_m mainly relies on the DNA length, salt/DNA concentration, and base composition (e.g., GC content). For example, the longer the DNA length, the higher the T_m and the more stable the structure. Similarly, a high composition of GC bases (with low AT content) in the structure increases stability. In biological methods, numerous molecular toolboxes are used for exquisite control of DNA structures. DNA can be easily connected, shortened, elongated, and modified by selecting an appropriate enzyme type among ligases, restriction enzymes, polymerases, and so on. In chemical methods, various bioconjugation techniques can be performed to incorporate functions onto DNA structures. For this, chemical moieties such as thiols, amines, biotins, and alkynes must first be conjugated to DNA strands in specifically assigned regions. These chemical-conjugated DNA can be further modified by additional substrates through either covalently or non-covalently to form functionalized materials. Different bifunctional cross-linkers, which are further explained in next section, are also available for this purpose. Taken together, these tools for DNA engineering provide a broad range of controllability and flexibility to generate precisely controlled branched DNA structures with functional moieties.

3.1.3 DNA Structure Purification and Characterization

Assembled DNA structures must be purified and evaluated before further usage. Gel electrophoresis, high-performance liquid chromatography (HPLC), purification columns, and dialysis are representative methods to achieve a high purity of DNA structures. Moreover there are several ways to characterize the formation of DNA structures. The most representative ways to evaluate multivalent DNA structures are electrophoresis using agarose or polyacrylamide gels. Other characterization methods to visualize DNA structures include high-resolution microscopy techniques, such as scanning electron microscopy (SEM), transmission electron microscopy (TEM), atomic force microscopy (AFM), and confocal/ fluorescence microscopy. It is important to use the appropriate instruments to resolve the final DNA products in terms of their sizes and structures.

3.2 Synthesis of Dendrimer-Like DNA

Recently, branched DNAs with diverse shapes such as Y-DNA, X-DNA, and DL-DNA have been designed [4]. The monodisperse branched DNA is able to be synthesized as relatively of high yield and purity, and in particular, it is highly suitable for biological applications because inherent DNA is both biocompatible and biodegradable. To construct branched DNA, ssDNA should be designed to have complementary sequences to matching parts on the other ssDNAs. For instance, Y-DNA as a basic building block can be self-assembled from three ssDNAs following base pairing interactions via hydrogen bonding (hybridization). In addition, these basic DNA building blocks can be further connected to form highly ordered and intricate structures, such as DL-DNA (*see* Fig. 1). In more detail, DL-DNA can be made from a non-palindromic core Y-DNA building block (G_0), surrounded by other Y-DNAs with sticky ends complementary to the core Y-DNA. These surrounding Y-DNAs are ligated to the initial core to form the first generation (G_1) of DL-DNA. Additional generations (G_2, G_3, G_4, etc.) are created by repeatedly ligating Y-DNAs to the sticky ends of the previous generation. Detailed procedures to generate DL-DNA with two different synthesis approaches—a solution phase and a solid phase—are explained in the following subsections. The example DNA sequences for DL-DNA construction using both methods are listed in Table 2.

3.2.1 Branched DNA Synthesis: Y-DNA

A Y-DNA building block is synthesized by annealing of ssDNAs (Y_{na}, Y_{nb}, Y_{nc}) with a one-pot approach.

1. Dissolve each ssDNA in an annealing buffer (10 mM Tris, pH 8.0, 1 mM EDTA).
2. Mix the three ssDNAs in an equal molar ratio (1:1:1).
3. Increase the temperature to 95 °C for 2 min to denature the ssDNA.

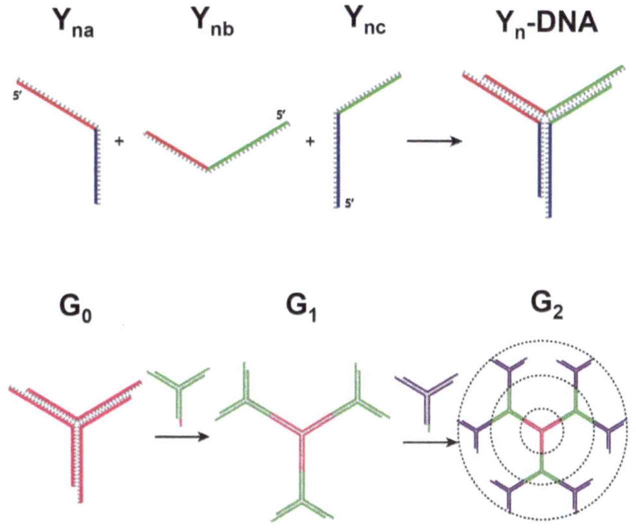

Fig. 1 Schematic illustration of the general dendrimer-like DNA in solution-phase synthesis [9]. Reproduced by permission of the Royal Society of Chemistry

4. Cool ssDNAs at 65 °C for 2 min.
5. Anneal ssDNAs at 60 °C for 5 min.
6. Subsequently, further anneal at 60 °C with a continuous temperature decrease at a rate of 1 °C per min.
7. Repeat the annealing steps total 40 times.
8. Y-DNA will be hybridized. Then, Y_0-DNA is also called the zero-generation DL-DNA (G_0, core Y-DNA), and Y_1, Y_2, Y_3, Y_4-DNAs are additional building blocks to grow further generation DL-DNA.
9. Confirm if hybridization of Y-DNAs was successful by gel electrophoresis (100 V for 60 min in 3 % agarose gel).
10. Store Y-DNAs at 4 °C.

3.2.2 DL-DNA Synthesis: A Solution-Phase Approach

1. To synthesize the first-generation DL-DNA (G_2), ligate the Y_1-DNA to the Y_0-DNA (G_0) at a 3:1 M ratio ($G_0 + 3Y_1 - \text{DNA} \rightarrow G_1$).
2. Add 10× T4 DNA ligase buffer and T4 DNA ligase, and mix well (*see* **Note 1**).
3. Incubate at room temperature for 16 h.
4. Confirm if ligation of first-generation DL-DNA (G_1) was successful by gel electrophoresis (100 V for 60 min in 3 % agarose gel).
5. To synthesize the second-generation DL-DNA (G_2), ligate the Y_2 DNA to the G_1 DL-DNA at a 6:1 M ratio ($G_1 + 6Y_2 - \text{DNA} \rightarrow G_2$).

Table 2
DNA sequences for the dendrimer-like DNA construction

Solution-phase synthesis			
Strand		**Segment 1**	**Segment 2**
Y_0	Y_{0a}	5′-p-TGAC	TGGATCCGCATGACATTCGCCGTAAG-3′
	Y_{0b}	5′-p-TGAC	AGGCTGATTCGGTTCATGCGGATCCA-3′
	Y_{0c}	5′-p-TGAC	CTTACGGCGAATGACCGAATCAGCCT-3′
Y_1	Y_{1a}	5′-p-GTCA	TGGATCCGCATGACATTCGCCGTAAG-3′
	Y_{1b}	5′-p-CAGT	AGGCTGATTCGGTTCATGCGGATCCA-3′
	Y_{1c}	5′-p-CAGT	CTTACGGCGAATGACCGAATCAGCCT-3′
Y_2	Y_{2a}	5′-p-ACTG	TGGATCCGCATGACATTCGCCGTAAG-3′
	Y_{2b}	5′-p-GCAT	AGGCTGATTCGGTTCATGCGGATCCA-3′
	Y_{2c}	5′-p-GCAT	CTTACGGCGAATGACCGAATCAGCCT-3′
Y_3	Y_{3a}	5′-p-ATGC	TGGATCCGCATGACATTCGCCGTAAG-3′
	Y_{3b}	5′-p-TGTC	AGGCTGATTCGGTTCATGCGGATCCA-3′
	Y_{3c}	5′-p-TGTC	CTTACGGCGAATGACCGAATCAGCCT-3′
Y_4	Y_{4a}	5′-p-GACA	TGGATCCGCATGACATTCGCCGTAAG-3′
	Y_{4b}	5′-p-GGAT	AGGCTGATTCGGTTCATGCGGATCCA-3′
	Y_{4c}	5′-p-GGAT	CTTACGGCGAATGACCGAATCAGCCT-3′

Solid-phase synthesis			
Strand		**Segment 1**	**Segment 2**
Spacer 1		Biotin-5′-p	CCGGATAAGGCGCAGCGGTCGGCTGAA TTCAGGGTTCGTGGCAGGCCAGCACAC TTGGAGACCGAAGCTTACCGGACTCCT AAC-3′
Spacer 2		5′-p-TCA	GTTAGGAGTCCGGTAAGCTTCGGTCTC CAAGTGTGCTGGCCTGCCACGAACCCT GAATTCAGCCGACCGCTGCGCCTTATC CGG-3′
Y_0	Y_{0a}	5′-p-TGA	GTGATCCGCATGACATTCGCCGTAAG-3′
	Y_{0b}	5′-p-TGAC	AGGCTGATTCGGTTCATGCGGATCAC-3′
	Y_{0c}	5′-p-TGAC	CTTACGGCGAATGACCGAATCAGCCT-3′
Y_1	Y_{1a}	5′-p-GTCA	TGGATCCGCATGACATTCGCCGTAAG-3′
	Y_{1b}	5′-p-CAGT	AGGCTGATTCGGTTCATGCGGATCCA-3′
	Y_{1c}	5′-p-CAGT	CTTACGGCGAATGACCGAATCAGCCT-3′
Y_2	Y_{2a}	5′-p-ACTG	TGGATCCGCATGACATTCGCCGTAAG-3′
	Y_{2b}	5′-p-GCAT	AGGCTGATTCGGTTCATGCGGATCCA-3′
	Y_{2c}	5′-p-GCAT	CTTACGGCGAATGACCGAATCAGCCT-3′

(continued)

Table 2 (continued)

Solid-phase synthesis			
Strand		**Segment 1**	**Segment 2**
Y_3	Y_{3a}	5′-p-ATGC	TGGATCCGCATGACATTCGCCGTAAG-3′
	Y_{3b}	5′-p-TGTC	AGGCTGATTCGGTTCATGCGGATCCA-3′
	Y_{3c}	5′-p-TGTC	CTTACGGCGAATGACCGAATCAGCCT-3′
Y_4	Y_{4a}	5′-p-GACA	TGGATCCGCATGACATTCGCCGTAAG-3′
	Y_{4b}	5′-p-GGAT	AGGCTGATTCGGTTCATGCGGATCCA-3′
	Y_{4c}	5′-p-GGAT	CTTACGGCGAATGACCGAATCAGCCT-3′

Adapted by permission from Macmillan Publishers Ltd: Nature Protocols [11], copyright (2006). 5′-p indicates the phosphate in the 5′ prime end of the DNA sequence

6. Repeat ligation procedure as described in G_1 DL-DNA synthesis to create higher generation DL-DNA (*see* **Note 1**).
7. Follow combination rule of molar ratio between G_{n-1} and Y*n*-DNA to generate higher generation DL-DNA [$G_{n-1} + 6 \times 2^{(n-2)} Y_n - \text{DNA} \rightarrow G_n$ (here, $n \geq 2$)]; hence, the *nth-generation* DL-DNA is formed by ligating new Y_n-DNA to the previous generation [(n−1)th-generation] at a $6 \times 2^{(n-1)}$:1 M ratio.
8. Generate the third-generation DL-DNA ($G_2 + 12Y_3\text{-DNA} \rightarrow G_3$).
9. Generate the fourth-generation DL-DNA ($G_3 + 24Y_4\text{-DNA} \rightarrow G_4$).
10. Generate the fifth-generation DL-DNA ($G_4 + 48Y_5\text{-DNA} \rightarrow G_5$).
11. Confirm if the assembly of each *n*th-generation DL-DNA (G_n, $n \geq 2$) was successful by gel electrophoresis (80 V for 60 min in 1 % agarose gel).
12. Store DL-DNA at 4 °C for further usages.

3.2.3 DL-DNA Synthesis: A Solid-Phase Approach

A solid-phase approach of DL-DNA provides more purity and yield than solution-phase synthesis because, in the solid phase, there is a synchronized process to assemble and purify final product. Here, the spacer is necessary as a linker that attaches the DL-DNA to a solid bead surface, and it is designed to have a complementarity with one arm of branched DNA. The scheme of a solid-phase approach for DL-DNA synthesis is illustrated in Fig. 2.

1. Prepare commercially synthesized two ssDNAs as spacers (SP): SP1 is 5′-biotin-modified 84 bases, and SP2 is 87 bases with 5′-sticky end which is complementary to one arm of Y_0-DNA.

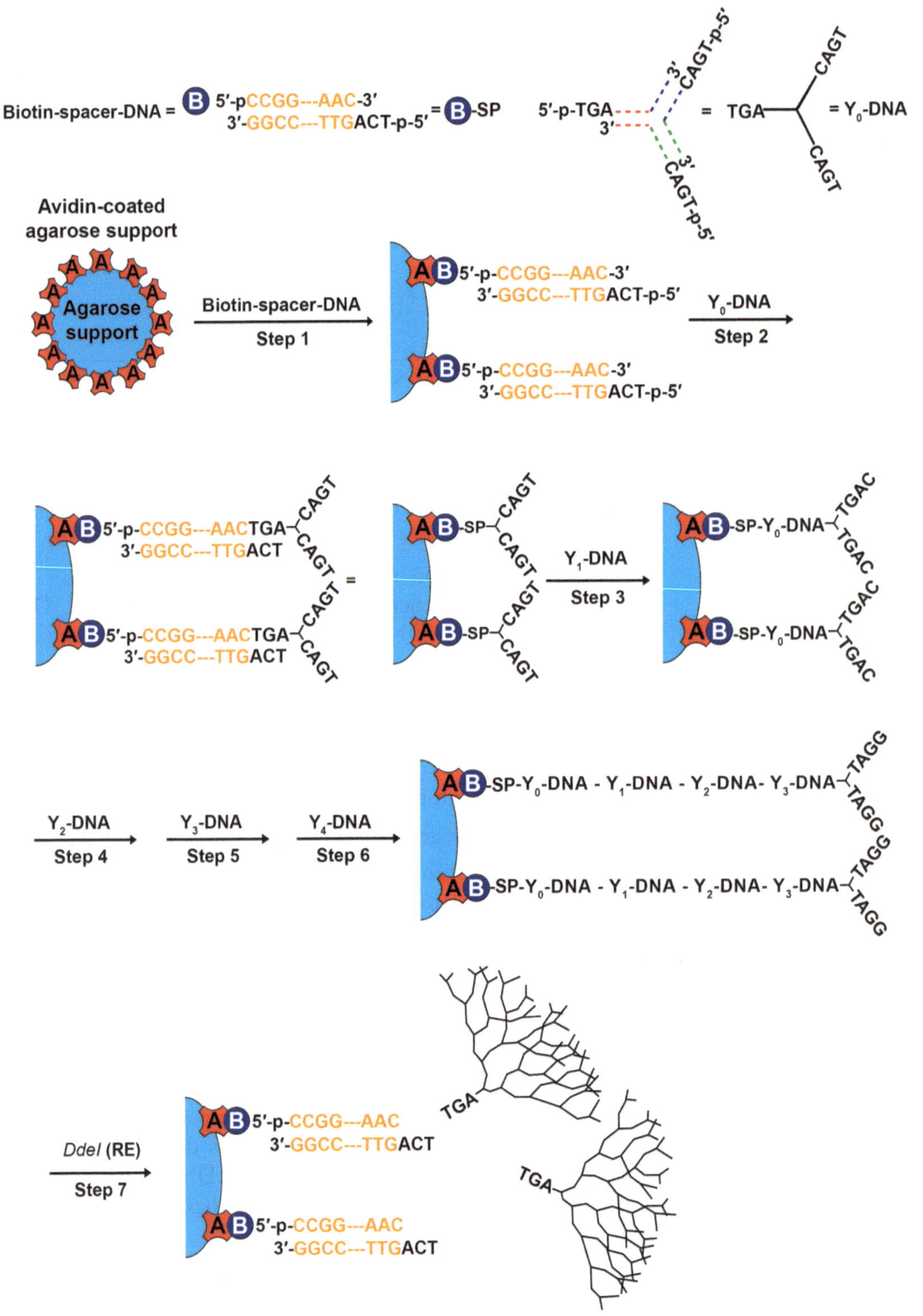

Fig. 2 Schematic diagram of the process for the solid-phase synthesis of dendrimer-like DNA. Reprinted by permission from Macmillan Publishers Ltd: Nature Protocols [11], copyright (2006)

2. Combine the spacer ssDNAs in an equal molar ratio (1:1) and hybridize as per the following methods: denaturation to 94 °C for 4 min, annealing at 80 °C for 2 min, and a continuous temperature decrease as a rate of 0.5 °C/min for 1 h.
3. Pretreat the avidin-coated agarose solid bead within 0.1 % sodium dodecyl sulfate (SDS) surfactant to reduce nonspecific binding of DNA on the solid surface (*see* **Note 2**).
4. Add 150 μl (8.2 nmole) of biotin-modified spacer to 100 μl of avidin-coated agarose support and incubate at room temperature overnight for conjugation.
5. Centrifuge the solid support at 2,500 g for 10 min, and discard the supernatant. Additionally, repeat the three times washing step with 1× PBS buffer.
6. Attach Y_0-DNA (G_0 DL-DNA) on spacer-modified bead using T4 DNA ligase (*see* **Note 1**).
7. The first-generation process (ligation) of DL-DNA in solid-phase synthesis is identical with it in solution-phase synthesis; however, combination rule of molar ratio in solid-phase synthesis is different from it in solution-phase synthesis; here G_1 is formed by ligating two Y_1 with one G_0 ($G_0 + 2Y_1$-DNA $\rightarrow G_1$).
8. The higher generation of DL-DNA on solid surface is constructed using the same ligation process and combination rule of molar ratio in the first generation [$G_{n-1} + 2^n Y_n$-DNA $\rightarrow G_n$ (here, $n \geq 1$)]; hence, the nth-generation DL-DNA is formed by ligating new Y_n-DNA to the previous generation [$(n-1)$ th-generation] at a 2^n:1 M ratio.
9. Generate the second-generation DL-DNA ($G_1 + 4Y_2 -$ DNA $\rightarrow G_2$).
10. Generate the third-generation DL-DNA ($G_2 + 8Y_3 -$ DNA $\rightarrow G_3$).
11. Generate the fourth-generation DL-DNA ($G_3 + 16Y_4 -$ DNA $\rightarrow G_4$).
12. Generate the fifth-generation DL-DNA ($G_4 + 32Y_5 -$ DNA $\rightarrow G_5$).
13. After ligation, DL-DNA is cleaved off from the solid phase by *DdeI* restriction enzyme (restriction site: 5′-C/TNAG-3′), which recognized a specific site on the spacer. The enzyme solution contained 100 units of *DdeI* and 4 μl BSA in restriction buffer D with 60 mM Tris–HCl, pH 7.9, 1.5 M NaCl, 60 mM $MgCl_2$, and 10 mM dithiothreitol (DTT).
14. Evaluate the restriction efficiency of the synthesis of the DL-DNA using PicoGreen dsDNA quantification assay kit.
15. Confirm if the assembly of each DL-DNA generation was successful by gel electrophoresis (80 V for 60 min in 1 % agarose gel).
16. Store DL-DNA at 4 °C for further usages.

3.3 Anisotropicity and Multifunctionality of Dendrimer-Like DNA

Unlike other chemical dendrimers, DL-DNAs can also be designed to be anisotropic due to their sequence complementary specificity and selectivity. Such anisotropy arises from the ability to conjugate different functional groups to different arms of the branched DNA monomers, and the diversity of these functional groups also gives it considerable potential multifunctionality. Anisotropic, branched, and cross-linkable DNA structures (ABC monomers) are one example [3]. These versatile DNA-based building blocks are made of highly branched DNA, forming structures with unique geometries by design. Each arm of the branched DNA structures can be conjugated or designed to have a particular functional moiety, allowing multiple functions within one single structure. The basic structure of ABC monomers consists of an X-DNA acceptor surrounded by up to three Y-DNA branches, each of which can be conjugated to different functional moieties. Then, two types of ABC monomers can be hybridized to form ABC dimers by the complementary linker DNA. After synthesis of the ABC monomers and dimers, these DNA building blocks can be further polymerized at adjacent PEGA groups using a UV-induced free radical reaction to form multifunctional ABC polymers (*see* Fig. 3). The detailed procedures for each step are described in the following subsection.

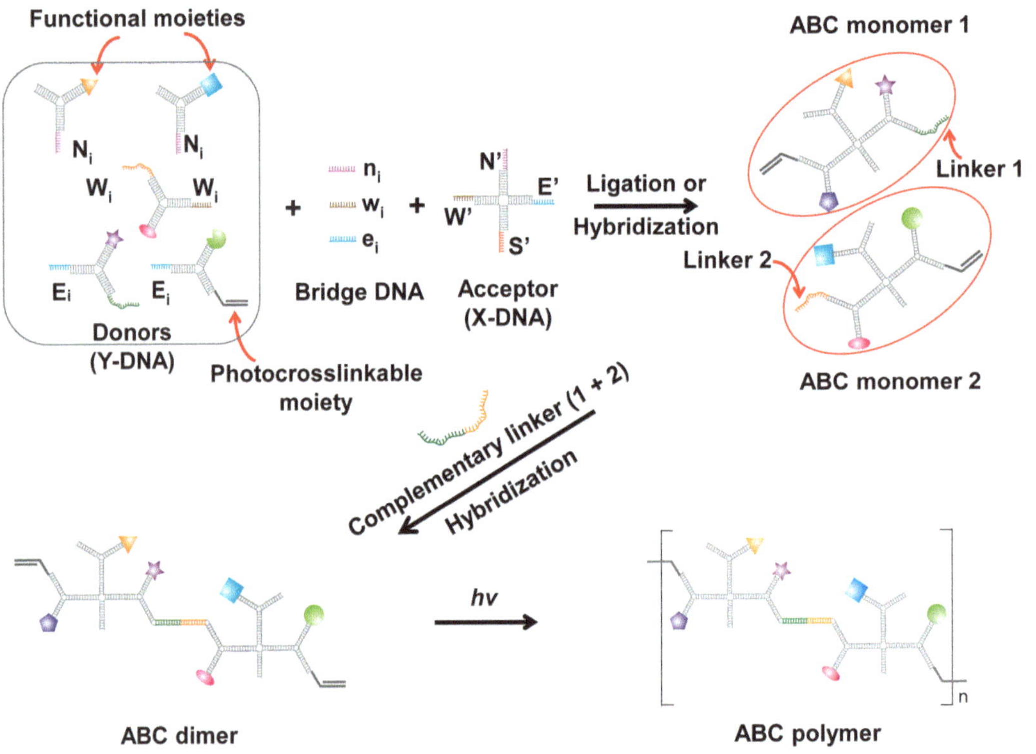

Fig. 3 Schematic illustration of the creation for anisotropic branched cross-linkable DNA structures. Reprinted by permission from Macmillan Publishers Ltd: Nature Nanotechnology [3], copyright (2009)

3.3.1 Construction of ABC Monomers and Dimers

1. Design acceptor X-DNA and donor Y-DNA sequences. For simplicity, each of the three Y-DNA donor and the corresponding arms of the X-DNA acceptor will be designated north, south, east, and west. Both X-DNA and Y-DNA arms must have a single-stranded sticky end region. The sticky ends on the donor and acceptor do not have to be complementary—a unique bridge DNA bearing regions complementary to both sticky ends may be used as a connector. This allows one to easily control the placement of particular Y-DNAs with specific functions.
2. Conjugate Y-DNA to desired functional groups. Each arm of the Y-DNA may be conjugated to one functional moiety (*see* **Notes 3** and **4**).
3. Hybridize donor and acceptor DNA with bridge DNA. Incubate Y-DNAs (north, east, west), bridge DNA, and X-DNA in equal molar ratios at 30 °C for 1 h.
4. Confirm the quality of ABC monomers using gel electrophoresis (80 V for 60 min in 1 % agarose gel).
5. Make ABC dimers by additional hybridization with a linear strand of linker DNA bearing sticky ends complementary to those on the "east" and "west" Y-DNA branches that will be connected (*see* **Note 5**).

3.3.2 Construction of Multifunctional ABC Polymers

Poly(ethylene glycol) acrylate-functionalized ABC dimers can be polymerized using a photocrosslinking method to form multifunctional ABC polymers (*see* **Notes 6** and 7).

1. Warm up the UV cross-linker instrument.
2. Plate various concentrations of the ABC dimer mixture into a UV transparent flat-well plate or Eppendorf tube.
3. Add a photoinitiator (e.g., 1-[4-(2-hydroxyethoxy) phenyl]-2-hydroxy-2-phenylpropan-1-one, 0.05 % w/v).
4. Place plate or tube in UV cross-linker and expose to UV radiation (365 nm) at 8 mW/cm^2 for 10 min (*see* **Note 8**).
5. Remove plate, and mix the final products thoroughly.
6. Confirm the final products by microscopy or gel electrophoresis (100 V for 60 min in 3 % agarose gel).

3.4 Conjugations of Multivalent DNA Structures with Functional Moieties

Multivalent DNA structures can be tailored by conjugating various functional moieties as desired. In general, precisely controlled conjugation is required to connect functional moieties onto multivalent DNA vectors in a specific region with a preassigned ratio. These moieties can be a variety of peptides, polymers, dyes, ligands, lipids, as well as functional genes of interest (e.g., plasmid DNA) for DNA vaccination. These functional moieties may also be mixed and matched to make additional combinations with unique

Table 3
Functional cross-linkers for DNA conjugation [9]

	Functional group of reactant A	Bifunctional cross-linker	Functional group of reactant B
Homo	Amino	NHS ester 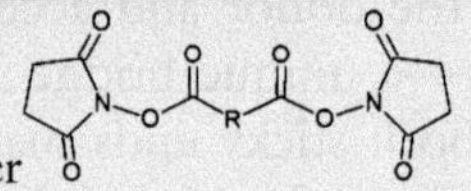	Amino
	Amino	Imidoester 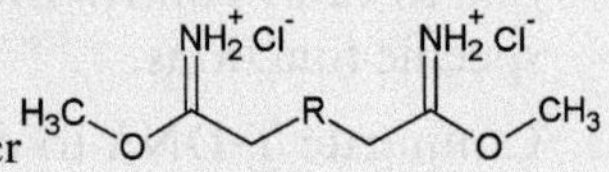	Amino
	Thiol	Maleimide 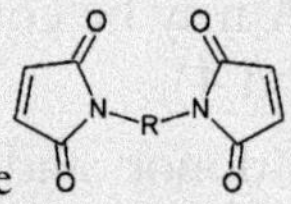	Thiol
	Thiol	Pyridyldithiol 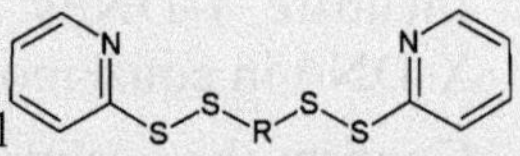	Thiol
Hetero	Amino	NHS ester/maleimide 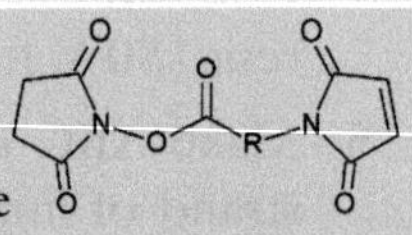	Thiol
	Amino	NHS ester/pyridyldithiol 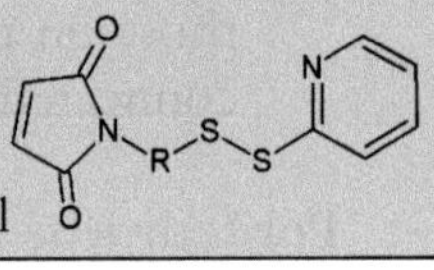	Thiol

properties. Different types of homo and hetero bifunctional cross-linkers have been utilized for chemical conjugation and modification. Representative cross-linkers that have been employed for nucleic acid conjugations between an amine ($-NH^{2+}$) and thiol (-SH) group are listed in Table 3. The detailed examples of conjugation procedures with peptide, polymer, lipid, and plasmid DNA are described as follows:

3.4.1 Conjugation of Multivalent DNA with Peptide

Succinimidyl 4-(*N*-maleimidomethyl) cyclohexane-1-carboxylate (SMCC) has an NHS ester and a maleimide group, which result in primary amine and sulfhydryl reactivity. The cyclohexane composition, in particular, makes the maleimide particularly stable.

1. Select an appropriate peptide candidate (*see* **Note 9**).
2. Resuspend the amine-modified ssDNA or branched DNA in PBS, pH 7.5.
3. Dissolve SMCC in dimethylformamide (DMF).
4. To activate DNA, mix with a 40 molar excess of SMCC in DMF.
5. Incubate the reaction solution at room temperature for 2 h.

6. Remove the excess of unreactive SMCC from activated DNA via filtration (e.g., Sephadex™ G-25 column, Bio-Rad, Hercules, CA) or centrifugation.
7. Concentrate the SMCC-activated DNA by using either a Microcon Y-3 (Millipore, Bedford, MA) or a freeze drying instrument.
8. Mix the SMCC-activated DNA with an eight molar excess of peptide, and adjust the reaction solution in 1× PBS.
9. Incubate the reaction mixture overnight at room temperature with gentle shaking.
10. Store the DNA–peptide conjugate at −20 °C for further usage.

3.4.2 Conjugation of Multivalent DNA with Polymer

The conjugation of multivalent DNA with a polymer involves a reaction between the amine-modified groups of DNA and the ester groups of the polymer.

1. Dissolve the amine-modified Y-DNA in annealing buffer to a concentration of 0.2 mM.
2. Prepare an appropriate polymer, for example polyethylene glycol (PEG) in 1× PBS solution.
3. *N*-hydroxysuccinimide-functionalized PEG can be conjugated to the selected Y-DNA arm by gentle mixing with a five molar excess of PEG.
4. Incubate the reaction mixture for 4 h at room temperature.
5. HPLC can be used for the removal of unreacted products and impurities (*see* **Note 10**).
6. Gel electrophoresis of HPLC fractions may be performed to confirm the conjugated products (90 V for 60 min in 3 % agarose gel).
7. Store the Y-DNA–PEG conjugate at −20 °C for further usage.

3.4.3 Conjugation of Multivalent DNA with Lipid

Y-DNA–lipid conjugation is achieved by reacting the amine-modified groups of Y-DNA with the activated ester groups of the lipid molecules. In the Y-DNA, one ssDNA is functionalized with a lipid molecule, and the other two ssDNAs can be used as annealing sites for multivalent DNA. Finally, self-assembly of the lipid molecules in the Y-DNA forms DNAsomes, liposome-like core–shell structures.

1. Dissolve dried *N*-glutaryl phosphatidylethanolamine (NGPE, 0.3 mg) with 0.5 ml of 0.16 M octylglucoside in MES buffer (pH 5.5).
2. Add EDC (2 mM) and NHS (5 mM) and incubate for 10 min at room temperature.
3. Centrifuge at 2,500 ×*g* for 60 s, and mix with 20 μl of 0.1 M NaOH (pH 8.0). Repeat shaking and centrifugation.

8. Combine the prepared Y-DNA products, which are hybridized with the amine-modified ssDNA, and incubate the reaction mixture with a 10 M excess of lipid for 4 h at room temperature.
4. Purify Y-DNA–lipid conjugation by HPLC (*see* **Note 10**).
5. Dialyze Y-DNA–lipid amphiphiles against 100 ml of aqueous solution for 48 h using a cellulose membrane bag to prepare DNAsomes. After dialysis, DNAsomes are collected and frozen using a freeze dry system.

3.4.4 Conjugation of Linearized Plasmid DNA and Multivalent DNA

The linearized plasmid DNA (pDNA) and multivalent DNA can be ligated or hybridized to one another at restriction sites that form complementary sticky ends [10].

1. To clone a target gene into a plasmid vector, for example, digest pIVEX1.3WG plasmid with *NcoI* and *SmaI* restriction enzymes, and ligate a target gene, also digested by *NcoI* and *SmaI*, at the respective restriction sites on the pIVEX1.3WG plasmid.
2. Transform the pIVEX1.3 plasmid into *E. coli*.
3. Purify pIVEX1.3 by Maxiprep (Qiagen, Valencia, CA) as outlined in the manufacturer's protocol.
4. Linearize the amplified pIVEX1.3 plasmid by *ApaI* restriction enzyme (restriction site: 5′-G***GGCC***/C-3′ and 3′-C/***CCGG***G-5′).
5. Design arms of multivalent DNA, for example, two ssDNAs in Y_{final}-DNA of DL-DNA can have complementary sequence of the restriction site (5′-***GGCC***-3′).
6. Combine pDNA with multivalent DNA in an equal molar ratio (1:1).
7. Add 10× T4 DNA ligase buffer and T4 DNA ligase, and mix well (*see* **Note 1**).
8. Incubate at room temperature for 16 h for enzyme ligation.
9. Then gel electrophoresis can be performed to confirm the conjugated products (80 V for 80 min in 1 % agarose gel).
10. Store the pDNA-multivalent DNA product at −20 °C for future use.

3.5 Evaluating the Efficiency of Multivalent DNA-Based Platforms for DNA Vaccine Delivery

The multivalent DNA-based platforms described thus far are natural vehicles for DNA vaccine delivery. It is important, however, to assess the delivery efficiency of each DNA construct before drawing conclusions about the functional effects of the DNA vaccine. The delivery efficiency of DNA vaccine-vehicle constructs can be evaluated at different levels using direct fluorescence visualization, quantification of antigen or reporter protein expression, and immunogenicity assays. Fluorescence visualization is ideal for tracking specific delivery or confirming uptake. Protein expression and mRNA quantification allow one to assess the functional delivery of the loaded sequences. Immunogenicity assays directly detect a desirable property of DNA vaccines in general. These methods can be applied to assess the delivery of a wide range of multivalent DNA-based delivery platforms, including DL-DNA, ABC polymers, DNA nanospheres, and DNAsomes. Each of these basic assay methods for these multivalent DNA-based platforms is briefly described below:

3.5.1 Visualization by Fluorescence Microscopy

Direct fluorescence imaging allows one to visualize the delivery of DNA constructs into the cell (*see* **Note 11**).

1. Plate cells from culture on an 8-well slide plate (*see* **Note 12**) at a density that will give 75–80 % confluence after growing overnight. Mammalian epithelial cells are recommended for testing DNA delivery, but other cell types may be used.
2. Incubate cells overnight at 37 °C, 5 % CO_2.
3. Prepare multivalent DNA vehicle under sterile conditions.
4. Load DNA vaccine onto multivalent DNA vehicle as described previously in Subheading 3.4.4 to generate the DNA vaccine-vehicle construct.
5. Add DNA vaccine-vehicle construct to 100 μl of Opti-MEM media (*see* **Note 13**).
6. Aspirate cell supernatant and replace with Opti-MEM media with DNA vaccine-vehicle construct.
7. Incubate for 2 h at 37 °C, 5 % CO_2.
8. Wash cells with 2× PBS to remove adsorbed DNA not actually taken up by the cell.
9. Stain cell nuclei and actin according to the manufacturer's protocols (Invitrogen, Life Technologies) to localize the cells and visualize the boundaries of the cell. We typically use DAPI for the nuclear stain and green fluorescent phalloidin for the actin stain, while the DNA vehicle is stained red (e.g., AlexaFluor 594).
10. After staining and fixing the cells, detach the wells and dry the slide.
11. Apply one or two drops of a signal-enhancer fluid to the slide or the cover glass, and seal to prevent evaporation.
12. Image the slide using fluorescence or confocal microscopy to confirm the cellular uptake of DNA vaccine-vehicle construct.

3.5.2 Protein Expression Assay

This assay can be used to test whether the DNA vaccine can escape the endosome after delivery, be released from the DNA carrier, be localized to the nucleus, and be expressed in the target cell (*see* **Note 14**).

1. Plate cells on a 6-well plate to yield 75–80 % confluence after overnight culture.
2. Incubate cells at 37 °C, 5 % CO_2 overnight.
3. Prepare DNA delivery construct, and load linearized pDNA by ligation or hybridization under sterile conditions.
4. Add pDNA-DNA delivery construct to Opti-MEM media.
5. Remove culture medium and replace with 2 mL Opti-MEM media containing pDNA-DNA delivery construct. Then incubate cells at 37 °C for 2 h.
6. Replace the medium with full 10 % FBS media.
7. Incubate cells for 48 h post-transfection to allow protein expression to occur.
8. Assay protein expression efficiency by qRT-PCR (to assess mRNA levels) or Western blot (to assess protein levels).

3.5.3 Immunogenicity Assay

The ability of a DNA vaccine to elicit an immune response is necessary for optimal function. Thus, it is of interest to measure the immunogenicity of the construct overall (*see* **Notes 15** and **16**).

1. Plate macrophages or other immune cell mixtures on a 48- or a 96-well plate.
2. Prepare the DNA vaccine construct, and load the vaccine pDNA on the multivalent DNA platform (*see* Subheading 3.4.4).
3. Incubate macrophages or other immune cells with the DNA vaccine construct for 4 days at 37 °C, 5 % CO_2.
4. Assess the production of inflammatory cytokines (TNF-α, IL-6) in cells using RT-PCR or in culture supernatant using enzyme-linked immunosorbent assay (ELISA) or parallel methods.

4 Notes

1. Avoid high temperate during ligation; T4 DNA ligase buffer contains ATP, which provides energy for enzyme activity, and high temperature may reduce the activity of the ligase due to fast degradation of ATP into ADP. Additionally, do not use more than 10 % glycerol because the high concentration of glycerol may inhibit the activity of the ligase [11].
2. SDS is used to block nonspecific binding of DNA.
3. For example, conjugation to PEGA groups allows the monomer to be photocrosslinked, while conjugation to fluorophores permits direct visualization and tracing. DNA-binding drugs may also be loaded by intercalation between the bases of

the DNA building blocks. Functional DNA or RNA (genes, mRNA, etc.) likewise may be loaded by designing a sticky end carrier on the Y-DNA branches.

4. The numbers of functions of a DNA building block can be manipulated by the generation number of the DL-DNA backbone. For example, the fifth generation of DL-DNA theoretically may contain up to 48 functional moieties. The types of functions also can be changed by designing sticky end sequences of donor/acceptor branched DNA and incorporating different functional moieties in unique combinations with desired ratios.
5. For application to DNA vaccination, the linear linker DNA may be replaced with pDNA or mRNA coding for the antigen to be expressed.
6. It is possible to vary the degree of photocrosslinking and functional diversity. For example, **DNA nanospheres** are formed by photocrosslinking X-DNA with PEGA functional groups on all four arms [5]. Some of these PEGA groups may be substituted with other functional groups to increase the multifunctionality of the construct.
7. Another type of anisotropic and multifunctional DNA-based vectors, named **DNAsomes**, is formed from amphiphilic monomers made of a simplified Y-DNA. The Y-DNA monomer contains a lipid functional moiety for self-assembly, a fluorescence dye for imaging, and sticky ends for carrying nucleic acid drugs on each arm. The liposome-like, layered spherical DNA structure assembles spontaneously, driven by the hydrophobic interactions of the lipid groups [6].
8. Reactive oxygen species and other free radicals induced by UV radiation have been observed to cause damage to the DNA building blocks. Addition of sterically hindering molecules such as PEG to DNA building block arm ends helps to reduce DNA damage. Decreasing the amount of photoinitiator used during cross-linking, UV light intensity, and time of exposure, as well as increasing the wavelength of the UV light and monomer concentration, also helps to minimize DNA damage. It also, however, decreases the photocrosslinking efficiency, which appears tied to the level of DNA damage [12].
9. There are basic criteria for selecting peptides to overcome various cellular uptake barriers. Peptides should be selected from different types of viruses to eliminate reassembly to an original viral capsid. An additional amino acid like Cys is necessary at the C-terminal to improve conjugation efficiency. The following peptides are good candidates: SV40 NLS, HIV TAT, Adno mu, and artificial condensing peptide [13–16].

10. For HPLC, a C18 column is equipped with a photodiode array detector for UV detection at 260 nm (Waters Corporation, Milford, MA). A gradient of 0–50 % acetonitrile in 0.1 M triethylammonium acetate (TEAA, pH 7.0) is used as the mobile phase with a flow rate of 1.0 mL/min.
11. Fluorescence microscopy can be done directly after transfection, 4 h after transfection to allow cells more time to take up the construct, or over multiple time points to assess uptake kinetics. To use this method, the DNA vehicle must include at least one fluorescent dye moiety. The slide preparation portion of the assay (after transfection) does not need to be performed under sterile conditions.
12. Use of a slide plate with removable culture wells allows the cells to be cultured and imaged directly without being transferred.
13. For transfection using DNA-based delivery platforms, use of a serum-free, low-nutrient medium like Opti-MEM is important because protein in serum can bind to the DNA construct and reduce delivery efficiency.
14. This is particularly needed when the DNA vaccine encodes an antigen designed to prime or stimulate the immune system. For testing purposes, a reporter gene (e.g., luciferase or GFP) may be used as the linear pDNA loaded onto the multivalent DNA platform.
15. The multivalent DNA structures themselves can be designed and utilized as an immune modulator for DNA vaccine therapy. In instance, an immunostimulating CpG motif may be incorporated into the backbone of the multivalent DNA platform [17]. The position of the DNA motifs is flexible, and the number of inserted motifs may be varied. Such modified DL-DNA platforms overall possess vaccine, carrier, and adjuvant functions.
16. The immunogenicity assay does not necessarily give a quantitative measurement of the amount of DNA vaccine construct that enters the cell. Degrees of cytokine production may vary with the antigen or the platform used.

References

1. Luo D, Saltzman WM (2000) Synthetic DNA delivery systems. Nat Biotechnol 18:33–37
2. Luo D (2003) The road from biology to materials. Mater Today 6:38–43
3. Lee JB, Roh YH, Um SH et al (2009) Multifunctional nanoarchitectures from DNA-based ABC monomers. Nat Biotechnol 4:430–436
4. Li Y, Tseng YD, Kwon SY et al (2004) Controlled assembly of dendrimer-like DNA. Nat Mater 3:38–42
5. Roh YH, Lee JB, Tan SJ et al (2010) Photocrosslinked DNA nanospheres for drug delivery. Macromol Rapid Commun 31: 1207–1211

6. Roh YH, Lee JB, Kiatwuthinon P et al (2011) DNAsomes: multifunctional DNA-based nanocarriers. Small 7:74–78
7. Zuker M (2003) Mfold web server for nucleic acid folding and hybridization prediction. Nucleic Acids Res 31:3406–3015
8. UNAFold: http://mfold.rit.albany.edu, Integrated DNA Technologies (OligoAnalyzer): http://www.idtdna.com/analyzer/applications/oligoanalyzer/, Sigma-Aldrich (OligoEvaluator): http://www.sigmaaldrich.com/life-science/custom-oligos/custom-dna/learning-center/calculator.html
9. Roh YH, Ruiz RCH, Peng S et al (2011) Engineering DNA-based functional materials. Chem Soc Rev 40:5730–5744
10. Park N, Kahn J, Rice EJ et al (2009) High-yield cell-free protein production from P-gel. Nat Protoc 4:1759–1770
11. Um SH, Lee JB, Kwon SY et al (2006) Dendrimer-like DNA-based fluorescence nanobarcodes. Nat Protoc 1:995–1000
12. Roh YH, Park JH, Ye JJ et al (2012) Systematic studies of UV stability and photopolymerization efficiency of DNA-based nanomaterials. ChemPhysChem 13:2517–2521
13. Zanta MA, Belguise-Valladier P, Behr JP (1999) Gene delivery: a single nuclear localization signal peptide is sufficient to carry DNA to the cell nucleus. Proc Natl Acad Sci U S A 96:91–96
14. Bonny C, Oberson A, Negri S et al (2001) Cell-permeable peptide inhibitors of JNK: novel blockers of beta-cell death. Diabetes 50:77–82
15. Keller M, Tagawa T, Preuss M et al (2002) Biophysical characterization of the DNA binding and condensing properties of adenoviral core peptide mu. Biochemistry 41:652–659
16. Corbel SY, Rossi FM (2002) Latest developments and in vivo use of the Tet system: ex vivo and in vivo delivery of tetracycline-regulated genes. Curr Opin Biotechnol 13:448–452
17. Mohri K, Nishikawa M, Takahashi N et al (2012) Design and development of nanosized DNA assemblies in polypod-like structures as efficient vehicles for immunostimulatory CpG motifs to immune cells. ACS Nano 6:5931–5940

Chapter 12

Superparamagnetic Nanoparticle Delivery of DNA Vaccine

Fatin Nawwab Al-Deen, Cordelia Selomulya, Charles Ma, and Ross L. Coppel

Abstract

The efficiency of delivery of DNA vaccines is often relatively low compared to protein vaccines. The use of superparamagnetic iron oxide nanoparticles (SPIONs) to deliver genes via magnetofection shows promise in improving the efficiency of gene delivery both in vitro and in vivo. In particular, the duration for gene transfection especially for in vitro application can be significantly reduced by magnetofection compared to the time required to achieve high gene transfection with standard protocols. SPIONs that have been rendered stable in physiological conditions can be used as both therapeutic and diagnostic agents due to their unique magnetic characteristics. Valuable features of iron oxide nanoparticles in bioapplications include a tight control over their size distribution, magnetic properties of these particles, and the ability to carry particular biomolecules to specific targets. The internalization and half-life of the particles within the body depend upon the method of synthesis. Numerous synthesis methods have been used to produce magnetic nanoparticles for bioapplications with different sizes and surface charges. The most common method for synthesizing nanometer-sized magnetite Fe_3O_4 particles in solution is by chemical coprecipitation of iron salts. The coprecipitation method is an effective technique for preparing a stable aqueous dispersions of iron oxide nanoparticles. We describe the production of Fe_3O_4-based SPIONs with high magnetization values (70 emu/g) under 15 kOe of the applied magnetic field at room temperature, with 0.01 emu/g remanence via a coprecipitation method in the presence of trisodium citrate as a stabilizer. Naked SPIONs often lack sufficient stability, hydrophilicity, and the capacity to be functionalized. In order to overcome these limitations, polycationic polymer was anchored on the surface of freshly prepared SPIONs by a direct electrostatic attraction between the negatively charged SPIONs (due to the presence of carboxylic groups) and the positively charged polymer. Polyethylenimine was chosen to modify the surface of SPIONs to assist the delivery of plasmid DNA into mammalian cells due to the polymer's extensive buffering capacity through the "proton sponge" effect.

Key words Superparamagnetic iron oxide nanoparticles, SPION, Polyethylenimine, PEI, DNA vaccine, Magnetofection

1 Introduction

Superparamagnetic iron oxide nanoparticles (SPIONs) have attracted significant attention in gene delivery applications because of their relatively low toxicity, low cost of production,

Monica Rinaldi et al. (eds.), *DNA Vaccines: Methods and Protocols*, Methods in Molecular Biology, vol. 1143, DOI 10.1007/978-1-4939-0410-5_12, ability to immobilize biological materials on their surfaces, and potential for direct targeting using external magnets. Magnetic particle-assisted gene delivery, also known as magnetic transfection or magnetofection, has been shown to improve both the efficiency of gene delivery and the rapidity of uptake in different tissues in vitro [1]. Magnetofection originated from the concept of magnetic drug delivery in the late 1970s, with the technique demonstrating applicability to gene delivery with viral and non-viral vectors [2]. Magnetic particles appear to be generally useable with any gene delivery vector, and the duration of the transfection process can be significantly reduced down to 10 min, compared to 4-h incubation usual with standard protocols [2]. Magnetofection is an appropriate tool for rapid and specific gene transfection needing only low doses in vitro and allowing site-specific in vivo applications [3, 4].

In biotechnology, the critical characteristics of magnetic nanoparticles are their nanoscale dimensions, magnetic properties, and ability to bind particular biomolecules and deliver them to specific targets. Studies performed over the last decade have used several types of iron oxides, among them maghemite, γ-Fe_2O_3, or magnetite, Fe_3O_4, which consist of a single domain of about 5–20 nm in diameter [5, 6]. Magnetite, Fe_3O_4, is the most common magnetic iron oxide candidate because its biocompatibility in biological systems has already been proved [7]. This form of iron oxide is stable in water or physiological saline under neutral pH conditions. It has a large surface area that can be modified to attach biological agents [8]. Nanoparticles of this composition with suitable surface coating materials can disperse widely in suitable solvents to produce a homogenous suspension called ferrofluid that permits further biochemical functionalization. Numerous synthesis methods have been used to produce magnetic nanoparticles for bioapplications including coprecipitation, microemulsions, polyols, sol–gel synthesis, sonochemical synthesis, hydrothermal, hydrolysis, thermolysis of organic precursors, flow injection, and electrospray [5, 9]. These methods have been used to prepare magnetic particles with homogeneous composition and narrow size distribution. However, the most common method for synthesizing magnetite particles in solution within the nanometer range is chemical coprecipitation of iron salts. The technique is probably the simplest and most efficient wet chemical route to obtain magnetic particles for biomedical applications [10].

The stabilization of iron oxide nanoparticles is an important feature in obtaining ferrofluid colloids that do not aggregate in both biological media and magnetic field. The hydrophobic surface of magnetic particles means that in the absence of coating materials, these particles tend to interact with each other to form large clusters, resulting in the increase of aggregate size [10]. Coating layers not only provide stability to nanoparticles in solution but also help to bind various biological ligands to the particle surface for various biomedical applications. Various materials have been used as protective coatings for magnetic nanoparticles.

Polyethyleneimine (PEI) is one of the most efficient cationic compounds for delivery of plasmid DNA into mammalian cells due to its extensive buffering capacity through the "proton sponge" effects [11, 12]. PEI polymer is known to form cationic complexes with SPIONs that then interact nonspecifically with negatively charged DNA and enter the cell via endocytosis [13]. In contrast to other cationic polymers, PEI has high transfection efficiencies even in the absence of endosomolytic agents such as fusogenic peptides or chloroquine which facilitates cellular uptake [14]. Many types of linkages have been used to couple magnetic nanoparticles to nucleic acids, and the simplest one is a physical method based on electrostatic interaction between positively charged magnetic particles with a cationic polymer coating layer and negatively charged nucleic acids. Different factors have been examined for their effects on magnetic gene complex preparation such as molecular weight and different structures (branched and linear structure) of PEI as well as charge density and charge-to-mass ratio of the polymer and DNA molecules [13]. For instance, our previous work showed that SPIONs/branched PEI complexes at pH 4.0 showed a better binding capability for DNA than at a neutral pH, despite negligible differences in the size and surface charge of the complexes [15]. This finding might be a result of protonation and mutual charge repulsion between PEI amine groups in acidic conditions, expanding the polymeric network to increase the amount of entrapped genetic material and consequently increasing gene expression upon injection. In contrast, the stiff stable structure of the polymer's six-membered rings under neutral conditions would decrease the particle's ability to entrap more DNA molecules, subsequently decreasing DNA dosage (*see* Fig. 1) [15].

In this chapter, we describe a coprecipitation method to produce SPIONs. This method is exceptionally useful as it is able to produce magnetic particles of a specific size within the nanometer range and with good magnetic properties. The produced particles can then be coated with PEI polymers as an example of useful cationic polymers that can form complexes with DNA molecules for gene delivery.

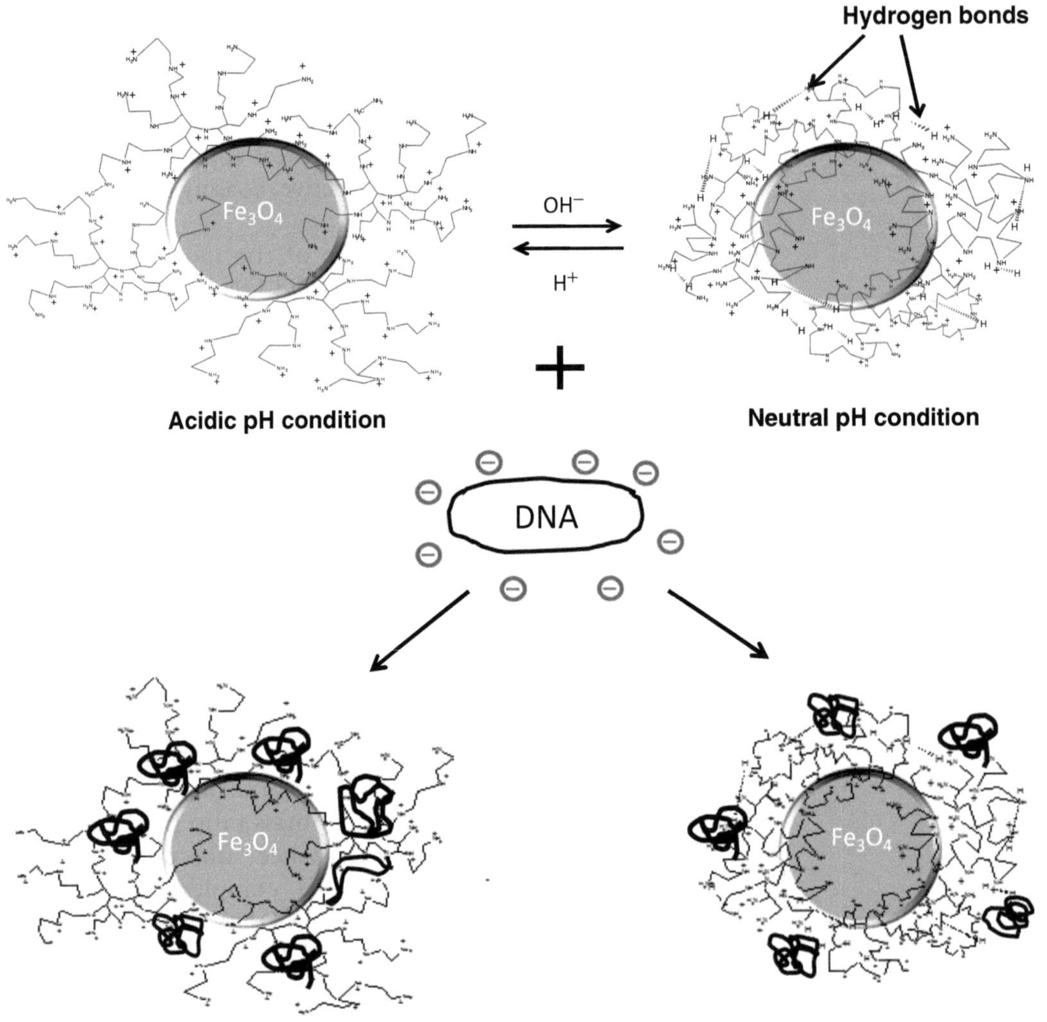

Fig. 1 A schematic demonstrating PEI structures under acidic and neutral pH conditions, showing a relatively branched structure due to mutual charge repulsion between amine groups under acidic condition and a stiff structure under neutral pH condition, with DNA likely to be entrapped within the respective structures. Adapted from Al-Deen et al. [15], with permission from American Chemical Society (ACS) publications of Langmuir

2 Materials

Prepare all solutions using ultrapure water (prepared by purifying deionized water dd H_2O to attain a sensitivity of 18 MΩ cm at 25 °C) and analytical grade reagents. Prepare and store all reagents at room temperature (unless indicated otherwise).

2.1 Iron Oxide Nanoparticle Preparation from Iron Salts

1. Fe (III) chloride ($FeCl_3.6H_2O$) and Fe (II) chloride ($FeCl_2.7H_2O$) (from Ajax Finechem and Ajax Chemicals, respectively).

2. Trisodium citrate dihydrate ($C_6H_5Na_3O_7$. $2H_2O$) (Sigma Aldrich).
3. Sodium hydroxide ACS reagent, ≥97.0 % in pellets (Sigma Aldrich).
4. Cooking oil for oil bath.
5. Zetasizer Nano ZS (Malvern Instruments Ltd., UK).
6. 1140 PW diffractometer with nickel-filtered Cu Kα radiation ($\lambda = 1.5405$ Å) (Philips).
7. Vibrating sample magnetometer (VSM) (Riken Denshi).
8. Transmission electron microscope (TEM) CM20 (Philips).

3 SPION/PEI Complexes

1. PEI solution: 10 % PEI in water (w/v), pH 7.9. Weigh 10 g of PEI (molecular weight of 25 kDa branched, Sigma Aldrich) and dissolve in 75 ml of H_2O. Adjust pH to 7.9 with concentrated HCL and add water to a volume of 100 ml. Filter the PEI solution through a 0.22 μm nitrocellulose filter. Store the solution at 4 °C (*see* **Note 1**).
2. 0.5 M HCl.
3. 0.5 M NaOH.
4. Zetasizer Nano ZS (Malvern Instruments Ltd., UK).
5. Dialysis tubing Spectra/Por® membranes (MWCO = 12,000–14,000) (Spectrum Medical Industries, Inc., Los Angeles, CA).

3.1 SPION/PEI/DNA Polyplexes

1. Endotoxin-free plasmid DNA: 10 μg/ml.
2. 1× PBS, pH 7.4.

3.2 Agarose Gel Electrophoresis

1. Agarose.
2. 6× Sample loading buffer: Weigh ~0.05 mg bromophenol blue and transfer to a 2 ml tube with 1 ml sterile H_2O and 1 ml glycerol. Add enough bromophenol blue to make the buffer deep blue. For long-term storage, keep the sample loading buffer frozen.
3. Ethidium bromide (EtBr) stock solution (10 mg/ml): 0.02 g in 1 ml sterile H_2O.
4. DNA ladder standard.
5. 50× tris–acetate–EDTA (TAE) buffer: 242 g tris base, 100 ml of 0.5 M EDTA solution, 57.1 ml glacial acetic acid, pH 8.5. Add 800 ml water to a 1 l graduated cylinder. Weigh 242 g tris base and transfer to the cylinder. Add 100 ml of 0.5 M EDTA and 57.1 ml glacial acetic acid, mix, and adjust the pH to 8.5

using KOH. Add up to 1 l with H_2O. Store the buffer at room temperature (*see* **Note 2**).

6. Electrophoresis chamber.
7. Power supply.
8. Gel casting tray and combs.

3.3 Reflux System

1. 100 ml three-necked round-bottom flask.
2. Dropping funnel.
3. Air evacuation vacuum pump.
4. Ultrasonic bath (Power Sonic 405, 40 kHz and 350 W).
5. Probe sonicator (Sonics vibra cell, 40 kHz and130 W).
6. N_2 gas cylinder.
7. Water-cooled condenser.
8. Temperature controller.
9. Heating magnetic stirrer.
10. Stir bar.

4 Methods

4.1 Synthesis and Characterization of SPIONs

This method involves coprecipitation of ferrous and ferric salts in an alkaline solution by the addition of a base such as concentrated ammonium hydroxide (NH_4OH) or sodium hydroxide (NaOH) in a non-oxidizing environment (N_2 gas atmosphere) with the following chemical reaction [16, 17]:

$$Fe^{2+} + 2Fe^{3+} + 8OH^- \rightarrow Fe_3O_4 + 4H_2O.$$

Control over size and shape of nanoparticles depends on the Fe^{2+} and Fe^{3+} ratio, the type of salts (e.g., sulphate, nitrate, chloride), and the pH of the reaction media [10].

1. Weigh 1.35 g (0.005 mol) of Fe (III) chloride ($FeCl_3.6H_20$) and 0.70 g (0.0025 mol) of Fe (II) chloride ($FeCl_2.7H_2O$) dihydrate (1:2 M ratios), and dissolve them in 20 ml of H_2O in the first beaker [1].
2. Weigh 1.2 g (1.5 mol) of NaOH and 1.47 g (0.005 mol) of trisodium citrate dihydrate, and dissolve them in 20 ml of Milli-Q H_2O in the second beaker [2] (*see* **Note 3**).
3. Sonicate these beakers in an ultrasonic bath with the homogenization shaking rate (3,600–9,000 rpm) for 10–15 min.
4. Transfer the solution in the first beaker [1] into 100 ml three-necked flask, and place a small magnetic bar inside the flask. Place the flask in the oil bath which has already been placed on the magnetic stirrer, and set the stirring rate to 1,000–1,500 rpm (*see* **Note 4**). Heat the oil bath to 80 °C (*see* Fig. 2).

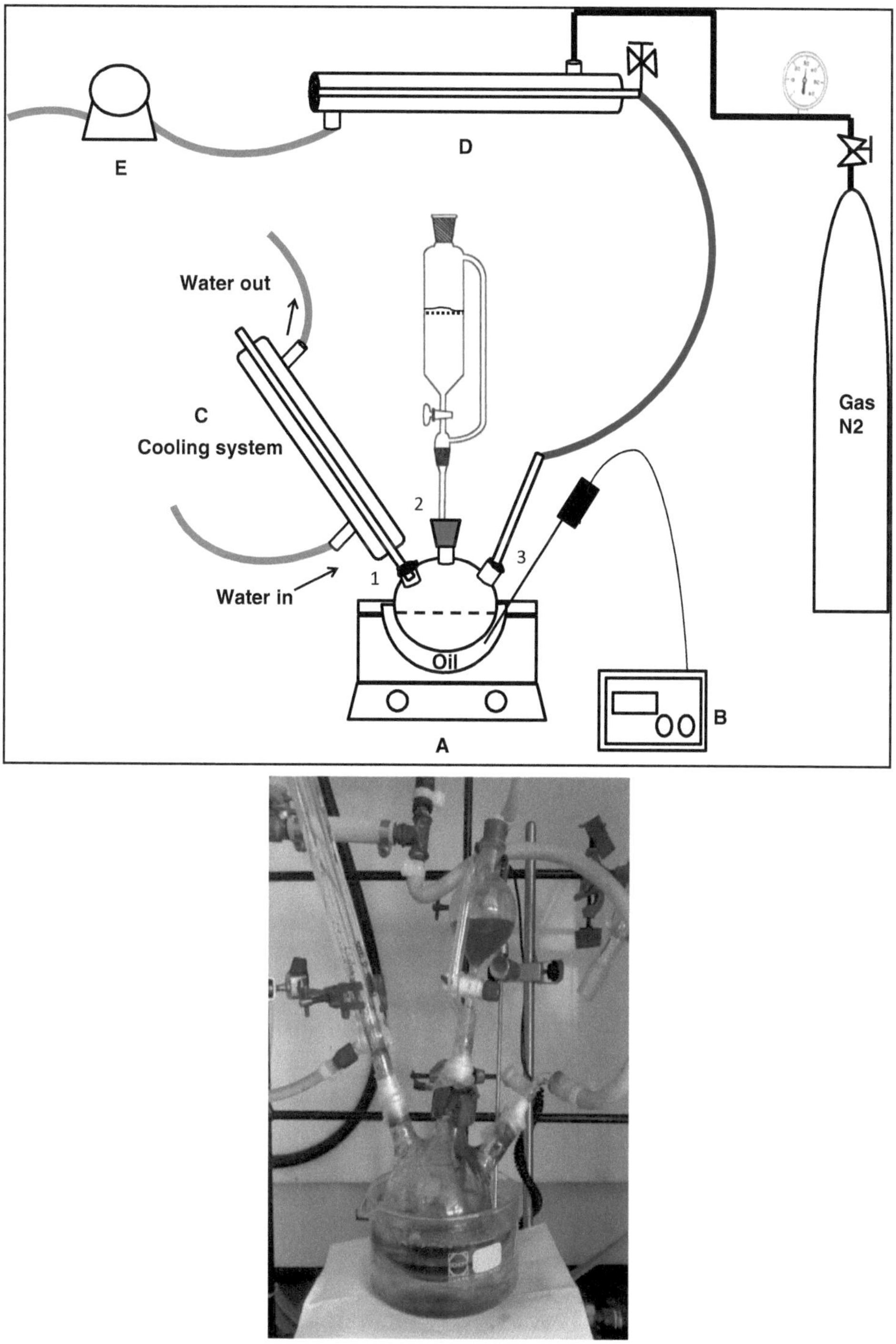

Fig. 2 A reflux system for synthesis of superparamagnetic iron oxide nanoparticles (SPIONs). (*A*) Heating magnetic stirrer, (*B*) temperature controller, (*C*) water-cooled condenser, (*D*) water-cooled condenser, (*E*) air evacuation vacuum pump

5. Transfer the solution in the second beaker [2] into a dropping funnel, and connect the funnel with the three-necked round-bottomed flask via neck [2]; make sure that its stopcock is closed (*see* **Note 5**) (*see* Fig. 2).
6. Introduce a separate hose of the nitrogen stream to the three-necked round-bottomed flask via neck 3 (*see* Fig. 2). Separate hose connected to the condenser which has been connected to N2 gas cylinder via a gas-trap arrangement connected to the top of the condenser.
7. Connect the water-cooled condenser to the three-necked round-bottomed flask via neck 1 (*see* Fig. 2) (*see* **Note 6**). Start water circulation through turning on a water tap. Be sure that cold water is flowing through the condenser in moderate water flow rates.
8. Wrap the connection of equipment in the reflux system with each other by a thin strip of paraffin film to avoid any outside air entering the system.
9. Remove the air out of the system by using a vacuum pump for 4–5 min. Open the cylinder tap cautiously to allow N_2 gas to enter the system for 4–5 min at a steady but controlled rate until 18.2 g to provide a nitrogen blanket to the reaction. Flowing N_2 gas through the reaction medium during the synthesis reaction can afford protection to the produced iron oxide particles from oxidation. The three-necked flask is kept under a positive nitrogen pressure by means of a gas-trap arrangement connected to the top of the condenser.
10. Once the system temperature reaches 80 °C, turn the funnel stopcock partially to start the aqueous coprecipitation of the iron salt solution with NaOH and trisodium citrate dihydrate solution.
11. After 1 h of reaction, collect the resulting black precipitates and remove them from solution by applying an external magnet. Then wash the precipitates four times, firstly with ddH_2O, then twice with ethanol, and finally with deionized (DI) water to remove excess ions and salts from the suspension. The concentration of the solution will be about 8 mg/ml.
12. Disperse the washed precipitate in 20 ml DI water. Zetasizer Nano ZS is used to determine the hydrodynamic diameter and zeta potential of these particles in suspension, while TEM CM20 is used to confirm the size and morphology of dry particles. X-ray powder diffraction (XRD), by means of diffractometer with nickel-filtered Cu Kα radiation ($\lambda = 1.5405$ Å), is used to determine the crystallinity and phase of iron oxide particles. Magnetic saturation is measured using a VSM under a magnetic field of up to 15 kOe at room temperature.

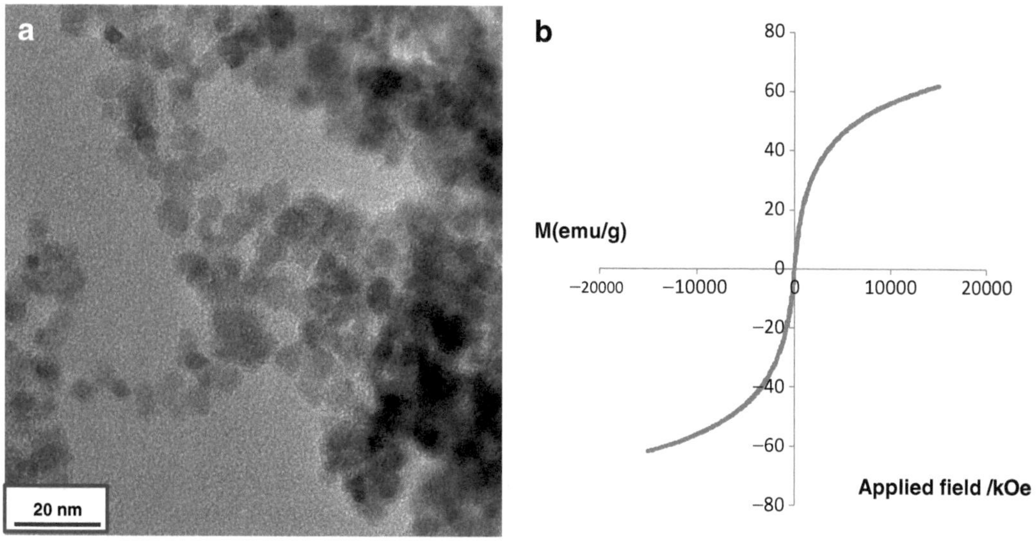

Fig. 3 (**a**) A TEM image of as-synthesized SPIONs, (**b**) VSM data for SPIONs, with *X*- and *Y*-axes in the graph indicating the applied field (kOe) and magnetization (emu/g), respectively

An example of size and morphology of prepared SPIONs under TEM and magnetic saturation using a VSM is shown in Fig. 3.

4.2 Coating SPIONs with PEI Polymer

In magnetofection, magnetic nanoparticles need appropriate surface coatings to form gene complexes, which also increase their stability in solution. The stability of magnetic nanoparticles in biological fluid can be improved by modifying their surface using materials including inorganic and polymeric materials to increase repulsive forces between particles, thus balancing magnetic and van der Waals attractive forces [18] (*see* **Note 7**).

1. Mix the prepared iron oxide suspension (0.1 mg/ml) with 10 % (w/v) PEI solution (25 kDa branched PEI), at PEI/Fe mass ratios of (R) = 10, while sonicating using a probe sonicator using a Sonics vibra cell 130 W apparatus at 40 kHz for 5 min.
2. Dialyze the produced SPION/PEI complexes using Spectra/Por membranes (MWCO = 12,000–14,000) against deionized water for 3 days to remove any unbound/excess PEI.
3. Acidify the mixture of SPION/PEI complex to pH 2.0 using 0.5 M HCl, and retain at this pH for 10 min to stabilize the complexes.
4. Divide each sample into two aliquots: increase the pH of the first part to 4.0 (referred to as SPION/PEI-A), while the other part is neutralized to pH 7.0 (referred to as SPION/PEI-N) using 0.5 M NaOH.

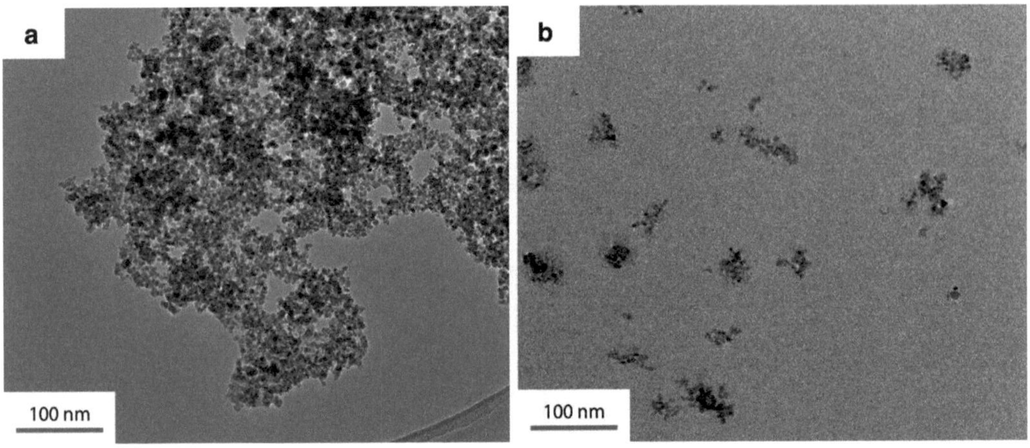

Fig. 4 TEM images of (**a**) as-synthesized SPIONs and (**b**) SPIONs/PEI (ratio = 10) at pH 4 displaying better dispersion. Adapted from Al-Deen et al. [15], with permission from American Chemical Society (ACS) publications of Langmuir

Zetasizer Nano ZS (Malvern Instruments Ltd., UK) is used to determine the hydrodynamic diameter and zeta potential of SPIONs/PEI in suspension, while TEM CM20 is used to confirm the size and morphology of dry particles. An example of the small aggregate size of prepared SPIONs/PEI at pH 4.0 compared with bare SPIONs under TEM is shown in Fig. 4.

4.3 Preparation of SPION/PEI/DNA Polyplexes

1. Mix plasmid DNA at a concentration of 10 μg/ml in PBS (pH 7.4) with SPION/PEI complexes at R of 10 at different N/P ratios (i.e., the molar ratio of PEI nitrogen to DNA phosphate) (*see* **Note 8** for details of calculating different N/P ratios).

4.4 DNA Retardation Assay

The DNA binding capabilities of SPION/PEI/DNA polyplexes are determined using 1 % agarose gel electrophoresis. SPION/PEI complexes with plasmid DNA were formed at N/P ratios of 0.5–30.

4.5 Agarose Gel Electrophoresis of SPION/PEI/DNA Polyplexes

1. Measure out 1 g of agarose.
2. Pour agarose powder into flask along with 100 ml of 1×TAE.
3. Boil with swirling the agarose solution on a heater (until all of the small translucent agarose particles are dissolved, the solution becomes clear, and there is a nice rolling boil) (approximately 10 min).
4. Allow the agarose solution cool to about 50–55 °C, swirling the flask occasionally to cool regularly.

5. Add EtBr to a final concentration of approximately 0.5 μg/ml (usually about 2–3 μl of stock solution per 100 ml gel) (*see* **Note 9**).
6. Seal the ends of the casting tray with two layers of tape.
7. Place the combs in the gel-casting tray.
8. Pour the melted agarose solution into the casting tray and let cool until it is solid.
9. Carefully pull out the combs, and remove the tape. Place the gel in the electrophoresis chamber.
10. Add enough TAE buffer so that there is about 2–3 mm of buffer over the gel (*see* **Note 10**).

4.6 Loading the Gel

1. For each N/P ratio, mix the appropriate amount of SPIONs/PEI with 0.5 μg plasmid DNA in the 25 μl PBS.
2. Incubate all SPION/PEI/DNA polyplex solutions at 37 °C for 30 min.
3. Add 5–6 μl of 6× sample loading buffer to each 25 μl SPION/PEI/DNA polyplexes.
4. Carefully pipette 20 μl of each sample/loading buffer mixture into separate wells in the gel.
5. Pipette 10 μl of the DNA ladder standard into at least one well of each row on the gel.

4.7 Running the Gel

Carry out the electrophoresis at 60 V for 90 min, and then visualize the DNA bands using a UV illuminator.

An example of agarose gel of gel electrophoresis of SPION/PEI/DNA polyplexes at different N/P ratios is shown in Fig. 5.

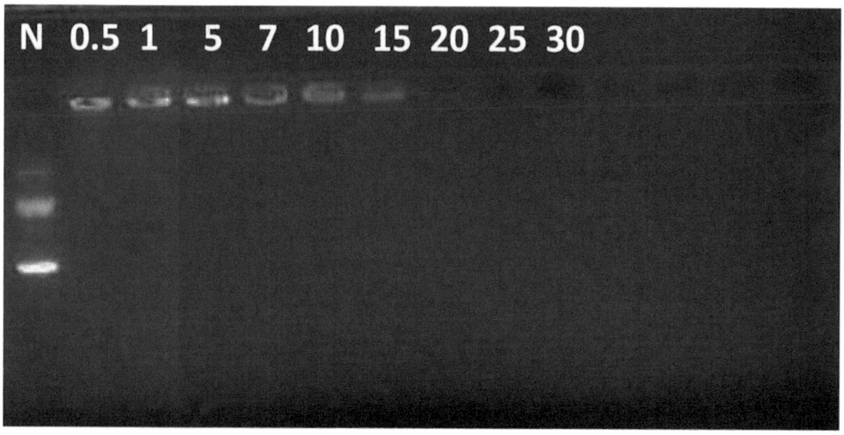

Fig. 5 Agarose gel electrophoresis of SPION/PEI/DNA polyplexes. Lane N: Plasmid DNA (naked). Lanes 0.5–30 correspond to SPION/PEI/DNA polyplexes at different N/P ratios

5 Notes

1. The PEI 10 % (w/v) solution is the working PEI reagent solution that should be stable for a long period of time at 4 °C. If a 6 % PEI solution is needed, add 6 ml of 10 % of PEI to 4 ml of H_2O.
2. For convenience, a concentrated stock of TAE buffer (either 10× or 50×) is often made ahead of time and diluted with water to 1× concentration prior to use.
3. The presence of trisodium citrate ($C_6H_5Na_3O_7.2H_2O$) on the magnetic nanoparticles works as an electrostatic stabilizer [15]. Trisodium citrate has three carboxyl groups which promote their adsorption onto the iron oxide particles, whilst ionization of carboxyl groups supplies the coated magnetic particles with a negative charge to give a stable dispersion in water due to a strong mutual electrostatic repulsion. Moreover, these negatively charged particles are able to adsorb cationic polyelectrolytes such as PEI.
4. High stirring rate (1,500 rpm) might play effective roles during particle nucleation in preventing the nanocrystals from growing further into large single crystals. Furthermore, the size of nanoparticles becomes smaller when the stirring rate increases due to the increasing amount of energy that is transferred to the suspension. SPIONs with this narrow size range could be used for the delivery of gene vector because the small size of nanoparticles has been shown to influence the rate of their uptake as well as their cytotoxicity [19].
5. Ensure that the dropping funnel stopcock is completely closed to prevent the coprecipitation reaction from starting while contents are being added to the three-necked flask at this stage.
6. A condenser is attached to the heated three-necked flask, and cooling water is circulated to condense the vapor, returning it back to the flask as a liquid.
7. A highly positively charged coating agent for magnetic nanoparticles such as PEI cationic polymer has advantages over other polycations in that it not only increases repulsive forces between the particles but also readily associate with negatively charged plasmid DNA accompanied by intrinsic endosomolytic activity [20].
8. The calculation of the N/P ratio for the SPION/PEI/DNA complexes is defined as the molar relation of primary amine groups in the PEI cationic molecule (secondary and tertiary amines are neglected in this calculation due to their lower pK_a values), which represent the positive charges, to phosphate groups in the DNA, which represent the negative charges. The

calculation of the N/P ratio was based on the assumption that one repeating unit of PEI containing one nitrogen (N) corresponds to 43.1 g/mol and one repeating unit of DNA containing one phosphate group (P) corresponds to 330 g/mol [21].

For example, if we need to prepare SPION/PEI/DNA polyplexes containing 10 μg DNA at N/P ratio of 2:

For DNA, 330 g/mol corresponds to one phosphate atom.

1 μg × 330 g/1 μg = 1 mol phosphate.

1 μg DNA = 1 mol phosphate × 10 − 6/330.

1 μg DNA = 3.03 × 10 − 9 mol phosphate.

10 μg DNA = 10 × 3.03 × 10 − 9 mol phosphate.

10 μg DNA = 30.3 × 10 − 9 mol phosphate.

For PEI, the number of N atom in 25 KDa of PEI = 25,000/43.1 = 580.0464 ≈ 580.05.

If we need to prepare SPION/PEI/DNA polyplexes with an N/P ratio of 2:

N/P = 2.

N = 2 × P.

N = 2 × 30.3 × 10 − 9 = 60.6 × 10 − 9.

(1 mol of PEI = 580.05 mol of N.)

How many moles of PEI are in 60.6 × 10 − 9 mol of N?

60.6 × 10 − 9/580.05 = 0.10 × 10 − 9 mol of PEI.

PEI mass = 0.10 × 10 − 9 mol × 25,000 g/mol = 2.61 × 10 − 6 g = 2.61 μg.

Alternatively, we can use the ratio to calculate the amounts of PEI and DNA required.

The g/mol ratio for one N (PEI) to one P (DNA) is N:P = 43.1:330. For N/P = 2, the g/mol ratio is 86.2:330.

Thus, mass of PEI:mass of DNA = 86.2:330.

For 10 μg DNA, the ratio becomes

Mass of PEI:10 μg = 86.2:330 or, in the form of division,

Mass of PEI/10 μg = 86.2/330.

Rearranging, mass of PEI = 86.2/330 × 10 μg = 2.61 μg.

Since we previously coated SPION with PEI at PEI/Fe mass ratios of (R) = 10 (*see* Subheading 3.2), the mass of SPION could be calculated depending on the mass of PEI, assuming that all SPIONs are coated completely with PEI polymer:

R (10) = PEI/SPIONs.

Mass of SPIONs = 2.61 μg/10 = 0.261 μg.

9. EtBr is an intercalating agent commonly used as a fluorescent tag (nucleic acid stain) that binds to the DNA and allows the visualization of DNA under ultraviolet (UV) light.
10. Gels can be made several days prior to use and sealed in a plastic wrap (without combs).

References

1. Krotz F, Sohn HY, Gloe T et al (2003) Magnetofection potentiates gene delivery to cultured endothelial cells. J Vasc Res 40: 425–434
2. Scherer F, Anton M, Schillinger U et al (2002) Magnetofection: enhancing and targeting gene delivery by magnetic force in vitro and in vivo. Gene Ther 9:102–109
3. Plank C, Scherer F, Schillinger U et al (2003) Magnetofection: enhancing and targeting gene delivery with superparamagnetic nanoparticles and magnetic fields. J Liposome Res 13:29–32
4. Dobso J (2006) Gene therapy progress and prospects: magnetic nanoparticle-based gene delivery. Gene Ther 13:283–287
5. Tartaj P, Morales MD, Veintemillas-Verdaguer S et al (2003) The preparation of magnetic nanoparticles for applications in biomedicine. J Phys D Appl Phys 36:R182–R197
6. Roca AG, Costo R, Rebolledo AF et al (2009) Progress in the preparation of magnetic nanoparticles for applications in biomedicine. J Phys D Appl Phys. doi:10.1088/0022-3727/42/22/224002
7. Schwertmann U, Cornell RM (1991) Iron oxides in the laboratory: preparation and characterization. Wiley-VCH, Weinheim, NY
8. Ankamwar B, Lai TC, Huang JH et al (2010) Biocompatibility of Fe_3O_4 nanoparticles evaluated by in vitro cytotoxicity assays using normal, glia and breast cancer cells. Nanotechnology. doi:10.1088/0957-4484/ 21/7/075102
9. Laurent S, Forge D, Port M et al (2008) Magnetic iron oxide nanoparticles: synthesis, stabilization, vectorization, physicochemical characterizations, and biological applications. Chem Rev 108:2064–2110
10. Gupta AK, Gupta M (2005) Synthesis and surface engineering of iron oxide nanoparticles for biomedical applications. Biomaterials 26: 3995–4021
11. Boussif O, Lezoualch F, Zanta MA et al (1995) A versatile vector for gene and oligonucleotide transfer into cells and in-vivo-polyethylenimine. Proc Natl Acad Sci U S A 92:7297–7301
12. Behr JP (1997) The proton sponge: a trick to enter cells the viruses did not exploit. Chimia 51:34–36
13. Godbey WT, Wu KK, Mikos AG (1999) Poly(ethylenimine) and its role in gene delivery. J Control Release 60:149–160
14. Kichler A, Leborgne C, Coeytaux E et al (2001) Polyethylenimine-mediated gene delivery: a mechanistic study. J Gene Med 3: 135–144
15. Al-Deen FN, Ho J, Selomulya C et al (2011) Superparamagnetic nanoparticles for effective delivery of malaria DNA vaccine. Langmuir 27:3703–3712
16. Massart R (1981) Preparation of aqueous magnetic liquids in alkaline and acidic media. IEEE Trans Magn 17:1247–1248
17. Massart R, Dubois E, Cabuil V et al (1995) Preparation and properties of monodisperse magnetic fluids. J Magn Magn Mater 149:1–5
18. Wu W, He QG, Jiang CZ (2008) Magnetic iron oxide nanoparticles: synthesis and surface functionalization strategies. Nanoscale Res Lett 3:397–415
19. Steitz B, Hofmann H, Kamau SW et al (2007) Characterization of PEI-coated superparamagnetic iron oxide nanoparticles for transfection: Size distribution, colloidal properties and DNA interaction. J Magn Magn Mater 311: 300–305
20. Lungwitz U, Breunig M, Blunk T et al (2005) Polyethylenimine-based non-viral gene delivery systems. Eur J Pharm Biopharm 60: 247–266
21. Geisse S, Jordan M, Wurm FM (2005) Large-scale transient expression of therapeutic proteins in mammalian cells. Methods Mol Biol 308:87–98

Part III

Production, Purification, and Quality

Chapter 13

Plasmid Fermentation Process for DNA Immunization Applications

Aaron E. Carnes and James A. Williams

Abstract

Plasmid DNA for immunization applications must be of the highest purity and quality. The ability of downstream purification to efficiently produce a pure final product is directly influenced by the performance of the upstream fermentation process. While several clinical manufacturing facilities already have validated fermentation processes in place to manufacture plasmid DNA for use in humans, a simple and inexpensive laboratory-scale fermentation process can be valuable for in-house production of plasmid DNA for use in animal efficacy studies.

This chapter describes a simple fed-batch fermentation process for producing bacterial cell paste enriched with high-quality plasmid DNA. A constant feeding strategy results in a medium cell density culture with continuously increasing plasmid amplification towards the end of the process. Cell banking and seed culture preparation protocols, which can dramatically influence final product yield and quality, are also described. These protocols are suitable for production of research-grade plasmid DNA at the 100 mg-to-1.5 g scale from a typical 10 L laboratory benchtop fermentor.

Key words Plasmid, DNA vaccine, Fermentation, Fed-batch, *E. coli*

1 Introduction

The ability to consistently manufacture large amounts of highly purified, high-quality plasmid DNA is becoming increasingly important as DNA vaccine technology advances. Plasmid DNA for use in DNA vaccination and gene therapy is produced by cultivation of *Escherichia coli* followed by lysis of the bacterial cell paste, cell debris removal, chromatography (e.g., anion exchange, hydrophobic interaction, size exclusion), buffer exchange, and final product formulation (reviewed in refs. 1–3). Shake flask cultivation is widely used for milligram-scale plasmid production but is not practical to produce the quantity and quality of plasmid required for preclinical and clinical studies and, ultimately, industrial scale production. Ideally, the process used to produce plasmid DNA for preclinical GLP toxicology studies and clinical studies will closely

Monica Rinaldi et al. (eds.), *DNA Vaccines: Methods and Protocols*, Methods in Molecular Biology, vol. 1143,
DOI 10.1007/978-1-4939-0410-5_13, © Springer Science+Business Media New York 2014

resemble the process which would be used for industrial scale production by fermentation in well-controlled bioreactors.

A plasmid fermentation process should result in high biomass yields but should also be designed to maximize the amount of product relative to the biomass, which enhances purification efficiency [4, 5]. This is especially important in the production of DNA vaccines because of the relatively large doses that may be required [2, 6]. Fed-batch fermentation, the preferred methodology for high-cell-density cultivation [7], is also useful for achieving much higher specific plasmid yields than can be achieved by batch cultivation in either shake flasks or fermentors, because the specific growth rate of the cells can be restricted by controlling the nutrient feed rate. Generally, lower specific growth rates result in increased plasmid copy numbers [8–11].

It is also critical that the fermentation produces high-quality supercoiled monomer plasmid. The FDA recognizes that open circle, linear, and nicked forms may be less effective therapeutically than the supercoiled form [12]. These other forms of plasmid can be very difficult to separate from the supercoiled plasmid during purification; therefore, the fermentation process should be optimized to retain a high percentage of supercoiled plasmid.

All reagents used in the production of clinical grade plasmid DNA should be sourced from non-animal origins. Laboratory media commonly used for plasmid DNA production (e.g., LB medium, Terrific Broth) contain tryptone, which is not acceptable for production of plasmid DNA for use in humans due to its animal origins. Soytone, a papaic digest of soy flour, may be used in place of tryptone as a non-animal-origin reagent (referred to in this chapter as "animal product-free"). Information regarding the source, manufacture, and characterization for all biologically sourced materials, or materials that may have used biological materials during manufacture, should be well documented [13, 14].

We have previously described a patented, scalable, fed-batch fermentation process (i.e., HyperGRO™, Nature Technology Corporation, NTC) [3–5, 15–18] that meets the above requirements and is ideal for production of clinical grade plasmid DNA. The HyperGRO™ process uses control parameters that reduce plasmid-associated metabolic burden by utilizing conditions that both restrict growth rate and maintain reduced plasmid copy number during a biomass accumulation phase prior to a plasmid accumulation phase immediately before harvest (with temperature-sensitive pUC origin-type plasmids). Yields up to 2.6 g plasmid DNA/L, with cell densities of around 100 OD_{600}, have been achieved in this process with DNA vaccine plasmids [19]. The HyperGRO™ process also results in improved yield and quality with unstable plasmid constructs and in severe cases enables production of plasmids which cannot otherwise be produced at practical levels using conventional 37 °C processes [18] and thus serves as a platform clinical grade fermentation process that may be used "generically" for very-high-yield plasmid production as well as for production of plasmids containing inserts which are unpredictably "toxic" or otherwise low yielding. Clinical grade plasmid manufacturing under current good manufacturing practice (cGMP) by the HyperGRO™ process is available through a commercial license from NTC or through several licensed cGMP facilities in the United States and Europe.

However, an inexpensive, simple, fed-batch fermentation process for production of research-grade plasmid DNA for use in non-GLP studies that are not used to support an Investigational New Drug (IND) application would be valuable. This chapter describes fermentation media (*see* Table 1) and detailed protocols for production of research-grade plasmid DNA using a simple fed-batch process (Fig. 1) (Subheading 3.3), sufficient to produce up to approximately 1.5 g plasmid DNA (final purified product yield) from a 10 L benchtop fermentor, when using high-copy-number plasmids that do not contain any unstable or toxic sequences.

Table 1
Fermentation media

Batch-phase medium	
Component	Concentration
Yeast extract	24.0 g/L
Soytone	12.0 g/L
Potassium phosphate monobasic, KH_2PO_4	3.0 g/L
Sodium phosphate dibasic anhydrous, Na_2HPO_4	6.78 g/L
Sodium chloride, NaCl	0.5 g/L
Ammonium chloride, NH_4Cl	1.0 g/L
Antifoam 204	0.5 mL/L
Glycerol*	10 g/L
Magnesium sulfate heptahydrate, $MgSO_4{\cdot}7H_2O$*	0.49 g/L
Thiamine HCl*	0.005 g/L
Trace metal solution (see below)*	1 mL/L
*Added post-sterilization	
Feed medium (prepare 0.25 L per L batch volume)	
Component	Concentration
Glycerol	600 g/L
Magnesium sulfate heptahydrate, $MgSO4{\cdot}7H_2O$	8 g/L
Trace metal solution	
Component	Concentration
Ferric chloride hexahydrate, $FeCl_3{\cdot}6H_2O$	1.6 g/L
Cobalt chloride hexahydrate, $CoCl_2{\cdot}6H_2O$	0.366 g/L
Copper sulfate pentahydrate, $CuSO_4{\cdot}5H_2O$	0.2 g/L
Zinc chloride, $ZnCl_2$	0.13 g/L
Sodium molybdate dihydrate, $Na_2MoO_4{\cdot}2H_2O$	0.2 g/L
Boric acid, H_3BO_3	0.05 g/L
Hydrochloric acid, HCL (37 %)	10 mL/L

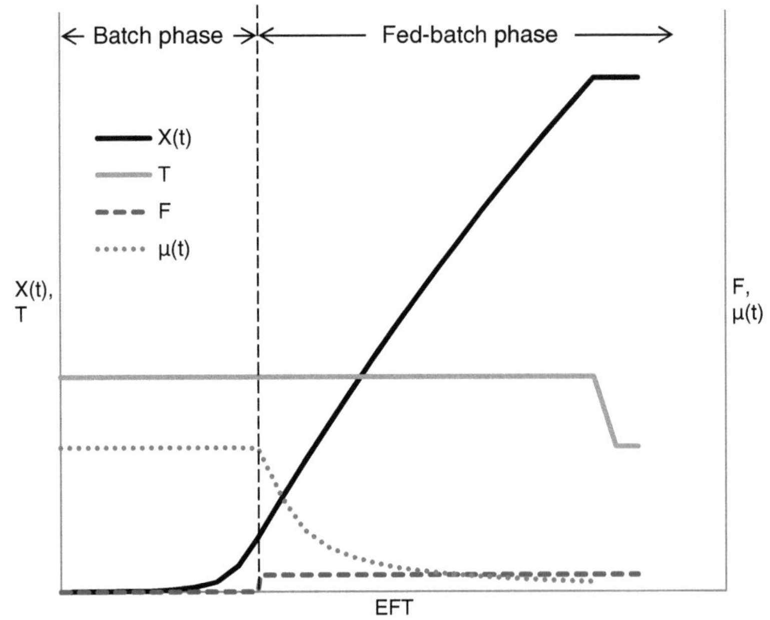

Fig. 1 The general trends of biomass ($X(t)$), temperature (T), feed medium feed rate (F), and specific growth rate ($\mu(t)$) over the course of the fed-batch fermentation process described in Subheading 3.3. The *vertical dashed line* indicates the transition from the batch phase to the fed-batch phase. Exponential growth at a specific growth rate near the cell line's μ_{max} is observed during the batch phase, followed by approximately linear growth during fed-batch phase with a constantly decreasing specific growth rate

Optimized protocols for cell banking (Subheading 3.1), seed culture preparation (Subheading 3.2), and fermentation analytical methods (Subheadings 3.4 and 3.5) are also included.

Results with a 3.7 kb DNA vaccine plasmid, pVAX1-EGFP, produced according to the protocols described herein at the 10 L fermentation scale are included as an example. This fermentation yielded 356 mg plasmid DNA/L (Fig. 2, Table 2) of predominately supercoiled monomer plasmid (Fig. 3).

2 Materials

2.1 Cell Banking

1. Purified sample of plasmid to be transformed, ≥1 μg.
2. Sterile ultrapure water (type 1 reagent grade).
3. Electrocompetent *E. coli* cells (strain selection is dependent on specific requirements, *see* **Note 1**).
4. 1.5 mL microcentrifuge tubes, sterile.

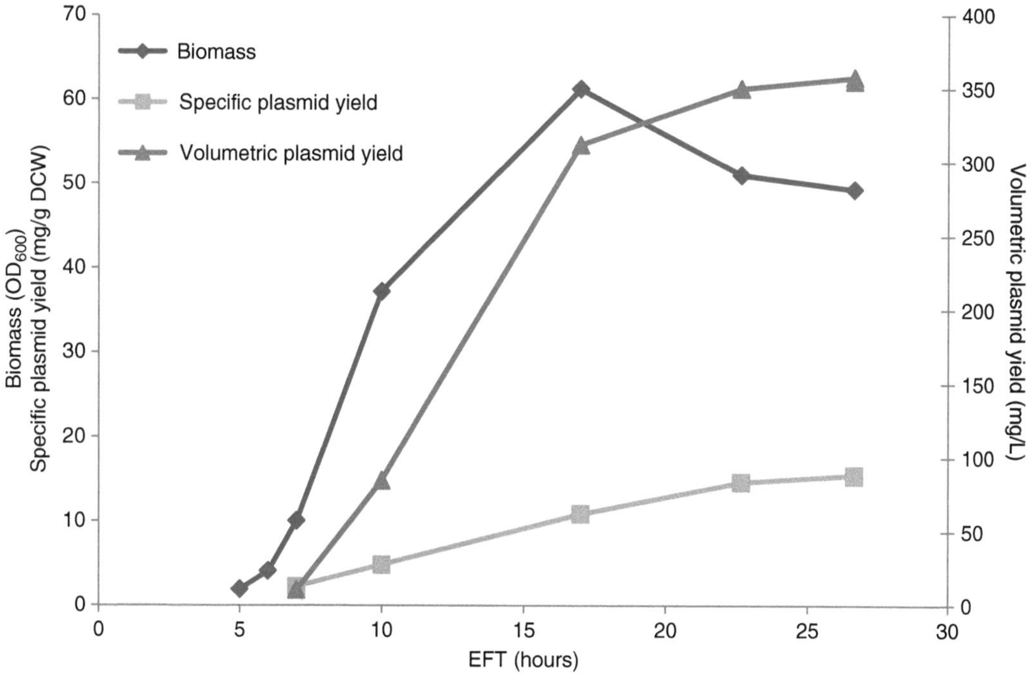

Fig 2 Biomass and plasmid yield profile from a 10 L fed-batch fermentation of *E. coli* DH5α producing a 3.7 kb pVAX1-EGFP DNA vaccine plasmid. The fed-batch phase was started at 7 h using a constant feed rate calculated based on a batch phase $\mu_{max} = 0.83\ h^{-1}$ (*see* Subheading 3.3, **step 7**). The fermentation process was completed at EFT 26.7 h, yielding 356 mg plasmid DNA/L and 15 mg plasmid DNA/g DCW

Table 2
Fed-batch fermentation results with the 3.7 kb pVAX1-EGFP DNA vaccine plasmid

Parameter	Result at harvest
OD_{600}	49.3
Wet cell weight (WCW)	129 g/L
Dry cell weight (DCW)[a]	23 g/L
Volumetric plasmid yield	356 mg/L
Specific plasmid yield (based on OD_{600})	7.2 mg/L/OD_{600}
Specific plasmid yield (based on DCW)	15 mg pDNA/g DCW
Total cultivation time	26.7 h

[a]DCW calculated based on a factor of 0.47 g/L DCW per 1 OD_{600} determined previously for DH5α cultures performed using the protocols of this chapter (*see* **Note 22**)

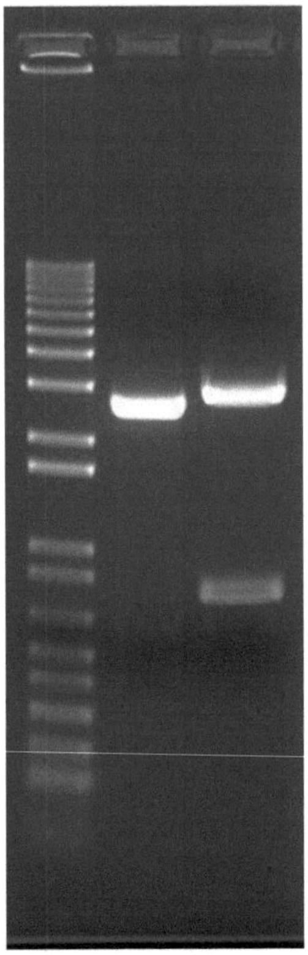

Fig. 3 Agarose gel electrophoresis analysis of the miniprep purified DNA from the final time point of the 3.7 kb pVAX1-EGFP vaccine plasmid fermentation. *Lanes*: (*1*) 1 kb Plus DNA Ladder (Invitrogen); (*2*) 1 μg of undigested plasmid; (*3*) 1 μg of PmeI-digested plasmid with expected fragments at 2,897 and 792 bp

5. Electroporation cuvettes, 0.1 cm gap, sterile.
6. Electroporator.
7. Animal product-free SOC medium: 20 g soytone, 5 g yeast extract, 2 mL 5 M NaCl, 2.5 mL 1 M KCl, 10 mL 1 M $MgCl_2$, 10 mL 1 M $MgSO_4$, 20 mL 1 M glucose, adjust volume with ultrapure water to 1.0 L and sterilize by autoclaving (*see* **Note 2**).
8. 14 mL round-bottom snap-cap tubes, sterile.
9. Sterile disposable inoculation loops.
10. Animal product-free LB plates with appropriate selection agent (*see* **Note 3**): 10 g soytone, 5 g yeast extract, 10 g NaCl, 15 g agar, adjust volume with ultrapure water to 1.0 L and sterilize by autoclaving. Add appropriate selection agent once cooled to 55 °C and pour into sterile Petri dishes (*see* **Note 4**).
11. UV/visible spectrophotometer, 1 cm pathlength cuvettes, and UV-transparent 96-well plates (*see* **Note 5**).

12. Qiagen QIAprep Spin Miniprep Kit, and included solutions (i.e., Buffer P1, RNase A, Buffer P2, Buffer N3, Buffer PE, and Buffer EB).
13. 50 % (w/v) glycerol solution in ultrapure water, sterile.
14. Sterile cryogenic vials.
15. Appropriate antibiotic/selection agent stock solution.
16. Agarose gel electrophoresis (AGE) system: For 150 mL of 1.0 % agarose gel, use 1.5 g of ultrapure agarose (electrophoresis grade) with 150 mL of 1× TAE. Prepare 1 L of 10× TAE stock solution in ultrapure water with 48.4 g of Tris base and 3.72 g disodium EDTA and adjust to pH 8.5 with glacial acetic acid.
17. Restriction enzymes, appropriate for the plasmid DNA being produced.
18. 1 kb DNA ladder.

2.2 Seed Culture

1. Animal product-free LB plates with appropriate selection agent (*see* Subheading 2.1, **item 10**) (*see* **Note 4**).
2. Animal product-free LB liquid medium: 10 g soytone, 5 g yeast extract, 10 g NaCl, adjust volume with ultrapure water to 1.0 L (*see* **Note 4**).
3. Appropriate antibiotic/selection agent stock solution.
4. Sterile disposable inoculation loops.
5. Incubator set to 30 °C.
6. Phosphate-buffered saline (PBS) solution: 0.01 M phosphate buffer, 0.0027 M KCl, 0.137 M NaCl, pH 7.4.
7. UV/visible spectrophotometer and 1 cm pathlength cuvettes (*see* **Note 5**).

2.3 Fed-Batch Fermentation

1. Fermentation system, including pH probe, dissolved oxygen (DO) probe, gas flow and temperature control, a calibrated peristaltic addition pump (for addition of feed medium), and, optionally, a computer with supervisory control and data logging software linked to the bioreactor controller.
2. Air and pure oxygen supply, regulated to 69 kPa/10 psi (gauge pressure).
3. Batch-phase medium (per liter of fermentation starting volume): 24.0 g yeast extract, 12.0 g soytone, 3.0 g KH_2PO_4, 6.78 g Na_2HOP_4, 0.5 g NaCl, 1.0 g NH_4Cl, and 0.5 mL antifoam 204, dissolved in ultrapure water to a total volume of 978 mL (*see* **Note 6**).
4. Carbon/Mg source solution: 500 g/L glycerol, 24.6 g/L $MgSO_4 \cdot 7H_2O$, in ultrapure water. Sterilize by autoclaving at 121 °C.

5. 5 mg/mL thiamine HCl in ultrapure water, filter sterilized through a 0.2 μm sterile filter. Protect from light and store at 4 °C.
6. Trace metal solution: 1.6 g/L $FeCl_3 \cdot 6H_2O$, 0.366 g/L $CoCl_2 \cdot 6H_2O$, 0.2 g/L $CuSO_4 \cdot 5H_2O$, 0.13 g/L $ZnCl_2$, 0.2 g/L $Na_2MoO_4 \cdot 2H_2O$, 0.05 g/L H_3BO_3, and 10 mL/L 37 % HCl, in ultrapure water, filter sterilized through a 0.2 μm sterile filter. Store at 4 °C.
7. Feed medium: 600 g/L glycerol, and 8 g/L $MgSO_4 \cdot 7H_2O$, in ultrapure water. Sterilize by autoclaving at 121 °C.
8. Ammonium hydroxide, 28–30 w/w % (base for pH control).
9. Phosphoric acid, 25 % (acid for pH control).
10. Acid- and base-resistant addition bottles and tubing assemblies for the feed medium, ammonium hydroxide, and phosphoric acid.
11. Appropriate antibiotic/selection agent stock solution.
12. UV/visible spectrophotometer and 1 cm pathlength cuvettes (*see* **Note 5**).
13. Centrifuge, continuous or batch.

2.4 Wet Cell Weight and Dry Cell Weight

1. Microcentrifuge.
2. 1.5 mL microcentrifuge tubes.
3. Analytical balance.
4. Heated vacuum concentrator compatible with 1.5 mL microcentrifuge tubes (e.g., Savant DNA SpeedVac).

2.5 Plasmid Yield and Quality Analysis

1. Qiagen QIAprep Spin Miniprep Kit, and included solutions (i.e., Buffer P1, RNase A, Buffer P2, Buffer N3, Buffer PE, and Buffer EB).
2. Microcentrifuge.
3. 1.5 mL microcentrifuge tubes.
4. UV/visible spectrophotometer and UV-transparent 96-well plates (*see* **Note 5**).
5. AGE system: For 150 mL of 1.0 % agarose gel, use 1.5 g of ultrapure agarose (electrophoresis grade) with 150 mL of 1× TAE. Prepare 1 L of 10× TAE stock solution in ultrapure water with 48.4 g of Tris base, 3.72 g disodium EDTA, and adjust to pH 8.5 with glacial acetic acid.
6. Restriction enzymes, appropriate for the plasmid DNA being produced.
7. 1 kb DNA ladder.

3 Methods

3.1 Cell Banking

The methods used during generation of cell bank glycerol stocks can have dramatic effects on fermentation plasmid yield, and the importance of this should not be overlooked. High-copy DNA vaccine vectors, unstable vector sequences (e.g., direct repeats, long terminal repeats, or palindromic hairpin inverted repeats), and some antigen sequences can result in poorly performing or unviable cell banks when the transformation incubation steps are carried out at 37 °C [18]. Performing transformation incubation steps at 30 °C works well for almost all host/plasmid combinations (including low- and high-copy vectors) and minimizes plasmid-mediated metabolic stress on the host cells to maximize health and stability of the transformed cell line.

All steps involving open tubes, bottles, plates, etc. should be performed using aseptic technique in a bio-safety cabinet.

1. Set the electroporator to deliver one pulse at 1.8 kV (*see* **Note 7**).
2. Dilute the plasmid sample to a concentration of approximately 5 ng/μL in sterile ultrapure water (*see* **Note 8**).
3. Remove a tube of electrocompetent cells from −80 °C storage and thaw on wet ice. While cells are thawing, chill one microcentrifuge tube and one electroporation cuvette on wet ice.
4. When the cells are thawed, tap gently to mix, and add 20 μL to the chilled microcentrifuge tube (*see* **Note 9**). Then add 1 μL of the 5 ng/μL plasmid solution.
5. Pipette the cell/plasmid mixture into the chilled electroporation cuvette.
6. Dry the outside of the cuvette with a paper towel. Insert the cuvette into the electroporator and electroporate with one pulse.
7. Immediately add 1 mL of animal product-free SOC medium to the cuvette, gently resuspend the cells with a pipette, and transfer into a 14 mL round-bottom snap-cap tube (*see* **Note 10**).
8. Perform outgrowth for 2 h by shaking at 30 °C (*see* **Note 11**).
9. Plate 10 μL of the outgrowth on animal product-free LB plates with appropriate selection and incubate at 30 °C for 18–24 h.
10. Choose a plate with well-isolated colonies; count and record the number of colonies. Using a sterile inoculation loop, pick a well-isolated colony and inoculate 6 mL of animal product-free LB (with the appropriate selection agent) in a 14 mL snap-cap tube, and then immediately use the same inoculation loop to streak an animal product-free LB plate (with the

appropriate selection agent) to get individual colonies. Repeat with a total of four well-isolated colonies. Incubate the 6 mL cultures at 30 °C with shaking at 225–300 rpm for 18–24 h, and incubate the plates at 30 °C in a plate incubator for 18–24 h.

11. Measure and record the OD_{600} (*see* **Note 12**) of each of the four saturated 6 mL cultures. Isolate plasmid DNA from the 6 mL cultures using the Qiagen QIAprep Spin Miniprep Kit, and determine the following for each culture: miniprep isolated plasmid concentration (mg/mL); plasmid-specific yield (mg/L/OD_{600}); identity (restriction mapping/AGE); and % plasmid monomer (undigested plasmid AGE). Identify the highest yielding plasmid monomer colony (*see* **Note 13**).
12. From this top yielding monomer isolate, pick one colony from the corresponding plate streaked in **step 10** to inoculate a new 6 mL animal product-free LB culture (with the appropriate selection agent) and incubate with shaking at 30 °C and 225–300 rpm for 18–24 h for glycerol stock creation.
13. Aseptically transfer 4.5 mL of the saturated culture from **step 12** to 1.1 mL of sterile 50 % glycerol in a sterile tube, for a final concentration of 10 % glycerol, and mix well by inversion.
14. Aseptically transfer 1 mL of this glycerol stock into each of the five sterile cryogenic vials. Label and store at −80 °C. The remaining ~0.6 mL may be used to make an additional glycerol stock that can be used to set additional cultures for evaluation of the cell bank's antibiotic resistance/sensitivity, plasmid identity and quality, etc.
15. The cell bank may be expanded by using one of these five glycerol stocks to inoculate a small shake flask containing animal product-free LB medium, with appropriate selection agent (e.g., a 250 mL shake flask with 40 mL medium), and incubating with shaking at 225–300 rpm and 30 °C until saturated. Add 1/4th volume of sterile 50 % glycerol, mix by swirling, make 1 mL aliquots in sterile cryogenic vials, and store at −80 °C.

3.2 Seed Culture

All seed culture propagation steps are performed at 30 °C to maintain maximum plasmid stability. Our preferred method of growing the seed culture is to streak LB-agar plates (with the appropriate selection agent) from glycerol stock. After overnight incubation the cells are resuspended from the plate(s) and used directly to inoculate the fermentor. This method is convenient because the cells are unlikely to become over-saturated on plates; thus, timing is not critical. By contrast, different plasmid–host combinations will give different growth rates, and a shake flask culture can quickly exceed the optimum target cell density if not carefully monitored.

Use of an overgrown shake flask seed can result in unpredictable growth and low yields. Nevertheless, shake flask seed cultures may be preferable in certain cGMP settings and at larger scales and will work just as well when care is taken to limit growth to mid-exponential phase (*see* **Note 14**).

1. Using a sterile inoculation loop, scrape some frozen cells from the top of a glycerol stock and use to streak an animal product-free LB-agar plate (with the appropriate selection agent). Spread the cells evenly across the entire surface of the plate. Incubate for 16–20 h at 30 °C. Verify confluent growth.
2. After incubation, resuspend the cells from the plate by adding 3 ml of animal product-free LB medium to the surface of the plate and gently scraping with a sterile inoculation loop. Pipette these cells into a sterile tube, and repeat with another 3 mL of medium to maximize cell recovery.
3. Vortex the tube to evenly resuspend the cells, and then measure the OD_{600} (a 1:50 dilution of the cells in PBS usually gives an absorbance within the linear absorbance range, which is typically 0.1–0.8 OD_{600} with a 1 cm cuvette).
4. Using this OD_{600} value, calculate the volume of resuspended cells required to obtain 5 OD_{600}·mL (*see* **Note 15**) per L of starting fermentor volume (*see* **Note 16**). This will give a calculated initial cell concentration in the fermentor of 0.005 OD_{600}.

3.3 Fed-Batch Fermentation Process

This protocol describes a simple fed-batch fermentation process that uses semi-defined media with a glycerol carbon source (Table 1). After an initial batch phase, during which cells are in exponential growth, the fed-batch phase is initiated and a nutrient feed medium is added at a constant rate. Because the feed does not require any control once started, supervisory software control is not necessary (although data logging is useful, if available). The constant feed rate is calculated based on the maximum specific growth rate of the culture, μ_{max}, during the batch phase, such that the feed medium will be supplied at a rate that supports growth at μ_{max} at the transition from the batch phase to the fed-batch phase. Because the feed rate is constant, the specific growth rate continually decreases as the fermentation progresses (Fig. 1). Thus, rapid growth early in the process allows for efficient biomass accumulation, and the decreasing growth rate late in the process results in increased plasmid copy numbers. This effect was demonstrated in a fermentation performed as described herein, with a 3.7 kb DNA vaccine plasmid, pVAX1-EGFP, as shown in Fig. 2.

The initial batch volume should be ≤75 % of the maximum working volume of the fermentor in order to allow volume for added feed medium, acid, and base. Follow the manufacturer's instructions for preparing the fermentation vessel for sterilization and pH and DO probe calibration. The pH probe is typically calibrated with pH 4 and pH 7 buffer standards prior to sterilization, and the calibration is adjusted after sterilization if necessary. The DO probe is typically calibrated after sterilization, by sparging nitrogen (to set the *zero* value to 0 % air saturation) and air (to set the *span* value to 100 % air saturation); polarographic DO probes must be connected to the controller for several hours in order to polarize before calibration can be performed.

1. Add the batch-phase medium to the fermentor, and prepare the fermentor for sterilization. Sterilize for 50 min at 121 °C. Reinstall the fermentor to the control systems after sterilization. Set the temperature set point to 37 °C. Aseptically connect the tubing from the feed medium, ammonium hydroxide, and phosphoric acid addition bottles to the fermentor, and install the tubing into the nutrient feed, base, and acid pumps, respectively.
2. Once the bioreactor has cooled to 37 °C, add the post-sterilization additions, consisting of 20 mL/L of the carbon/Mg source solution, 1 mL/L of the 5 mg/mL thiamine HCl solution, 1 mL/L of the trace metal solution, and the appropriate volume of the antibiotic/selection agent.
3. Aseptically remove a 5–10 mL sample of the batch-phase medium, and measure the pH using an external calibrated pH meter. Adjust the fermentor pH calibration to match the off-line pH measurement. Set the fermentor pH set point to 7.0. A deadband of ±0.1 pH is recommended to prevent "over-control" and excessive acid and base consumption.
4. Once the DO probe has been calibrated, set the aeration sparge rate to 1 vessel volume per minute (VVM, vessel volume refers to the initial fermentation liquid media volume). Set the DO set point to 30 % air saturation controlled by an automatic agitation/O_2 cascade.
5. The fermentor is now ready to be inoculated. Aseptically add the seed culture. This point is designated elapsed fermentation time (EFT) = 0 h.
6. Remove small samples of the culture periodically to measure the cell density (OD_{600}), and reserve samples for plasmid DNA yield analysis.
7. Growth should be evident by 5-h EFT (indicated by visual turbidity of the culture and decreasing dissolved oxygen concentration). The maximum specific growth rate, μ_{max}, can vary for different host/plasmid combinations. To determine μ_{max}, measure the cell density (OD_{600}) from at least two time points during log-phase growth of the batch phase (e.g., while the

cell density is between 0.5 and 10 OD_{600}), and calculate the specific growth rate, μ, using the equation

$$\mu = \frac{\ln \frac{OD_{t2}}{OD_{t1}}}{t_2 - t_1}$$

where OD_{t1} and OD_{t2} are the OD_{600} values at the different time points (t_1 and t_2, in hours), respectively. Because all nutrients are present in excess during the batch phase, this calculated μ is assumed to be equal to μ_{max}. This μ_{max} value will be used in the next step to calculate the fed-batch feed rate.

8. Begin constant feeding of the feed medium when the culture reaches a target cell density of 10 ± 2 OD_{600}, which typically occurs by about 7–10-h EFT. The feed rate is calculated by the following equation:

$$F = \frac{\mu_{max} X_B V_B}{S_f \Upsilon_{X/S}}$$

where F is the feed rate (L/h); μ_{max} is the maximum specific growth rate during the batch phase (h^{-1}); X_B is the biomass concentration at the end of the batch phase, which we assume to always be 4.7 g/L (dry cell weight (DCW) corresponding to 10 OD_{600}); V_B is the initial batch medium volume (L); S_f is the glycerol concentration in nutrient feed medium (600 g/L for this protocol); and $\Upsilon_{X/S}$ is the yield coefficient of biomass from glycerol (assumed constant at 0.4 g/g). Continue feeding for the rest of the fermentation (*see* **Note 17**).

9. Continue to remove small samples of the culture periodically (e.g., every 2–3 h) to measure the cell density (OD_{600}, *see* **Note 12**), and reserve samples for biomass and plasmid DNA yield analysis.
10. Once the cell density stops increasing, chill the culture to 25 °C, hold for 30 min, and then chill to 10–15 °C for harvest (*see* **Note 18**).
11. Take a final culture sample for analysis, and harvest the culture by centrifugation (*see* **Note 19**).

3.4 Wet Cell Weight and Dry Cell Weight Analysis

1. Weigh three microcentrifuge tubes using an analytical balance, and record each of their weights (tare weight).
2. Pipette 1.0 mL of well-mixed culture into each tube and centrifuge at 16,000–17,000 × *g* (*see* **Note 20**) for 2 min. Use a pipette to carefully remove as much supernatant as possible from the cell pellets (*see* **Note 21**).

3. Weigh each tube again on the analytical balance, and subtract its tare weight to calculate the wet cell weight (WCW).
4. Place the open tubes in the heated vacuum concentrator and dry on "high" overnight or until the tubes reach a constant weight.
5. Weigh each tube again on the analytical balance, and subtract its tare weight to calculate the DCW (*see* **Note 22**).

3.5 Plasmid Yield Analysis

Plasmid yield can be quickly determined from small culture samples taken throughout the course of a fermentation using a Qiagen QIAprep Spin Miniprep Kit, as described here. This protocol, which has been adapted from the QIAprep Miniprep Handbook [20], will also provide small amounts of purified plasmid that are useful for evaluating the quality of the plasmid DNA produced (e.g., restriction mapping and AGE). Yields determined using this protocol will represent the total amount of plasmid DNA contained within the cells. Maximum large-scale downstream purification recoveries of about 40 % can be expected [2] when manufacturing DNA vaccines for human use, which is consistent with gram-scale purification recoveries in our lab calculated based on the yields determined using this protocol. This protocol was used to determine the plasmid yields from the fermentation performed with the 3.7 kb pVAX1-EGFP DNA vaccine plasmid, as shown in Table 2 and Fig. 2. The plasmid DNA purified from the final fermentation sample using this protocol was further analyzed by restriction enzyme digestion and AGE (Fig. 3).

1. Pellet 0.5–4 OD_{600} mL of cells from a sample of fresh fermentation culture by centrifuging in 1.5 mL microcentrifuge tubes (*see* **Notes 23** and **24**). Remove supernatant completely, and proceed immediately or store at −80 to −20 °C for later yield determination.
2. Resuspend the cells in 250 μL Buffer P1 (ensure that RNase A has been added to Buffer P1). No cell clumps should be visible after resuspension of the pellet.
3. Add 250 μL Buffer P2 and mix gently but thoroughly by inverting the tube 4–6 times. Allow the lysis reaction to proceed for 5 min.
4. Add 350 μL Buffer N3 and mix immediately and thoroughly by inverting the tube 4–6 times. The lysate should become cloudy with a flakey, white precipitate.
5. Centrifuge for 10 min at maximum speed in a microcentrifuge. A compact white pellet will form.
6. Apply the supernatants from **step 5** to the QIAprep spin column by pipetting. Be sure to collect as much of the supernatant as possible without disturbing the pellet.

7. Centrifuge the spin columns for 60 s at 12,000 × *g*. Discard the flow-through.
8. Wash the spin columns by adding 0.75 mL Buffer PE and centrifuging for 60 s at 12,000 × *g*.
9. Discard the flow-through, and centrifuge for an additional 60 s at 12,000 × *g* to remove residual wash buffer.
10. Place the spin columns in new 1.5 mL microcentrifuge tubes. To elute DNA, add 50 μL Buffer EB to the center of each QIAprep spin column, let stand for 60 s, and centrifuge for 60 s at 12,000 × *g* (*see* **Note 25**). Repeat elution with another 50 μL of Buffer EB into the same tubes. The total elution volume for each spin column will be about 100 μL.
11. Determine the plasmid concentration and purity by triplicate measurement of A_{260} and A_{280}. Dilute the miniprep samples in TE buffer as necessary to obtain accurate A_{260} and A_{280} measurements (*see* **Notes 5** and **26**).
12. Make sure that the total amount of plasmid from each miniprep is 10–20 μg for the most accurate yield determination (*see* **Note 24**). If necessary, repeat the yield analysis with a different volume of culture.
13. The volumetric plasmid yield, in terms of mg plasmid DNA per L of culture (mg/L), is back calculated by dividing the amount of plasmid DNA obtained from the miniprep by the volume of culture that was used for the miniprep:

    mg pDNA/L culture = (100 μL) × (miniprep concentration, μg/μL)/(mL of culture pelleted).
14. The specific plasmid yield at each time point can be quickly calculated by dividing the volumetric plasmid yield (mg/L) by the cell density (OD_{600}) of the culture at the time the sample was taken to give a specific plasmid yield value in terms of mg/L/OD_{600} (as shown in Table 2).
15. The specific plasmid yield may also be expressed in terms of mg plasmid DNA per g DCW (mg pDNA/g DCW), once the DCW concentration has been determined (*see* Subheading 3.4). Divide the volumetric plasmid yield (mg/L) by the DCW concentration (g DCW/L) of the culture at the time the sample was taken to get a specific yield value in terms of mg pDNA/g DCW (as shown in Table 2).
16. Use the miniprep purified plasmid DNA from the fermentation harvest sample to evaluate plasmid quality. Analyze 1 μg of undigested plasmid and 1 μg of restriction enzyme-digested plasmid by AGE on a 1 % agarose gel next to a 1 kb Plus DNA ladder maker (as shown in Fig. 2). Compare to AGE the plasmid isolated in Subheading 3.1, **step 11**, from the colony used to create the cell bank. Specifically, check the undigested plasmid lane for multimerization problems that may have arose

during fermentation (*see* **Note 13**) and nicking, and check the undigested lane and restriction digest lane(s) for plasmid sequence deletion products or rearrangements (*see* **Note 27**).

4 Notes

1. The choice of *E. coli* host strain can have a significant effect on the plasmid yield and quality. The host strain should be derived from *E. coli* K-12 and should contain the *recA* mutation, which minimizes recombination of cloned DNA, and the *endA* mutation, which eliminates nonspecific digestion of plasmid during purification by a periplasmic endonuclease. The three host strains listed below all contain these mutations. DH5α (F^{-}φ80*lac*ZΔM15 Δ(*lac*ZYA-*arg*F) U169 *rec*A1 *end*A1 *hsd*R17 (r_k–, m_k+) *gal-pho*A *sup*E44 λ-*thi*-1 *gyr*A96 *rel*A1) performs very well as a general-purpose plasmid production host. XL1-Blue (*recA1 endA1 gyrA96 thi-1 hsdR17 supE44 relA1 lac* [F′ *proAB lac*Iq*ZΔM15* Tn*10* (Tetr)]) has been shown to give higher specific plasmid yields than DH5α with plasmids smaller than 3 kb [16]. NTC4862 (DH5*α* att*λ*:P5/6 6/6-RNA-IN-*Sac*B,*Cm*R) is derived from DH5α and is required when using NTC antibiotic-free sucrose selectable plasmid vectors [16, 21]. These three host strains also include the *rel*A mutation, which allows slow growth without induction of the stringent response.
2. The performance of complex media components such as yeast extract and soytone can vary among different suppliers. In choosing suppliers for these components, it is important to test several different lots of each component in order to establish consistency in raw material sources. We have identified BBL™ Yeast Extract (Becton, Dickinson and Company, Cat. No. 211931) and a soytone, Soy Peptone (Amresco, Cat. No. N454-500G), as reliable complex media components for use in our lab.
3. Generally, liquid media and plate recipes calling for tryptone can be made animal product-free by substituting soytone (papaic digest of soybean meal) in equal amounts in place of tryptone.
4. If using *E. coli* NTC4862 for sucrose selection of NTC antibiotic-free vectors, use a modification of the animal product-free LB plate and liquid media, which eliminates the NaCl:

   Animal product-free LB plates for sucrose selection: 10 g soytone, 5 g yeast extract, 15 g agar, adjust volume with ultra-pure water to 880 mL, and sterilize by autoclaving. When cooled to ~50 °C, add 120 mL 50 % sucrose (w/v) (0.2 μm filter sterilized), mix, and pour plates.

Animal product-free LB liquid medium for sucrose selection: 10 g soytone, 5 g yeast extract, 15 g agar, adjust volume with ultrapure water to 880 mL, and sterilize by autoclaving. When cooled to ~50 °C, add 120 mL 50 % sucrose (w/v) (0.2 μm filter sterilized) and mix.

The 50 % sucrose (w/v) should be sterilized by 0.2 μm filtration, rather than by autoclaving which may cause the sucrose to partially degrade into glucose and fructose.

5. The UV/visible spectrophotometer is used for measuring cell density at 600 nm in 1 cm cuvettes (OD_{600}) as well as DNA concentration and purity at 260 and 280 nm. Either quartz or disposable polystyrene cuvettes may be used for OD_{600} and will give nearly equivalent OD_{600} values (when measured against a reference blank value for the particular cuvettes used). For DNA quantitation at 260 and 280 nm, the use of a spectrophotometer with a plate reader and UV-transparent 96-well plates minimizes sample volume and allows multiple samples to be read simultaneously. Alternatively, DNA quantitation may be performed in a cuvette, in which case a quartz cuvette should be used because many disposable cuvettes are not fully UV transparent down to 260 nm.
6. For each liter of fermentation starting volume, the batch-phase medium is initially dissolved to a volume of 978 mL to account for the 22 mL of post-sterilization additions that will bring the volume to 1.0 L.
7. We use a Bio-Rad MicroPulser electroporator. The "Ec1" setting on the Bio-Rad MicroPulser will deliver one pulse 1.8 kV. If using an electroporator other than the Bio-Rad MicroPulser, follow the manufacturer's recommended settings for transformation of *E. coli* using 0.1 cm gap cuvettes.
8. When diluting samples for electroporation, water is used to achieve a low electrical resistance of the sample and prevent arcing during the electroporation pulse.
9. If using NTC *E. coli* NTC4862, use 25 μL of electrocompetent cells, instead of 20 μL.
10. Delays between applying the electroporation pulse and mixing the cells with the SOC medium will result in rapidly declining viability and transformation efficiency [22].
11. If transforming a plasmid that uses an ampicillin resistance marker (or another β-lactam-type resistance marker), perform the outgrowth at room temperature for only 30 min. However, this scenario should be unlikely because the use of β-lactam antibiotics for selection is generally not acceptable by the regulatory agencies due to potential hyper-reactivity to residual trace β-lactam antibiotics in the product.

12. Cell samples should be diluted as necessary with PBS to obtain an OD_{600} measurement within the linear range of the spectrophotometer being used (typically 0.1–0.8 OD_{600} with a 1 cm cuvette).

13. When creating a cell bank for use in the fermentation process outlined in this chapter, it is critically important to isolate a colony containing a very high fraction of the monomer plasmid isoform and very little dimer and higher multimer isoforms, in order to avoid a "dimer catastrophe," in which the fraction of plasmid dimer/multimer quickly overtakes the culture [23]. Unlike the HyperGRO™ process, which is inherently stabilizing due to the reduced temperature and growth rate during biomass accumulation [18], plasmids grown in batch and fed-batch processes performed at 37 °C (including the process described in this chapter) are more susceptible to rapid multimer accumulation, which imposes a severe metabolic burden on the host cells [24] and yields a mixed plasmid isoform population unsuitable for use.

14. The traditional LB shake flask seed culture will work just as well as the plate method if the following guidelines are followed: Use a growth temperature of 30 °C with sufficient shaking to ensure good aeration (≥300 rpm if possible). It is important that the shake flask culture is not allowed to near stationary phase; the best time to use the LB shake flask culture to inoculate the fermentor is when the cell density is between 0.3 and 0.7 OD_{600} (i.e., still in exponential growth). Using shake flask cultures that have become oversaturated may result in unpredictable growth and poor yields. A 1 % of batch volume seed culture is suggested, resulting in an initial cell density of approximately 0.005 OD_{600} in the fermentor.

15. OD_{600}·mL refers to the total number of cells in a specified volume at a specified cell density and is calculated by multiplying the OD_{600} of a cell suspension sample by the volume of the sample in mL. For example, the total number of cells in a 2 mL sample at a cell density of 5 OD_{600} can be expressed as 10 OD_{600}·mL; the same total number of cells would be present in a 10 mL sample at a cell density of 1 OD_{600}.

16. For example, after resuspending the cells from the plate, an OD_{600} of 0.192 is measured with a 50-fold dilution in PBS. The corrected OD_{600} is $50 \times 0.192\ OD_{600} = 9.6\ OD_{600}$. To inoculate a fermentor with a 6 L starting volume, the volume of seed culture is calculated as follows:

    Determine the total amount of cells required, in terms of OD_{600}·mL:

$$5\ OD_{600}\cdot\text{mL/L} \times 6\ \text{L} = 30\ OD_{600}\cdot\text{mL}.$$

Convert this figure into volume, according to the measured OD_{600}:

$$(30\ OD_{600}\cdot mL)/9.6\ OD_{600} = 3.13\ mL.$$

This is the volume of resuspended cells that should be used to inoculate the fermentor.

17. In order to simplify calculation of the feed rate when following this protocol, the values $X_B = 4.7$ g/L and $Y_{X/S} = 0.4$ g DCW/g glycerol can be successfully used with most host/plasmid combinations, even though the exact X_B and $Y_{X/S}$ values may vary somewhat from batch to batch and among different host/plasmid combinations. An estimate of $\mu_{max} = 0.8\ h^{-1}$ may be used instead of the actual μ_{max} if necessary (e.g., if the operator is not present to take batch-phase samples and perform the calculation according to Subheading 3.3, **step** 7). However, best results will likely be achieved if the μ_{max} value for the specific host/plasmid combination being produced is used in the feed rate calculation.

18. The preharvest hold at 25 °C, a temperature which is not permissive for initiation of new plasmid replication but may allow completion of partially replicated plasmid molecules, often results in increased plasmid yield and increased plasmid quality with high-copy-number plasmids [18].

19. We use a CEPA Z41 high speed continuous centrifuge (New Brunswick Scientific) with a 2 L clarifying bowl to harvest cell paste at the 10 L fermentation scale. The 2 L clarifying bowl will effectively retain about 1 kg of cell paste, which corresponds to the WCW in approximately 8 L of an *E. coli* DH5α culture at a cell density of 50 OD_{600}. Multiple centrifugation runs are used as required to keep the mass of expected cell paste less than 1 kg per run.

20. Use the maximum speed with Eppendorf 5418 centrifuges (16,873 × *g*).

21. With this method, we have found that it is not necessary to wash the cells with water before drying. However, if desired, the cell pellets may be washed by resuspending in 0.5–1.0 mL of water, centrifuging again, and removing the supernatant. In a comparison done in our lab, DCW measured from washed and unwashed cell pellets from the same culture were within 2 % of each other.

22. It can be useful to correlate OD_{600} to WCW and DCW. We have consistently measured about 0.47 g/L DCW per 1 OD_{600} with DH5α cultures.

23. 50 μL is about the minimum volume of a dense culture that can be accurately pipetted into an empty tube for pelleting. Accurate plasmid yield analysis will often require smaller culture volumes, especially near the end of fermentation when

cell density and plasmid copy number are both at their highest (such small volumes may be required to prevent column overloading with very high yielding fermentations). When smaller culture samples are required, we will pipette up to 20 μL of culture directly into the 250 μL of P1 used in **step 2** of Subheading 3.5 and proceed with lysis. Smaller volumes of culture can also be analyzed by using a fraction of a resuspended pellet obtained from a larger known culture volume.

24. The QIAprep Spin Miniprep columns have a capacity of up to 20 μg of DNA [20]. We have found that determination of fermentation plasmid yield using the QIAprep minipreps is most accurate when 10–20 μg of pDNA are recovered from a miniprep of a culture sample. If >20 μg of plasmid DNA is recovered, the spin column has been overloaded and plasmid yield will be underestimated. Generally, for the fed-batch fermentation process described in this chapter, 4 OD_{600}·mL of cells (e.g., 0.5 mL of a culture at OD_{600} 8.0) during the batch phase, down to 0.5–1 OD_{600}·mL of cells toward the end of the fermentation, will give miniprep yields in the range of 10–20 μg.
25. For large plasmids (>10 kb), heat the Buffer EB to 65 °C before applying to the miniprep column to aid in complete recovery of the plasmid.
26. To calculate the plasmid concentration in the miniprep elution by the A_{260} measurement (1 cm pathlength absorbance at 260 nm), first correct for the dilution and then multiply by the DNA concentration factor for A_{260} (1 A_{260} corresponds to 0.050 mg/mL double-stranded DNA). The ratio A_{260}/A_{280} should be between 1.8 and 2.0 for pure DNA [25].
27. For example, we have observed severe instability with retroviral plasmid vectors containing two long terminal repeats (LTRs, known to cause stability problems) when produced by fermentation at 37 °C, resulting in entire deletion of the sequence between the LTRs by the end of fermentation. Plasmid instability observed arising during fermentation might be remedied, in some cases, by performing the fermentation entirely at a temperature of 30 °C instead of 37 °C; doing so, however, will result in much lower plasmid yields.

Acknowledgments

We thank Sheryl Wright for performing the fermentation and plasmid yield analysis and Jeremy Luke for critical review of the cell banking protocol.

References

1. Prather KJ, Sagar S, Murphy J et al (2003) Industrial scale production of plasmid DNA for vaccine and gene therapy: plasmid design, production, and purification. Enzyme Microb Technol 33:865–883
2. Hoare M, Levy MS, Bracewell DG et al (2005) Bioprocess engineering issues that would be faced in producing a DNA vaccine at up to 100 m^3 fermentation scale for an influenza pandemic. Biotechnol Prog 21:1577–1592
3. Carnes AE, Williams JA (2007) Plasmid DNA manufacturing technology. Recent Pat Biotechnol 1:151–166
4. Carnes AE, Hodgson CP, Williams JA (2006) Inducible *Escherichia coli* fermentation for increased plasmid DNA production. Biotechnol Appl Biochem 45:155–166
5. Carnes AE (2005) Fermentation design for the manufacture of therapeutic plasmid DNA. BioProcess Intl 3:36–44
6. Brand E, Ralla K, Neubauer P (2012) Strategies for plasmid DNA production in *Escherichia coli*. In: Subramanian G (ed) Biopharmaceutical production technology, vol 1, 1st edn. Wiley-VCH, Weinheim, pp 3–41
7. Lee SY (1996) High cell-density culture of *Escherichia coli*. Trends Biotechnol 14:98–105
8. Lin-Chao S, Bremer H (1986) Effect of the bacterial growth rate on replication control of plasmid pBR322 in *Escherichia coli*. Mol Gen Genet 203:143–149
9. Seo JH, Bailey JE (1985) Effects of recombinant plasmid content on growth properties and cloned gene product formation in *Escherichia coli*. Biotechnol Bioeng 27: 1668–1674
10. Satyagal VN, Agrawal P (1989) A generalized model of plasmid replication. Biotechnol Bioeng 33:1135–1144
11. Rozkov A, Avignone-Rossa CA, Ertl PF et al (2006) Fed batch culture with declining specific growth rate for high-yielding production of a plasmid containing a gene therapy sequence in *Escherichia coli* DH1. Enzyme Microb Technol 39:47–50
12. FDA (2007) Guidance for industry: considerations for plasmid DNA vaccines for infectious disease indications. Rockville, MD
13. WHO (2005) Guidelines for assuring the quality and nonclinical safety evaluation of DNA vaccines. World Health Organization, Geneva
14. EMA (2001) Note for guidance on the quality, preclinical and clinical aspects of gene transfer medicinal products. European Medicines Agency, London
15. Carnes AE, Hodgson CP, Luke JM et al (2009) Plasmid DNA production combining antibiotic-free selection, inducible high yield fermentation, and novel autolytic purification. Biotechnol Bioeng 104:505–515
16. Carnes AE, Luke JM, Vincent JM et al (2010) Critical design criteria for minimal antibiotic-free plasmid vectors necessary to combine robust RNA Pol II and Pol III-mediated eukaryotic expression with high bacterial production yields. J Gene Med 12:818–831
17. Carnes AE, Williams JA (2011) US Patent 7943377
18. Williams JA, Luke J, Langtry S et al (2009) Generic plasmid DNA production platform incorporating low metabolic burden seed-stock and fed-batch fermentation processes. Biotechnol Bioeng 103:1129–1143
19. Carnes AE, Luke JM, Vincent JM et al (2011) Plasmid DNA fermentation strain and process-specific effects on vector yield, quality, and transgene expression. Biotechnol Bioeng 108: 354–363
20. Qiagen (2006) QIAprep® Miniprep Handbook
21. Luke JM, Vincent JM, Du SX et al (2011) Improved antibiotic-free plasmid vector design by incorporation of transient expression enhancers. Gene Ther 18:334–343
22. Dower WJ, Miller JF, Ragsdale CW (1988) High efficiency transformation of *E. coli* by high voltage electroporation. Nucleic Acids Res 16:6127–6145
23. Summers DK, Beton CW, Withers HL (1993) Multicopy plasmid instability: the dimer catastrophe hypothesis. Mol Microbiol 8: 1031–1038
24. Field CM, Summers DK (2011) Multicopy plasmid stability: revisiting the dimer catastrophe. J Theor Biol 291:119–127
25. Maniatis T, Fritsch EF, Sambrook J (1982) Molecular cloning : a laboratory manual. Cold Spring Harbor Laboratory, Cold Spring Harbor, NY, pp 468, 545

Chapter 14

Pharmaceutical Grade Large-Scale Plasmid DNA Manufacturing Process

Marco Schmeer and Martin Schleef

Abstract

For pharmaceutical applications of plasmid DNA, either direct or indirect, certain quality standards are required. Whereas for direct gene transfer into human "Good Manufacturing Practice" (GMP) grade is mandatory, for GMP production of, e.g., viral vectors (AAV, etc.) the plasmid DNA used needs not necessarily be produced under GMP.

Besides such regulatory aspects up-scaling of the plasmid DNA production process from research laboratory scale (up to a few milligrams) to industrial scales (milligram to gram scales) is an issue that is addressed here.

Key words High Quality Grade DNA, Plasmid DNA manufacturing, Plasmid quality control, Scale-up

1 Introduction

1.1 DNA Vectors: From Research Tools to Pharmaceutical Substances

Since plasmids were identified initially as elements carrying extrachromosomal genetic information of *Escherichia coli (E. coli)* bacteria [1], these typically circular DNA molecules became one of the most important tools in genetics and molecular biology and since the idea of gene therapy or genetic vaccination grew also in medicine. Current research focuses on technologies for curative and preventive applications of genetic material. In gene therapy, malfunction due to an incorrect genetic code is overcome by addition of the appropriate genetic information. Additional information is encoded on the therapeutic gene construct, e.g., to make malignant cells accessible for the autologous immune system.

DNA vaccine is mainly used preventively. The DNA encodes for an antigen—the cellular reaction is the same as after a viral infection. This is an important difference to classical vaccines using virus particles or synthetic peptides: the peptide produced by the target cell itself presents the same antigen as in a real infection.

The number of plasmid per *E. coli* cell (approx. 1,000 for high-copy plasmids) is quite high in comparison to the number of

Monica Rinaldi et al. (eds.), *DNA Vaccines: Methods and Protocols*, Methods in Molecular Biology, vol. 1143,
DOI 10.1007/978-1-4939-0410-5_14,

genomes per cell (just 1). This is why the optimization of the plasmid copy number has an economic impact on the use of these tools in pharmaceutical applications. Certain approaches have been made to increase the productivity of the "plasmid-producer cells" over the last decades [2–5]. The productivity is just an economical and important aspect if a DNA pharmaceutical should receive marketing authorization and permission to be used in human or veterinary vaccination or in cell and gene therapy. However, more important is the quality of the DNA, namely, the purity and the topology of plasmids. The different types of purity are presented within Subheading 2 of this chapter.

1.2 Regulatory Aspects of DNA Manufacturing

The use of plasmids, miniplasmids, or, in the future, minicircle DNA requires—as any other biological used for *direct or indirect* pharmaceutical application—the permission of the respective regulatory agency. The regulations are not identical worldwide but harmonized. The principle behind regulations is always to protect the patient (human or veterinary) from any potential risk associated with the use of such. It is important to distinguish between "direct" application of a DNA and those cases where the DNA is not finally administered to the patient. For example, the DNA vaccination is a direct application. The plasmid produced is formulated and filled and subsequently (after product release and certain storage) injected, chemically transferred, shot ("gene gun"), or "pushed" by electro-gene transfer into the target patient cell. This means that the DNA itself is later present within the body of the patient either in active form (expressed) or in inactivated and later degraded form. Any potential harmful element contained will as well be there, and this is why these drugs need to be subject to quality assurance in manufacturing and quality control at the end of manufacturing. The use of plasmid DNA produced under "Good Manufacturing Practice" (GMP) is necessary because an initial clinical trial introduces an investigational drug, here plasmid DNA, into human. GMP ensures that manufacturing, formulation, packaging, quality controls, etc. result in a product fitting for the intended application with respect to product safety, quality, and efficiency.

GMP facilities are regularly inspected by the national regulatory agencies making sure that processes and documentations are in compliance with *contemporary* regulatory requirements. This is why often the term cGMP for *current* Good Manufacturing Practice is used.

If the DNA itself is not directly applied to the patient ("indirect application") the quality assurance and control are extremely important as well, even though the DNA is not the active pharmaceutical ingredient (API) as in the case of directly administering plasmids in DNA vaccination. As an example, these plasmids are used to transfect a cell and such cell—now modified to produce a viral vector, a

RNA or an antibody or any other therapeutic protein—will not be transferred to the patient. Hence no plasmid is transferred to the patient. The product of the cell (viral vector, antibody, recombinant protein) is purified from the cell or the cell medium, and this is the latter API. The DNA for the cell modification is just a raw material. For example, this DNA is only used as a template for in vitro transcription for mRNA production. Such mRNA is the final API, while the DNA used to transcribe is not applied to the patient cells. PlasmidFactory developed a special manufacturing process to produce these raw materials called "High Quality Grade" that is presented within Subheading 2.3 of this chapter.

2 Materials: Quality Levels of Plasmid DNA

Depending on the intended application, plasmid DNA can be produced at different quality levels. The specifications of the produced plasmid DNA can be further fine-tuned to meet particular requirements, through additional optional procedures, as outlined below.

2.1 Research Grade

This production process comprises the following features:

1. Transformation of a characterized *E. coli* production strain using the intended plasmid.
2. Reproducible quality of plasmid DNA by establishing an individual cell bank.
3. Cultivation of cells without animal-derived complex media.
4. Individual DNA production in dedicated plasmid laboratories using standardized manufacturing processes including verified lipopolysaccharide (LPS) endotoxin removal technology.
5. Precisely defined quality, including certain quality control (QC) assays (standard QC, see below).
6. Concentration from 0.5 to 5.0 mg/mL upon individual requirements.
7. Filling of plasmid DNA aliquots, formulated in a buffer of choice in certified cryotubes.
8. Labelling according to the individual requirements.

The production of plasmid DNA in *Research Grade-animal free* (*AF*) quality is the same as *Research Grade* quality, but an enhanced quality is achieved using a recombinant RNase, such that no animal-derived substances are used throughout any part of the production process.

Plasmid DNA is typically obtained as covalently closed circular (ccc) supercoiled molecules when extracting them from *E. coli* cells through classic approaches such as the alkaline lysis initially presented by Birnboim and Doly [6]. Since then, alternative approaches have been tried without success for those cases where large amounts are the goal. The highest quality level is a plasmid DNA that contains a maximum of ccc monomers and no relaxed open circular (oc) or linear molecules in parallel. This can be achieved by producing ccc-Grade DNA (see next paragraph). However, in various plasmid DNA preparations the so-called dimers are present as a result of recombination between two identical plasmids. These can be of ccc Grade as well but may show a different transfection efficiency compared to the monomer. An overview is presented by Maucksch et al. [7].

2.2 ccc Grades

The production of the DNA *ccc Grades* is based on the steps provided for Subheadings 2.1 *(Research Grades)*. Through the use of a special (additional) chromatography step open circular and linear plasmid forms as well as bacterial chromosomal DNA (chrDNA) are advantageously removed. The high purity obtained for the product permits an increased gene expression (efficiency) and—more importantly—a lower toxicity as presented earlier [8]. No animal-derived substances are used in any part of the manufacturing process either by using a recombinant RNase or, preferably, by special chromatography procedures, in a completely enzyme-free process. "Enzyme-free" means that no enzymes for the removal of RNA, open circular or linear plasmids and the contaminating bacterial chrDNA are to be applied.

Two important technical analytical tools can be used to evaluate the plasmid DNA quality: the capillary gel electrophoresis (CGE) developed in our laboratory for the quantification of the various plasmid topologies and the atomic force microscopy (AFM), a way to visualize the plasmid DNA samples (e.g., from DNA vaccine manufacturing processes).

The CGE was initially presented for this application by Schmidt et al. [9] and used routinely for the evaluation of the DNA quality in production applications [10] and also to quantify the potential change of plasmid topology in long-term storage [11]. With AFM the DNA molecules can be visualized. This is a tool that shows and measures surface structures with unprecedented resolution and accuracy at the nm scale [12–14]. Two recent publications present data on analyzing minicircle and plasmid DNA [15, 16].

2.3 High Quality (HQ) Grade

Amongst other applications, plasmid DNA is used in the GMP-compliant production of recombinant viruses, antibodies, and RNA, where these (but not the plasmids) are the API used in clinical trials. In order to use this for such applications, it is not always necessary to produce the plasmid DNA under GMP as well. An alternative is the so-called *High Quality Grade* plasmid DNA, which is highly purified and well characterized in accordance with the EM(E)A guidelines CHMP/BWP/2458/03 and CPMP/BWP/3088/99 [17, 18] and, hence, meets the requirements of most regulating agencies.

High Quality Grade plasmid DNA is produced in a dedicated facility, based on a research cell bank (RCB) and the very effective *ccc Grade* DNA technology described above (PlasmidFactory, Bielefeld, Germany). A number of quality controls, both to the cell bank and to the plasmid DNA product, ensure that the final result is a product designed especially for the intended application and that this complies with the appropriate regulatory standards. To ensure product safety, substances of animal origin are not used at any stage of the entire process, guaranteeing maximum possible product purity by reliable exclusion of contaminants such as bacterial chrDNA or damaged plasmids. Only one plasmid is produced in each area—different plasmids are not produced in parallel in the same lab, minimizing the risk of cross-contaminations. Typically the plasmid DNA cultivation is completely separated from purification (chromatography) to ensure that the sensitive DNA production process is not affected by any living contamination. Together with a comprehensive documentation the High Quality Grade DNA fulfills the typical requirements for such a product. However, in all cases the respective regulatory authorities will decide on the application of the technology presented here.

2.4 GMP Grade

The GMP process for plasmid DNA production is related to the *Research Grade* process. The most important difference is the facility, which has to be a special, dedicated GMP laboratory, equipped with clean rooms. Second, the documentation of each process step including results obtained is much more comprehensive in GMP. Also, all steps of the GMP process are validated. A general overview is presented by Gengenbach [19].

Whether GMP is necessary clearly depends on the intended application. According to the guidelines mentioned above, for direct clinical applications of a biomolecule (e.g., a plasmid DNA in DNA vaccines) this has to be manufactured in a GMP facility starting from a GMP-produced master cell bank (MCB). However, for indirect use also HQ is an attractive option. Here it should be strongly recommended to discuss as early as possible the actual requirement with the responsible regulatory agency and, where applicable, with the responsible QA of the laboratory further processing the material.

As in any pharmaceutical manufacturing process, the product safety starts with the appropriate design of the facility, clean room design, and maintenance of this during production and—preferably—in between. Also the access of personal and the material used needs to be monitored and controlled.

A working quality assurance (QA) system with validated QC and in process control (IPC) is a minimum requirement and any raw material, intermediate, and product need to be tested and released. A comprehensive documentation of all this is necessary and basis for any product release. All working steps need to be carried out following standard operating procedures (SOP), and a manufacturing batch record compiles all documents generated before, during, or after a GMP process for plasmid DNA vaccine productions. All personnels involved need to be trained in all aspects of the manufacturing, monitoring, and QA/QC process. All equipment is calibrated and qualified.

Major aspects of this are also applied in HQ production runs as, e.g., the production of just one (and not multiple) DNA products per facility. Importantly, the rooms are dedicated to DNA productions in the HQ laboratory. This might be a special benefit due to the fact that GMP facilities frequently are multi-product facilities used for the manufacture of a protein in a first campaign, a life vaccine in a second run, and subsequently a DNA vaccine. Here it becomes obvious that the change over procedure is of an extremely high relevance—especially for GMP—since the type of potential contamination makes the correct choice of decontaminant necessary.

An overview on the production process is given in Fig. 1.

Specific important characteristics of *High Quality and GMP Grade* productions are summarized in Table 1.

3 Methods

3.1 Large-Scale DNA Vaccine Production

Independent from the quality grade, large-scale productions of plasmid DNA comprise the following steps:

1. *E. coli* cells used for plasmid DNA manufacturing were obtained from sources as DSMZ (German Collection of Microorganisms and Cell Cultures). The transformed cell clones are tested for plasmid identity by restriction analysis, and—if requested—further analysis is applied (*see* **Note 1**).
2. A pre-culture of approx. 50 mL is used to inoculate the main culture of 30 L. Cells are cultivated overnight at typically 37 °C or—if necessary to maintain unstable sequences or desired by customer—at lower temperatures. Cells are harvested in this scale by batch centrifugation (*see* **Notes 1** and **2**).
3. Alkaline lysis is performed as published earlier, by a rapid alkaline extraction procedure [6]. Significant modification is necessary to scale up this step, as discussed herein later. Removal of residual particles, cells, or cell fragments is performed by batch centrifugation and multiple filtration steps including a final 0.2 μm filtration of the "cleared lysate" (*see* **Notes 1** and **3**).

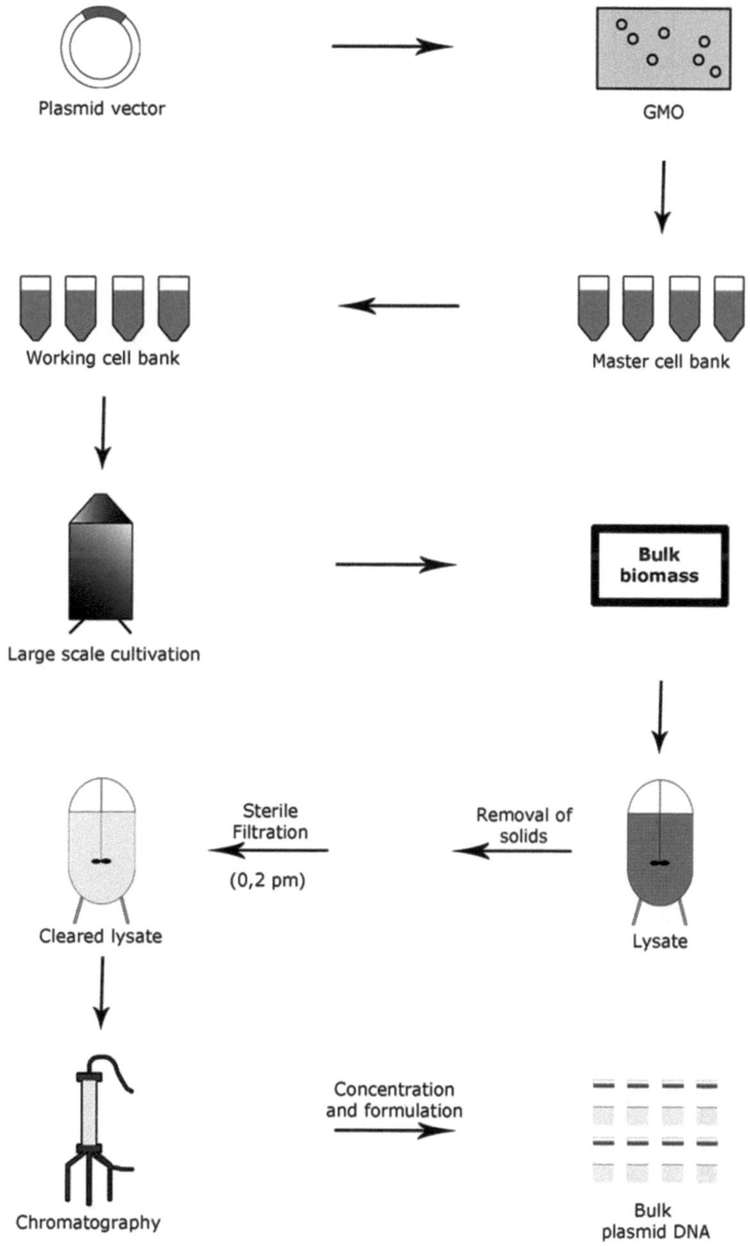

Fig. 1 Production steps in large-scale plasmid DNA manufacturing from plasmid DNA construct to final product

4. Chromatographic steps are performed to reduce the v*olume (chromatogra*phy I) and to remove (additional chromatography) contaminating LPS endotoxins, RNA, proteins, and undesired DNA contaminants (chromosomal bacterial DNA, non-supercoiled plasmid forms). The final eluate is filtered (0.2 μm), precipitated, washed, and reconstituted within buffer (*see* **Notes 1** and **3**).

Table 1
This table summarizes important aspects addressed during audits at the HQ and GMP facilities

Questions	HQ	GMP
Applicable regulations	Most relevant are EM(E)A guidelines CHMP/BWP/2458/03 and CPMP/BWP/3088/99	Any relevant regulatory GMP guidelines
PRODUCTION		
Is the production done in dedicated, separate, or classified clean rooms?	YES, dedicated separate *High Quality* Lab	YES, classified clean rooms
How is contamination or cross-contamination prevented?	– Exclusively one production per time in the HQ lab – Documented change over procedure between production runs	– Exclusively one production per time in GMP lab – Documented change over procedure between production runs
Where is open manipulation performed?	Laminar airflow	Laminar airflow
How are production rooms released?	Documented change over procedure between productions	Documented change over procedure between productions
Is personnel trained according to SOPs? Is the training documented?	YES	YES
Do operators follow gowning SOP to enter production room?	YES	YES
Is environmental monitoring EM performed?	NO	YES
How is equipment having product contact handled and cleaned?	Wherever possible dedicated equipment or disposables, fermenter, chromatography pumps, etc. with product contact will be cleaned (change over procedures) according to SOPs and "Hygieneplan," German GenTG	Wherever possible dedicated equipment or disposables, fermenter, chromatography pumps, etc. with product contact will be cleaned (change over procedures) according to SOPs and "Hygieneplan," German GenTG

How is material identified during production?	Unique manufacturing codes and lot numbers	Unique manufacturing codes and lot numbers
Are raw materials tested and released for production?	Documentation of certificates	YES
Is the production process completely free of animal-derived materials?	YES	Not necessarily
Is equipment cleaned, calibrated, and maintained according to SOP?	Yes, according to "Hygieneplan," German GenTG, and SOPs	YES, according to "Hygieneplan," German GenTG, and GMP SOPs
How is the production room cleaned?	According to our "Hygieneplan" and German GenTG	According to our "Hygieneplan," German GenTG, and SOPs
Are facility and equipment cleaning procedures validated?	Cleaning according to "Hygieneplan" and German GenTG; no validation according to GMP standards	Cleaning according to "Hygieneplan" and German GenTG; validated procedures according to GMP standards
Is a production batch record created?	YES, HQ documentation	YES, GMP compliant
How are manufacturing deviations handled?	Documentation of deviations; production will be repeated if necessary	YES, documentation of deviations; procedure according to SOPs
Who reviews production records?	Manufacturing operators and head of production	QA
How is the plasmid DNA stored to avoid contamination or confusion with other products?	Storage in separate, labeled box, labeled vials, comprehensive documentation; temperature monitoring, access control	Storage in separate, labeled box, labeled vials, comprehensive documentation; temperature monitoring, access control
QC		
Are QC tests performed according to SOPs?	YES	YES
Are QC assays qualified?	YES	YES, GMP compliant

(continued)

Table 1 (continued)

Questions	HQ	GMP
Are QC analysts trained according to SOP? Is this training documented?	YES	YES
How are OOS results handled?	Documentation of deviations; production will be repeated if necessary	Documentation of deviations; procedure according to SOPs
Who reviews testing records?	Head of production	Head of production, QC, and QA
Can stability studies be performed?	YES, e.g., by CGE, depends on customer's requirements	YES, e.g., by CGE, depends on customer's requirements
Who signs off the CofA?	Head of production	QA
Will client obtain a copy of CofA?	Comprehensive QC report with release statement, signed by head of production and QM	CofA signed by QA

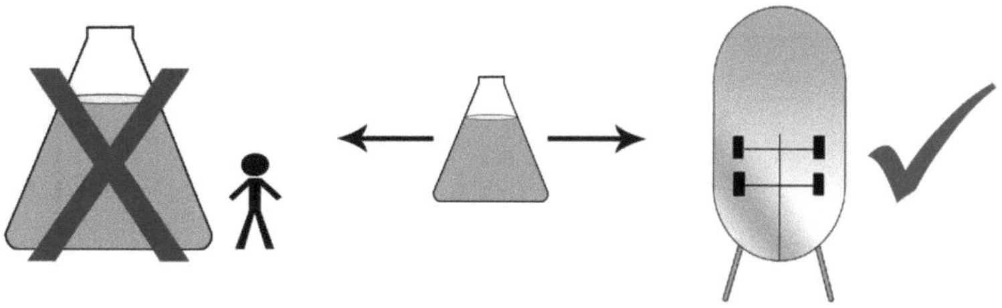

Fig. 2 Up-scaling is not just a question of using bigger devices but of carefully adjusting each process step in a way that can be reproducibly handled

5. According to the requested aliquots, certified cryotubes are manually filled within a laminar airflow. Vials are labeled and stored under quarantine until product release after QC at approx. –20 °C (*see* **Notes 1**and **3**).

3.2 Scale-Up of Cultivation

Before performing large-scale manufacturing of plasmid DNA, a sufficient amount of bacterial biomass is necessary as the source of plasmid DNA. The scale-up of cultivation between 1 and 10 mL or 100 mL is easy since a shaker flask replaces just a tube. The result seems to be sufficient if just the larger amount of biomass with more (plasmid) product is considered. The scale-up into larger scales is excluded since no senseful scale-up is possible by designing a 100 L shaker flask from glass (Fig.2).

However, if the cultivation is performed by fermentation this scale-up is a) on cultivation volume and in parallel b) on the productivity (cells per mL) but at the same time a completely different type of cultivation. The cell density can be maintained quite high, and still growth happens. The pH is controlled and corrected, and carbon source and any other important trace elements are supplied. The finished biomass needs to be harvested in a way keeping the product intact—and also the device for this needs to be adapted to the scale (no dead-end centrifugation is possible). Large-scale approaches apply fed-batch technology [20] (*see* **Notes 1** and **2**).

3.3 Scale-Up of Chromatography

Besides bacterial fermentation also lysate processing and final purification of the plasmid DNA have to be adjusted when shifting to large-scale production, especially for clinical applications. This has to be done with respect to the DNA amount to be produced (up-scaling) but also with respect to product purity and reproducibility of the process steps and the results achieved. Plasmid preparations using commercially available purification kits are not applicable for efficient multi-milligram- or gram-scale plasmid productions (Fig. 3). And, even more important, the resulting plasmid amount and purity are not at all reproducible which would be a prerequisite especially for GMP.

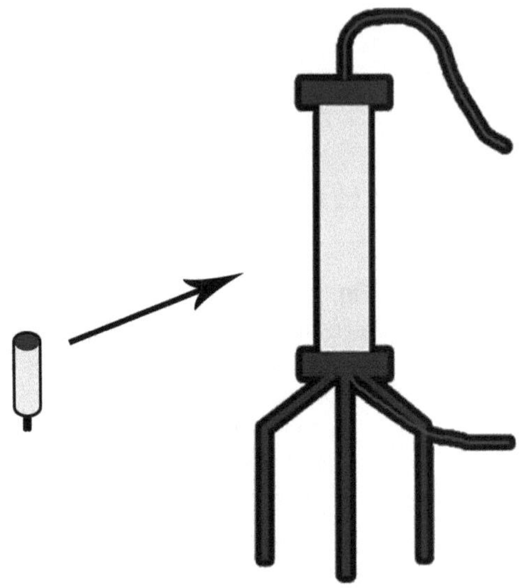

Fig. 3 Up-scaling of chromatography: Plasmid preparation kits frequently used in research labs cannot be used to produce reproducible amounts and qualities of the isolated plasmid DNA. For large-scale productions, especially for clinical applications, automated chromatography systems are used applying controlled, and hence reproducible, conditions

For an efficient purification process, a chromatography system that uses a gradient forming pump and includes UV detection for product sampling is used in order to have controlled conditions. The size of such device as well as size of the chromatography columns can be adjusted to the plasmid amount produced. Typically two or three chromatography steps are run to isolate the pure and (therapeutically) active ccc plasmid form [10] (*see* **Notes 1** and **3**).

3.4 Quality Controls

3.4.1 Quality Controls for E. coli Cell Banks (See **Note 2**)

1. Host identity (K12)
   The test identifies a suspension of *E. coli* as K12. The growth of an *E. coli* K strain differs from the growth of an *E. coli* B strain in the presence of valine and (iso)leucine.
2. Host identity (API-20E)
   API-20E is an identification system for Enterobacteriaceae and other gram-negative rods which uses 21 standardized and miniaturized biochemical tests and a database. The API-20E strip consists of 20 microtubes containing dehydrated substrates. These tests are inoculated with a bacterial suspension that reconstitutes the media. During incubation, metabolism produces color changes that are either spontaneous or revealed by the addition of reagents. The reactions are read according to the reading table, and the identification is obtained by using the identification software.

*3.4.2 Quality Controls for Plasmid DNA (See **Note 3**)*

1. Appearance
   Various plasmid DNA samples are checked visually and compared with the defaults.
   There must be a clear, particle-free liquid present.
2. Determining the concentration of a nucleic acid solution
   The concentration of nucleic acid in aqueous solutions is determined by dilution with TE buffer and photometric measurement at 260 nm. The following correlation is used for the conversion of the absorption value in a concentration: 50 μg/mL of double-stranded DNA is equivalent to an absorption value of 1.0 at 260 nm.
3. UV scan to determine purity
   The absorption of an aqueous plasmid DNA solution is measured and recorded in a single-beam photometer in the 220–320 nm wavelength range. The scan must have a maximum at 258 nm ± 2 %.
4. Restriction digestion of plasmid DNA
   With suitable enzymes (restriction endonucleases) under defined reaction conditions plasmid DNA can be split at certain points (specific sequence of bases) to produce various fragments of the original DNA, which can be made visible in agarose gel. The identity of the plasmid can then be clarified with the aid of plasmid maps by comparing the obtained fragment sizes with those calculated theoretically.
5. Visual checking for RNA in agarose gel
   A digital image of an agarose gel is examined on a PC for the presence or the absence of RNA contamination.
   The absence of RNA can be identified as follows:
   The DNA sample shows no "cloud-like form" in the agarose gel below 500 bp—i.e., there is no RNA in the preparation.
6. Visual checking for bacterial chrDNA in agarose gel
   A digital image of an agarose gel is examined on a PC for the presence or the absence of bacterial chrDNA.
   The absence of bacterial chrDNA can be identified as follows:
   The DNA sample in the relevant agarose gel trace shows neither a distinct band at a level higher than 10 kbp nor a stained "background smear" in the range between 15 and 5 kbp—i.e., there is no visually recognizable bacterial chrDNA in the preparation.
7. Quantification of bacterial chrDNA
   To determine the proportion of bacterial chrDNA in plasmid preparations, aliquots in suitable dilution are used as templates in quantitative real-time PCRs (qRTPCRs) so that the number of gene copies of the ribosomal operons can be determined.
   The quantity of bacterial chrDNA is calculated on this basis.

8. Homogeneity/densitometric agarose gel evaluation
   Plasmids exist in a ccc form, which is negatively superspiralized. Individual strand ruptures caused enzymatically or through mechanical stress on the DNA cause a loosening of the molecule. The resulting form is described as an oc form. Double-strand ruptures lead to linearization of the molecule.

   Errors in replication or homologous replication can also lead to multimeric plasmid forms.

   The various isoforms in a plasmid preparation can be made visible by agarose gel electrophoresis and subsequent staining with the fluorescence indicator ethidium bromide. The bands of the gels marked in this way are made visible on a transilluminator by UV stimulation (312 nm).

   Assignment of bands to plasmid forms entails some difficulties, as the electrophoretic mobility of plasmids with differing structures changes with operating conditions during electrophoresis. It can generally be said, however, that plasmids of the ccc form move most rapidly, followed by the oc monomer forms and ccc dimer forms. But the precise distinction of the ccc dimer forms and oc monomer forms is generally recognizable only in small plasmids.

   The quantitative evaluation of stain intensities, i.e., the quantitative measurement of stain density, is done with densitometers, which measure the light absorption in one band in the agarose gel.

9. Homogeneity/CGE analysis
   Plasmids exist in a *ccc* form, which is negatively superspiralized. Individual strand ruptures caused enzymatically or through mechanical stress on the DNA cause a loosening of the molecule. The resulting form is described as an *oc* form. Double-strand ruptures lead to linearization of the molecule.

   Errors in replication or homologous replication can also lead to multimeric plasmid forms.

   The plasmid topologies that occur can be distinguished and quantified by CGE. These are divided in a gel-filled capillary in an electrical field and then detected optically.

   The CGE analyses are performed using a P/AGE™ 2050 (Beckman Coulter, München) equipped with an LIF detector (488 nm/520 nm). Coated capillaries (DB-17; J&W Scientific, Folsom, CA) with an effective length of 30 cm, an inner diameter of 100 μm, and a coating thickness of 0.1 μm are used. Just prior to analysis, the intercalating dye YOYO (Molecular Probes, Eugene, OR) is added. After pre-staining with YOYO, the plasmid samples are introduced hydrodynamically into the capillary and electrophoresis is carried out at 100 V/cm and 30 °C.

10. Determination of endotoxins

The gram-negative bacterial endotoxin content in a plasmid DNA solution is determined by the kinetic chromogenic test "Kinetic—QCL® Kinetic Chromogenic Assay" (Lonza, Prod. No 50-650U).

LPS (endotoxins) located on the outer membrane of gram-negative bacteria cause the activation of a proenzyme in limulus amebocyte lysate (LAL). The activated enzyme in turn catalyzes the splitting of p-nitroaniline (pNA) from a chromogenic synthetic substrate, which leads to a yellow staining. The freeing of the pNA is measured photometrically over a specified incubation period at 405 nm by a microtitre plate reader with the relevant software. The increase in yellow staining correlates with the concentration of endotoxins. The reference value is provided by the pure standard LPS of *E. coli* 055:B5. The concentration of endotoxins in the sample can be calculated from the absorption values of a dilution series of the standard. The test has a high sensitivity (0.005–10.0 EU/mL) and linearity through the use of a chromogenic reagent.

11. Determination of proteins

The quantification of the total protein content in a plasmid DNA solution is determined by the chromogenic test "BCA™ Protein Assay Kit" (Pierce, Prod# 23227).

The BCA test used is based on the Biuret reaction. This method permits the colorimetric verification of a bicinchoninic acid/Cu^{+} color complex. Proteins reduce alkaline Cu(II) to Cu(I). Bicinchoninic acid forms an intensively violet complex with copper ions with an absorption maximum at 562 nm. The absorption is in turn directly proportional to the protein concentration. Bovine serum albumin (BSA) at a concentration of 10–250 μg/mL is used as a reference substance.

12. Bioburden

This relates to the microbiological testing of non-sterile products.

The test is for the total number of:

(a) Aerobic bacteria: Total aerobic microbial count (TAMC).

Medium: Casein-soya peptone (caso) agar medium.

Incubation conditions: 3–5 days at 30–35 °C.

(b) Yeasts and moulds: Total combined yeast/mould count (TYMC). Medium: Sabouraud-dextrose agar.

Incubation conditions: 5–7 days at 20–25 °C.

Anaerobic bacteria can also be tested. The procedure is similar to the TAMC test but without oxygen, which is however not described in Ph. Eur 2.6.12.

4 Notes

4.1 Discussion of Important Process Steps

1. Up-scaling
Some important aspects influencing the process steps summarized above are discussed in the following section:
Any plasmid manufacturing process starts at the laboratory or the pilot scale. Here, if applicable, one has to make sure that all components used are available in a quality suitable also for a GMP process. It has to be taken into account that up-scaling of a process is not just multiplying of relevant factors, but all steps to be performed have to be carefully evaluated, e.g., with respect to reproducibility and influences on product quality. Modification of temperature within a large fermenter requires a fast temperature transfer to all parts of the vessel or alkaline lysis is not just shaking a very large bottle instead of a small one. In-line lysis and mixing devices help to overcome such bottlenecks if applicable.

2. *E. coli* production strain, cultivation, and harvest
Plasmid DNA manufacturing starts with the transformation of the fully characterized plasmid into appropriate and characterized *E. coli* K12 host cells. The resulting "genetically modified organism" (GMO) has to be checked carefully for the expected characteristics. Subsequently, it will be transferred into the GMP environment for GMP-conformal processing. This includes the generation of an MCB and working cell bank (WCB), which are required for reproducible large-scale cultivation of the bacterial biomass. The cell banks have to be fully characterized to be of sufficient quality for a further manufacturing. Relevant QC assays for *E. coli* cell banks are described below.
The resulting transformed bacteria are used to prepare a pre-culture for inoculation of the bioreactor.
Applying proper cultivation parameters, fermentation overnight is sufficient to obtain large amounts of cellular biomass with a high amount of plasmid [10]. Different operating variables can be monitored and adjusted in the bioreactor like pH, dissolved oxygen, indirect determination of the specific growth rate, or limiting substrates [20–24].
Productivity and quality with respect to the content of plasmid forms very much depend on the host strain used and the cultivation conditions. With some plasmids the same process type may be used, and with others (typically those used for virus production likely caused by their repetitive elements such as ITR for AAV or LTR for lentiviral transfer plasmids) for each new plasmid a cultivation process development is required. After fermentation, the resulting biomass has to be separated

from the supernatant liquid of the culture. In lab scale this can be performed by batch centrifugation and in large scale by flow-through centrifugation or cell concentration with tangential flow filtration processes. The biomass is subject to QC tests for product content and absence of any contamination and will be processed, if released for manufacturing. Process elements and cultivation media are potential sources for such contamination. Today's technology for the generation of complex bacterial growth media uses soybean peptones to avoid animal-derived protein sources. Generally, in order to avoid BSE or TSE risk materials as recommended by regulatory guidelines the use of synthetic growth media should be favored [25].

Besides high biomass and product concentrations, homogeneity of the plasmid at the cultivation stage is important. Schmidt et al. [20] described a dissolved, oxygen-controlled fed-batch cultivation on a defined glycerol medium. A product concentration of 100 mg/dm^3 and a dry biomass concentration of 48 g/L were achieved, resulting in a yield of 2.1 mg/g.

The cultivation of *E. coli* to high cell densities for plasmid DNA production in a batch mode was described by Voss et al. [26]. By using a fully defined synthetic glycerol medium 45 mg/dm^3 plasmid DNA could be produced. Recent developments allow even higher productivity [27].

A cultivation process needs to be performed with identical results at least three times to ensure constant quality of biomass; this is also a regulatory requirement for GMP. After any up-scaling the quality of biomass should fulfil identical specifications.

Before purification the produced biomass is separated from the culture medium by centrifugation or microfiltration and can be stored at –20 °C or below.

3. Cell lysis, chromatography, and filling

Plasmids are released from host cells by alkaline lysis. In this procedure, suspended cells are destroyed by an alkaline extraction reagent. During neutralization with an acidic potassium acetate solution, proteins, chrDNA, and cell debris flocculate together with potassium dodecyl suphate. Plasmid DNA (nucleic acid) remains within solution and can be separated from the majority of the flocculated contaminants.

A further problem of the alkaline lysis is the separation of cellular components. Due to the high viscosity of the precipitate, this can only be performed by pre-filtration or centrifugation and subsequent clearing filtration. This approach is in any case time consuming and expensive. Furthermore the chrDNA within the precipitate is sensitive to shear forces [28]. An approach to mix cell lysate and neutralizing agent with low shear forces is necessary to avoid the contamination of the plasmid containing liquid with chrDNA that cannot be removed easily. This process step is critical because one major contamination source of plasmid DNA productions is the bacterial chrDNA [29]. The resulting DNA fragments will be present as a contamination in the subsequent purification steps as well. Hence, it is worth to avoid co-isolation of chrDNA fragments by keeping the bacterial high-molecular DNA intact and remove it together with bacterial debris during lysis separation and filtration. The resulting cleared lysate contains only about 3 % plasmid DNA and is sterilized by filtration for further downstream processing (DSP).

The initial alkaline lysis procedure was in fact never really changed. After certain reengineering approaches applied by manufacturers of plasmid DNA kits, the general idea is still the same. Same small-scale preparation protocols suggest a boiling technique in combination with detergents, but this is not scalable. Useful approaches were presented when applying in-line lysis techniques [30–33] as we did for our large-scale DNA production facilities [34] and shown in Fig. 4.

The design of the DSP depends on the required purity of the plasmid product. Chromatography separates the plasmid molecules from soluble biomolecules (e.g., host chrDNA, RNA, nucleotides, lipids, residual proteins, amino acids, saccharides),

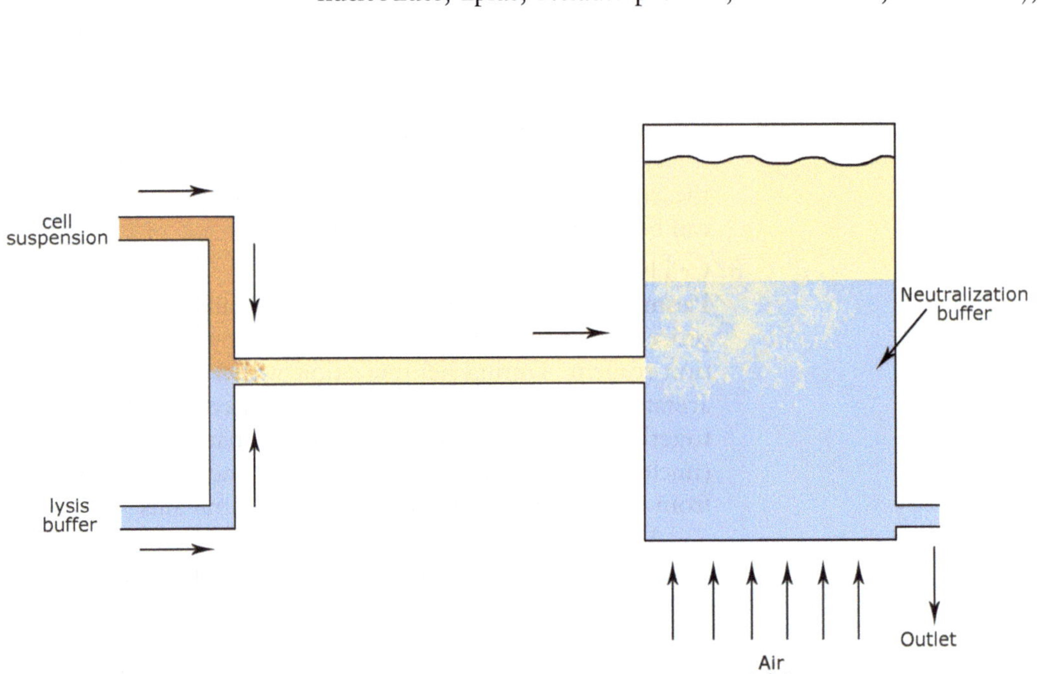

Fig. 4 In-line lysis process scheme as initially disclosed in Patent EP1593741 by PlasmidFactory. The cell resuspension is mixed with the lysis buffer resulting in a flocculation of the cell components and SDS. These flakes are pumped into a volume of neutralization buffer and separate by flotation, while the clear lysate can be taken from the outlet

salts, and buffer components. So far, only a few techniques are used in pharmaceutical grade processing of plasmids. Anion-exchange chromatography was initially described for certain processes [29, 33, 35–38].

Other approaches describe plasmid DNA purification systems for the selective removal of contaminants or improved binding capacity for the intended plasmid product (e.g., monolithic stationary phase) [39–41].

Recent manufacturing technology makes use of removing the undesired oc and linear plasmid forms as well as of fragmented chrDNA. The removal of damaged plasmid forms can be demonstrated by AGE and CGE and the removal of chrDNA by qPCR ([10], and PlasmidFactory, unpublished).

This DSP step of plasmid manufacturing ends with bulk purified plasmid DNA being formulated within the appropriate buffer or solution for further processing or storage and application.

5 Application and next-generation DNA vaccines

The DNA vaccines used so far were based on purified DNA molecules deriving from typically *E. coli* K12 cells. A majority of these were plasmid DNA and in some cases fragments thereof or mini-plasmids. Recently presented vectors reduced to almost the vaccine-encoding sequences are called minicircle, were published by our group (review ref. 42).

The significant advantage of these is their extremely small size resulting in an unsurpassed potency—due to their access to the target tissue and the option to have a significantly lower dose with equivalent (or even superior) activity. The second important aspect is a regulatory on: The minicircle DNA does not contain bacterial sequences and also no antibiotic resistance or any other selective marker.

At present the production of such DNA vectors is limited to certain mg in scale, but large-scale approaches are under development (PlasmidFactory, unpublished).

Plasmids used in gene therapy and vaccination are usually at minimum 2 kb in size although recently smaller—the so-called miniplasmids—have been successfully tested (for an overview *see* ref. 42). These are now approx. as little as 1,000 bp (without carrying a transgene for the "preventive" (vaccine) or the "therapeutic" (vaccine or modification of expression pattern)) in size, so these new plasmids consist exclusively of the "*ori*" (bacterial origin of replication) and a cloning site or—more convenient—a multiple cloning site (MCS) with certain restriction sites for the integration of a transgene into such plasmid [43]. The advantage of miniplasmids is that one prominent element typically used so far—the selection marker, typically an antibiotic resistance gene to be expressed in *E. coli*—is not present any more within this. These elements are approx. 1,000 bp in size and the reason for a minimum size of 2,000 bp for standard plasmids. The advantage is not only a size reduction but also the absence of resistance markers from these plasmids that in the past made these tools be suspected to spread resistance elements and cause resistances of pathogens. Especially in medical gene transfer this is a side effect that should be avoided, and also the use of such in biological food production should be avoided.

Approaches to even reduce the portion of a plasmid from earlier 2 kb plus transgene not only to 1 kb plus transgene (miniplasmids) but also to nearly zero (just a "circular transgene") were presented as well [42] with the so-called minicircles. These tools do not contain any *ori* nor do they contain any selective marker and for sure no antibiotic resistance elements. They have in between been tested in various applications and are just entering the field of pharmaceutical DNA application due to their extremely high safety profile.

All plasmid DNA products require—especially if manufactured in large scale and stored for subsequent use—an appropriate storage system. We did certain storage evaluation studies over years now with real-time data for HBV DNA vaccines [11]. Other studies were performed to evaluate the optimum storage conditions for a cat fibrosarcoma veterinary DNA vaccine [44] analyzing different types of formulation buffer, DNA concentration, filling and vial type, freeze and thaw techniques, and plasmid sizes. All these experiments (PlasmidFactory, data unpublished) were accompanied by the ccc quantification per CGE (see above) and demonstrated that DNA is a quite stable molecule and can even be lyophilized [45].

Acknowledgement

We thank the research team of PlasmidFactory, Bielefeld, Germany, for critical discussion and contributing work; Janine Conde-Lopez for support with figures; and the German Federal Ministry of Education and Research (BMBF) for grants BioChancePLUS (0313749) and Nano-4-Life (13N9063). Part of this work has also been supported by the CliniGene Network of Excellence funded by the European Commission FP6 Research Programme under contract LSHB-CT-2006-018933.

References

1. Lederberg J (1952) Cell genetics and hereditary symbiosis. Physiol Rev 32:403–430
2. Lahijani R, Hulley G, Soriano G et al (1996) High-yield production of pBR322-derived plasmids intended for human gene therapy by employing a temperature-controllable point mutation. Hum Gene Ther 7:1971–1980
3. Schmidt T, Friehs K, Flaschel E et al (1998) Method for the isolation of ccc plasmid DNA. WO 1999/061633
4. Carnes AE, Hodgson CP, Williams JA (2006) Inducible *Escherichia coli* fermentation for increased plasmid DNA production. Biotechnol Appl Biochem 45:155–166
5. Hoare M, Levy MS, Bracewell DG et al (2005) Bioprocess engineering issues that would be facing in producing a DNA vaccine at up to 100 m^3 fermentation scale for an influenza pandemic. Biotechnol Prog 2005(21): 1577–1592
6. Birnboim HC, Doly J (1979) A rapid alkaline extraction procedure for screening recombinant plasmid DNA. Nucleic Acids Res 7:1513–1523
7. Maucksch C, Connor B, Rudolph C (2013) Plasmid DNA concatemers: influence of plasmid structure on transfection efficiency. In: Schleef M (ed) Minicircle and miniplasmid DNA vectors—the future of non-viral and viral gene transfer. Wiley-Blackwell, Weinheim
8. Wooddell CI, Subbotin VM, Sebestyén MG et al (2011) Muscle damage after delivery of naked plasmid DNA into skeletal muscles is batch dependent. Hum Gene Ther 22: 225–235
9. Schmidt T, Friehs K, Schleef M, Voss K, Flaschel E (1999) Quantitative analysis of plasmid forms by agarose and capillary gel electrophoresis. Anal Biochem 274:235–240
10. Schleef M, Schmidt T (2004) Animal-free production of ccc-supercoiled plasmids for research and clinical applications. J Gene Med 6:S45–S53
11. Schleef M, Baier R, Walther W et al (2006) Long-term stability and topology analysis of plasmid DNA by capillary gel electrophoresis. Bioprocess Int 4:38–40
12. Janshoff A, Neitzert M, Oberdorfer Y et al (2000) Force spectroscopy of molecular systems-single molecule spectroscopy of polymers and biomolecules. Angew Chem Int Ed Engl 39(18):3212–3237
13. Binnig G, Quate CF, Gerber C (1986) Atomic force microscope. Phys Rev Lett 56(9):930–933
14. Drake B, Prater CB, Weisenhorn AL et al (1989) Imaging crystals, polymers, and processes in water with the atomic force microscope. Science 243(4898):1586–1589
15. Rischmüller A, Viefhues M, Dieding M et al (2013) Analytic tools in minicircle production. In: Schleef M (ed) Minicircle and miniplasmid DNA vectors—the future of non-viral and viral gene transfer. Wiley-Blackwell, Weinheim
16. Viefhues M, Wegener S, Rischmüller A, Schleef M, Anselmetti D (2013) Dielectrophoresis based continuous-flow nano sorter: fast quality control of gene vaccines. Lab Chip 13: 3111–3118
17. EMEA (2005) Guideline on development and manufacture of lentiviral vectors. CPMP/BWP/2458/03, London
18. EMEA (2001) Note for guidance on the quality, preclinical and clinical aspects of gene transfer medical products. CPMP/BWP/3088/99, London
19. Gengenbach RJ (2008) GMP-Qualifizierung und Validierung von Wirkstoffanlagen. Wiley-VCH, Weinheim
20. Schmidt T, Schleef M, Friehs K, Flaschel E (1999) Hochzelldichtefermentation zur Gewinnung von Plasmid-DNA für Gentherapie und genetische Impfung. BIOforum 22: 174–177
21. Lee SY, Chang HN (1994) High cell density cultivation of *Escherichia coli* using sucrose as a carbon source. Biotechnol Lett 15:971–974
22. Nakano K, Rischke M, Sato S, Märkl H (1997) Influence of acetic acid on the growth of *Escherichia coli* K12 during high-cell-density cultivation in a dialysis reactor. Appl Microbiol Biotechnol 48:597–601
23. Macaloney G, Hall JW, Rollins MJ et al (1997) The utility and performance of near-infrared spectroscopy in simultaneous monitoring multiple components in a high cell density recombinant *Escherichia coli* production process. Bioproc Eng 17:157–167
24. Paalme T, Tiisma K, Kahru A et al (1990) Glucose-limited fed-batch cultivation of *Escherichia coli* with computer-controlled fixed growth rate. Biotechnol Bioeng 35:312–319
25. EMEA (2001) Note for guidance on minimising the risk of transmitting animal spongiform encephalopathy agents via human and veterinary medicinal products. CPMP/410/01 rev 1, London
26. Voss C, Schmidt T, Schleef M et al (2004) Effect of ammonium chloride on plasmid DNA production in high cell density batch

culture for biopharmaceutical use. J Chem Technol Biotechnol 79:57–62
27. Urthaler J, Buchinger W, Necina R (2005) Improved downstream process for the production of plasmid DNA for gene therapy. Acta Biochim Pol 52:703–711
28. Levy MS, Collins IJ, Tsai JT et al (2000) Removal of contaminant nucleic acids by nitrocellulose filtration during pharmaceutical-grade plasmid DNA processing. J Biotechnol 76:197–205
29. Schleef M (1999) Issues of large-scale plasmid manufacturing. In: Rehm HJ, Reed G, Pühler A et al (eds) Biotechnology, vol 5a, Recombinant proteins, monoclonal antibodies and therapeutic genes. Wiley-VCH, Weinheim, pp 443–470
30. Lee AL, Sagar S (1999) A method for large scale plasmid purification. WO 96/36706
31. Schumacher I, Freitag R, Hilbrig F (2002) Method for treating biomass for producing cell lysate containing plasmid DNA. WO 02/057446 A2
32. Hebel H, Ramakrishnan S, Gonzales H et al (2004). Devices and methods for biomaterial production. WO2004/108260
33. Voß C, Schmidt T, Schleef M (2005) From bulk to delivery: plasmid m, manufacturing and storage. In: Schleef M (ed) DNA pharmaceuticals: formulation and delivery in gene therapy, DNA vaccination and immunotherapy. Wiley-VCH, Weinheim, pp 23–42
34. Voss C, Flaschel E (2010) Method for producing extra-chromosomal nucleic acid molecules. US 7,842,481 B2
35. Colpan M, Schorr J, Moritz P (1995) Process for producing endotoxin-free or endotoxin-poor nucleic acids and/or oligonucleotides for gene therapy. WO 95/21177
36. Thatcher DR, Hitchcock AG, Hanak JA et al (1997) Method of plasmid DNA production and purification. WO 97/29190
37. Bussey L, Adamson R, Atchley A (1998) Methods for purifying nucleic acids. WO 98/05673
38. Ferreira GNM, Prazeres DMF, Cabral JMS et al (2001) Plasmid manufacturing—an overview. In: Schleef M (ed) Plasmids for therapy and vaccination. Wiley-VCH, Weinheim, pp 193–236
39. Green AP (1999) Purification of supercoiled plasmid. In: Lowrie DB, Whalen RG (eds) DNA vaccines: methods and protocols. Humana, Totowa, NJ, pp 1–9
40. Lemmens R, Olsson U, Nyhammar T et al (2003) Supercoiled plasmid DNA: selective purification by thiophilic/aromatic adsorption. J Chromatogr B 784:291–300
41. Strancar A, Podgornik A, Barut M et al (2002) Short monolithic columns as stationary phases for biochromatography. Adv Biochem Eng Biotechnol 76:49–85
42. Schleef M (ed) (2013) Minicircle and miniplasmid DNA vectors—the future of non-viral and viral gene transfer. Wiley-Blackwell, Weinheim
43. Preston A (2003) Choosing a cloning vector. In Casali N, Preston A (eds) *E. coli* plasmid vectors. Humana Press, Totowa NJ, pp 19–26. Gene Ther 22:225-235
44. Schwarz B, Kempf T, Schillinger U et al (2007) Cloning of the feline cytokines IL-2, IFNy and GM-CSF for an adjuvant nonviral gene therapy of feline fibrosarcoma. Kleintierpraxis 52:569–578
45. Schleef M, Blaesen M (2009) Production of plasmid DNA as a pharmaceutical. In: Walther W, Stein US (eds) Methods in molecular biology, vol 542, Gene therapy of cancer., pp 471–495

Part IV

New Vaccine Applications

Chapter 15

Protective and Therapeutic DNA Vaccination Against Allergic Diseases

Almedina Isakovic, Richard Weiss, Josef Thalhamer, and Sandra Scheiblhofer

Abstract

DNA vaccines represent a novel approach for protective and therapeutic intervention against type I allergies. In contrast to classical subcutaneous immunotherapy, which relies on the injection of alum-adsorbed protein extracts, DNA vaccines do not suffer from side effects such as anaphylaxis or therapy-induced IgE antibodies. In animal models, DNA vaccines have been demonstrated to prevent TH2 sensitization or balance an existing TH2-mediated allergic immune response by induction of TH1 or regulatory T cells, rendering them promising candidates for prophylactic vaccination as well as therapy. In this chapter, we discuss methods relevant for evaluation of DNA expression vectors for targeting antigen to different cellular compartments for use as a vaccine in an asthma mouse model. Attaching signal sequences has proven to be a successful way to manipulate and boost the immune responses following DNA immunization and also creating hypoallergenic DNA vaccines.

Key words DNA vaccine, Genetic immunization, Immunotherapy, Ubiquitin, LIMP-II, Tissue plasminogen activator, Hypoallergen, TH1/TH2 responses, Allergy, BALB/c

1 Introduction

In this book chapter, we show how to specifically target antigen to different antigen-presenting pathways by attaching signal sequences using the major birch pollen allergen Bet v 1.0101 as an example. After applying the DNA vaccine modified with various regulatory sequences, the plasmid is taken up by antigen-presenting cells (APCs) or somatic bystander cells such as keratinocytes and transported into the nucleus, where the DNA is transcribed into mRNA and shuttled to the cytosol for protein translation. The first 63 bp of human tissue plasminogen activator (tPA) sequence encode a strong signal peptide for protein secretion, thereby mediating the release of antigen into the extracellular space of transfected cells from where it is taken up by APCs, resulting in presentation of

Monica Rinaldi et al. (eds.), *DNA Vaccines: Methods and Protocols*, Methods in Molecular Biology, vol. 1143,
DOI 10.1007/978-1-4939-0410-5_15, © Springer Science+Business Media New York 2014

Table 1
Expected immunological outcomes of different targeting routes

Construct	Humoral responses	CD4 responses	CD8 responses
w/o targeting	+	+	+
ER (tPA)	++ [1, 2]	++ [1]	+
Poteasomal (ubi)	- [3]	+ [3]	++ [4, 12]
Lysosomal (LIMP-II)	+/-	++ [7, 13]	+

Numbers in brackets refer to the references listed in the bibliography

Table 2
Effects of different targeting routes on protection from allergic sensitization

Construct	IgE suppression	TH2/TH22 suppression	Reduced lung inflammation
pCI-Bet	+++	+++	++
pCI-tPA-Bet	++	+	++
pCI-ubi-Bet	+++	++	+++
pCI-Bet-LIMP-II	+++	+++	+++

pCI-Bet: Plasmid expressing the major birch pollen allergen Bet v 1.0101 in the cytoplasm of the cell; pCI-ubi-Bet: plasmid targeting Bet v 1.0101 to the proteasome; pCI-tPA-Bet: plasmid targeting Bet v 1.0101 for secretion; pCI-LIMP-II: plasmid targeting Bet v 1.0101 into the endosome

antigenic peptides on MHC-II [1, 2]. However, antigen will also be presented on MHC-I molecules by cross-priming.

Conjugation of allergen to the 76-amino acid (AA)-long cellular protein ubiquitin (ubi) shuttles the antigen into the polyubiquitination pathway and the endoplasmic reticulum (ER), thereby promoting the presentation of antigenic peptides on MHC-I [3, 4]. In contrast, linking of allergen to the 20-AA C-terminal tail of lysosomal integral membrane protein-II (LIMP-II) peptide facilitates the antigenic transport from ER directly to lysosomes, making it accessible for MHC-II presentation [5–7]. Unmodified allergen, which lacks a signal sequence, is expressed as cytoplasmic protein and targets MHC-I and MHC-II presentation by autophagy.

Secretion of antigen will increase cellular as well as humoral immune responses. In contrast, targeting the proteasome or the endosome via ubiquitination and the LIMP-II targeting sequence (*see* **Note 1**) reduces the amount of available B-cell epitopes (via intracellular degradation) and consequently the IgE binding capacity of the antigen while maintaining the immunogenicity on the T cell level (*see* Tables 1 and 2).

Therefore, allergens targeted for intracellular degradation render the gene product hypoallergenic (i.e., it cannot cross-link cell-bound IgE) and represent the most promising candidates in a therapeutic setting.

2 Materials

2.1 Plasmids

1. pCI mammalian expression vector (Promega).
2. pCI-Bet: Plasmid expressing Bet v 1.0101 in the cytoplasm of the cell [8].
3. pCI-ubi-Bet: Plasmid targeting Bet v 1.0101 to the proteasome [3].
4. pCI-tPA-Bet: Plasmid targeting Bet v 1.0101 for secretion [1].
5. pCI-LIMP-II: Plasmid targeting Bet v 1.0101 into the endosome [9].
6. *Escherichia coli* XL1-blue competent cells.
7. Standard Luria Broth medium with ampicillin (LB-Amp): 1 % tryptone, 0.5 % yeast extract, 1 % NaCl, 100 μg/mL ampicillin.
8. Luria agarose plates with ampicillin (LA-Amp plates): LB-Amp, 1.5 % bacto agar.
9. Endotoxin-free plasmid preparation kit.
10. 37 °C bacterial incubator.
11. 37 °C bacterial incubator shaker.

2.2 Limulus Amebocyte Lysate Assay

1. Endotoxin-free H_2O.
2. 96-well V-bottom plate.
3. Pyrotell-T.
4. Purified plasmids.
5. Control standard endotoxin.
6. Tecan infinite M200Pro reader.

2.3 Vaccination and Sensitization of Mice

1. Female BALB/c mice (6–10 weeks of age).
2. 27-gauge needles.
3. 1-mL syringes.
4. Aluminum hydroxide gel (e.g., Alu-Gel-S).
5. Recombinant Bet v 1.0101 (Biomay, Austria).
6. Purified plasmid DNA.
7. Endotoxin-free phosphate-buffered saline (PBS): 137 mM NaCl, 2.7 mM KCl, 10 mM Na_2HPO_4, 2 mM KH_2PO_4, pH7.4
8. Endotoxin-free H_2O.

9. Curved forceps.
10. Rodent isoflurane anesthesia machine.
11. PARI BOY® SX nebulizer with PARI LL nebulizer head (median aerosol size 3.7 μm).
12. Acrylic glass nebulizing chamber (approx. 25 × 25 × 25 cm).

2.4 Lung Resistance/ Compliance (R/C) Measurement

1. 0.5 % Methacholine solution: 5 mg/mL methacholine, 0.9 % NaCl.
2. 1 % Methacholine solution: 10 mg/mL methacholine, 0.9 % NaCl.
3. 2 % Methacholine solution: 20 mg/mL methacholine, 0.9 % NaCl.
4. Anesthetic: 20 mg/mL ketamine-HCl, 2 mg/mL xylazine-HCl, 0.9 % NaCl.
5. 0.9 % NaCl.
6. Forceps and scissors.
7. 37 °C heating pad.
8. Ethilon II monofil suture.
9. Rodent Fine Pointe™ RC station.

2.5 Bronchoalveolar Lavage

1. Ice-cold PBS: 137 mM NaCl, 2.7 mM KCl, 10 mM Na_2HPO_4, 2 mM KH_2PO_4, pH7.4.
2. 40 μm cell strainer.
3. 96-well V-bottom plate.
4. Anti-mouse CD45-PE/Cy7, Clone 30-F11 (BD Pharmingen).
5. Anti-mouse Ly-6G (Gr1)-APC, Clone RB6-8C5 (eBioscience).
6. Anti-mouse CD4-APC-eFluor 780, Clone RM4-5 (eBioscience).
7. Anti-mouse CD8-FITC, Clone 53-6.7 (eBioscience).
8. Anti-mouse Siglec-F-PE, Clone E50-2440 (BD Pharmingen).
9. Anti-mouse FoxP3-PerCP/Cy5.5, Clone FJK-16s (eBioscience).
10. FACS buffer: PBS, 0.5 % BSA, 2 mM EDTA.
11. Lysing solution (e.g., BD FACS Lyse).

2.6 Basophil Activation Test

1. RPMI 1640 culture medium.
2. 96-well V-bottom plate.
3. Fresh whole blood.
4. Heparin.
5. Recombinant Bet v 1.0101 (Biomay, Austria).
6. Anti-mouse IgE, Clone 23G3 (eBioscience).
7. Anti-mouse IgE-FITC, Clone RME-1 (BioLegend).

8. Anti-mouse CD19-PE/Cy7, Clone 6D5 (BioLegend).
9. Anti-CD200R-APC, Clone OX110 (eBioscience).
10. Anti-mouse CD4-PerCp/Cy5.5, Clone GK1.5 (BioLegend).
11. FACS buffer: PBS, 0.5 % BSA, 2 mM EDTA.
12. Lysing solution (e.g., BD FACS Lyse).

2.7 CFSE Staining

1. 5- (and 6-) carboxyfluorescein diacetate succinimidyl ester (CFSE): 5 mM in DMSO.
2. Recombinant Bet 1.0101 (Biomay, Austria).
3. PBS: 137 mM NaCl, 2.7 mM KCl, 10 mM Na_2HPO_4, 2 mM KH_2PO_4, pH7.4.
4. Fetal bovine serum (FBS).
5. Proliferation medium: Earle's modified Eagle's medium (MEM), 1 % FCS, 1 % L-glutamine, 2.5 % penicillin/streptomycin, 1 mM sodium pyruvate, 2 % HEPES, 1 % nonessential amino acids, 2 μM 2-mercaptoethanol.
6. Ammonium, chloride, potassium (ACK) lysis buffer: 0.15 M NH_4CL, 10 mM $KHCO_3$, 0.1 mM EDTA, pH 7.2–7.4. Sterilize.
7. 24-well flat-bottom tissue culture plate.
8. FACS buffer: PBS, 0.5 % BSA, 2 mM EDTA.
9. 96-well V-bottom plate.
10. Anti-mouse CD4-APC-eFluor 780, Clone RM4-5 (eBioscience).
11. Anti-mouse CD25-PE/Cy7, Clone PC61.5 (eBioscience).
12. Anti-mouse FoxP3-PerCp/Cy5.5, Clone FJK-16s (eBioscience).
13. Foxp3/Transcription Factor Staining Buffer Set (eBioscience).

3 Methods

3.1 Plasmids

The pCI mammalian expression vector (GenBank®/EMBL accession number: U47119) has a strong human cytomegalovirus (CMV) major immediate–early gene enhancer/promoter region, which drives constitutive high protein expression levels in cells [10]. This feature promotes a transient and stable expression of cloned DNA inserts. Furthermore, a late SV40 polyadenylation site is included, which is very efficient in increasing the steady-state level of RNA, stability, and translation [11].

By attaching different signal sequences to the cDNA cloned into the expression vector, different cellular compartments can be targeted. In this chapter, we use pCI-Bet, which encodes the major birch pollen allergen Bet v 1.0101 for cytosolic expression, and its derivatives pCI-ubi-Bet (targeting the proteasome), pCI-tPA-Bet (facilitating secretion), and pCI-Bet-LIMP-II (promoting endosomal processing) (Fig. 1).

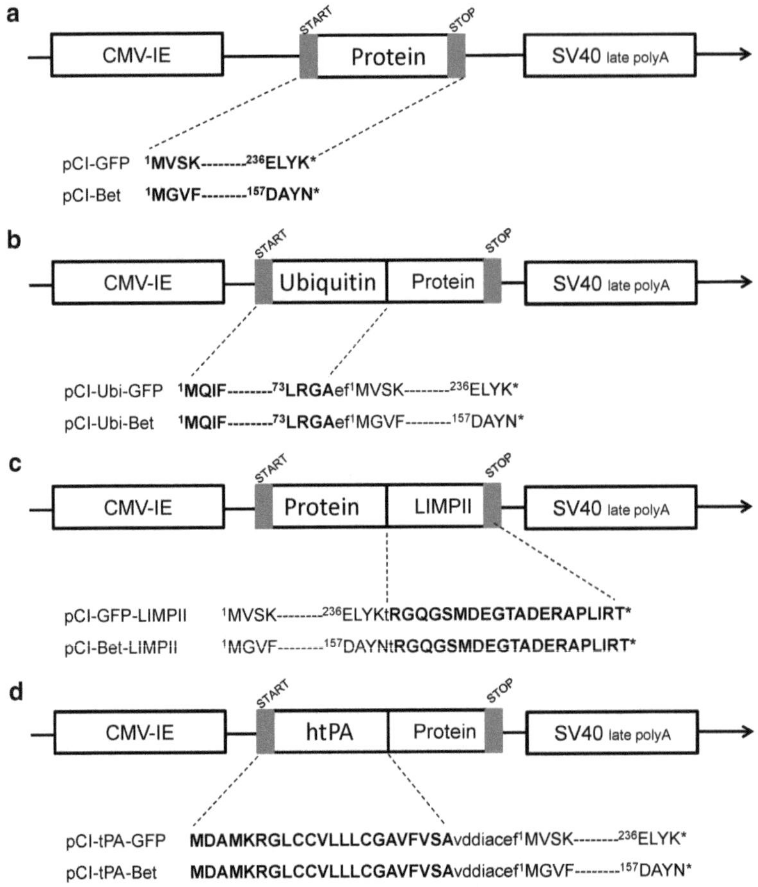

Fig. 1 Vector maps and AA sequences are shown for each modified construct. The superscripted numbers indicate the position in the full-length protein of the adjacent AA. Any additional AAs inserted as a result of the cloning process are shown in lowercase. The start and stop codons are shown as a *grey box*. The *asterisk* in the AA sequences represents a stop codon. The vector encoding the unmodified protein is shown in panel (**a**). Ubiquitin is 76AAs in length (**b**). LIMP-II comprises 20 AAs from the C terminus of LIMP-II full-length protein (**c**) and human tPA is 22 AAs in size (**d**). CMV-IE = Cytomegalovirus immediate–early enhancer/promoter region; SV40 late poly A = SV40 late polyadenylation signal

3.1.1 Preparation of Plasmid DNA

1. Transform XL-1 blue competent cells with plasmid DNA, and inoculate LA-Amp plates overnight at 37 °C.
2. Pick a single colony, and inoculate a starter culture of 2 mL of LB-Amp for 6–8 h at 37 °C, at 200 rpm.
3. Inoculate 1,000 mL LB-Amp with 2 mL starter culture and incubate for 16–20 h at 37 °C at 200 rpm.
4. Prepare plasmid DNA using a commercially available endotoxin-free plasmid DNA preparation kit.

3.2 LAL-Assay

Limulus amebocyte lysate (LAL) is very sensitive for detection and quantification of lipopolysaccharides (LPS), a cell membrane component of Gram-negative bacteria, also known as endotoxin. The assay is used to determine the LPS content in DNA plasmid samples after purification, because LPS is released during the purification process. Removal of endotoxins is recommended in order to avoid false interpretations of experimental results, as the LPS content can influence the immunogenicity of a plasmid. When incubating LAL with a sample, the turbidity of the lysate increases over time, proportionally to the LPS content. LPS content of samples is estimated by comparing the time required to reach half-maximal turbidity for different standard curves with that of the sample curve.

1. First dilute an endotoxin control standard with endotoxin-free H_2O to different concentrations (ranging from 250 to 3.90 pg per well) in duplicates.
2. Add 50 μL per well of each dilution to a 96-well V-bottom plate. Endotoxin-free water serves as negative control.
3. Apply 10 μg of purified DNA sample in 50 μL ddH_2O per well to the 96-well V-bottom plate.
4. Reconstitute Pyrotell-T with endofree water as indicated on the vial label. The lyophilized LAL pellet will go into solution within a few minutes (*see* **Note 2**).
5. Add 50 μL LAL to each sample, and measure the OD at 405 nm during a 60-min vibration-free incubation at 37 °C with a Tecan infinite M200Pro reader.

3.3 Mice and Experimental Setup

Female BALB/c mice are chosen for allergy models, because they show a strong TH2-biased immune response with increased levels of IL-4 and IL-5 and also high amounts of IgE. For protective approaches, animals are intradermally pre-vaccinated with DNA plasmids three times in weekly intervals, followed by subcutaneous sensitization with alum-adjuvanted recombinant allergen 2–4 weeks later. 1–2 weeks after the final sensitization, mice are challenged via the airways on three consecutive days. Sham-immunized mice that received pCI vector without an insert served as controls. In a therapeutic setting, mice are sensitized and challenged via the airways similar to the prophylactic approach. 2–4 weeks after the airway challenge, mice are treated by weekly injections of plasmid DNA for 3–6 weeks. 2–4 weeks after the final treatment, mice are re-challenged by another 3-day round of aerosol exposure to assess treatment efficacy compared to sham-treated controls.

For intradermal DNA immunization different plasmids are used, namely, pCI-Bet v 1.0101, pCI-tPA-Bet v 1.0101, pCI-Ubi-Bet v 1.0101, pCI-Bet v 1.0101-LIMP-II, and an empty pCI plasmid (sham immunization). In addition, some cohort-matched naïve animals serve as negative controls for immunological assays.

24 h after the last aerosol challenge, lung resistance and compliance are analyzed, BAL is performed, and splenocytes are pre-

pared and in vitro restimulated with and without rBet v 1.0101 to evaluate proliferation via CFSE staining. Cell-bound IgE in serum is detected by a basophil activation test (BAT).

3.3.1 Intradermal DNA Immunization

1. Prepare 100 μg plasmid DNA in a volume of 200 μL endotoxin-free PBS per mouse.
2. Shave mice at the back with an electric clipper.
3. For anesthesia use isoflurane with an appropriate rodent anesthesia machine.
4. Place the mice into a pre-filled anesthesia chamber until they sleep.
5. Keep mice under anesthesia during the treatment by placing the snout into an inhalation tube.
6. Use forceps to raise a small fold in the skin.
7. Inject the plasmid DNA very superficially into the dermis, and distribute the solution into 6–8 injection sites at the shaved back (*see* **Note 3**).

3.3.2 Subcutaneous Sensitization

1. Mix 5 μg of recombinant Bet v 1.0101 with 100 μL Alum and 20 μL endotoxin-free 10 × PBS, and fill up with endotoxin-free water to a total volume of 200 μL per mouse.
2. Shake the solution for some hours at room temperature to allow adjuvant–protein complex formation.
3. Inject the solution subcutaneously into two sites on the shaved back by lifting up the skin with the fingers or using forceps.
4. Administer a second sensitization after 2 weeks (*see* **Note 4**).

3.3.3 Airway Challenge

1–2 weeks after the final sensitization, mice undergo an airway challenge by means of evaporation of 1 mg/mL of recombinant Bet v 1.0101 allergen in endotoxin-free PBS three times in daily intervals. Airway challenge by means of evaporation is carried out using a PARI BOY® SX nebulizer with PARI LL nebulizer head (*see* **Note 5**).

1. Fill the nebulizer with 5 mL of 1 mg/mL rBet v 1.0101 protein in endotoxin-free PBS.
2. Place the mice into a chamber which is connected with the nebulizer head during the treatment.
3. Evaporation of 5 mL will take approximately 20–30 min.

3.4 Lung Resistance (R_L) and Dynamic Compliance (C_{dyn}) Measurement

1. Anesthetize the mouse with 140 μL/25 g body weight with an i.p. injection of ketamine and xylazine.
2. Keep mouse on a 37 °C heating pad during surgery.
3. As soon as the mouse is in deep anesthesia (*see* **Note 6**) you can expose the trachea.

4. Use two forceps to reach behind the trachea, and pull a suture of ~7 cm length under the trachea to form a loose ligature. Do not tighten the knot yet.
5. Make a small incision into the trachea, and insert the tracheal needle.
6. Fix the needle with the ligature.
7. Move the animal to the chamber, and insert the esophageal tube down to the stomach to measure the transpulmonary pressure (*see* **Note 7**).
8. Immediately connect the tracheal needle with the ventilator; otherwise, the mouse would suffocate.
9. Slowly pull out the tracheal tubing while watching the pressure signal. Place the tube where you see the maximal pressure deflection and minimize heart artifacts. Aim for 3–5 cm H_2O deflection.
10. If the flow and pressure signal are consistent, you can run the experiment (*see* **Note 8**). Naïve BALB/c mice usually show resistance values of 1.5–2.0 and compliance values of 35–50.
11. Record the baseline and add 10 μL of increasing concentrations of methacholine (5, 10, 20 mg/mL) to the nebulizer head following the instructions of the software.
12. After each run the nebulizer head should be rinsed with 0.9 % NaCl to avoid contamination.
13. Values for each dosage are given as percentage of baseline values.

3.5 Bronchoalveolar Lavage and Intracellular FoxP3 Staining

Bronchoalveolar lavages (BAL) are taken in order to analyze cellular lung infiltrates via FACS analysis (*see* **Note 9**).

1. Sacrifice mice after R/C measurement by cervical dislocation, and insert a flexible tubing into the already existing hole of the trachea to flush the animal's lungs twice with 1 mL ice-cold PBS.
2. Flush the lavages through a 40 μm cell strainer and centrifuge at 1,200 rpm (259 × *g*) for 10 min at 4 °C (*see* **Note 10**).
3. Resuspend the pellets in 100 μL FACS buffer, and transfer the samples to a 96-well V-bottom plate.
4. Prepare staining mix containing anti-mouse CD45-PE/Cy7, anti-mouse Ly-6G (Gr1)-APC, anti-mouse CD4-APC-eFluor 780, anti-mouse CD8-FITC, and anti-mouse Siglec-F-PE, each diluted 1:100 in FACS buffer.
5. Centrifuge plate at 1,200 rpm (259 × *g*) for 5 min at 4 °C.
6. Resuspend the cell pellets in 20 μL of staining mix and incubate for 15 min at 4 °C in the dark.

7. Add 150 μL FACS buffer and centrifuge at 1,200 rpm (259 × *g*) for 5 min at 4 °C.
8. Resuspend pellets in 150 μL cold 1 × fix/perm buffer and incubate for 30 min at 4 °C. Centrifuge samples at 1,600 rpm (460 × *g*) for 5 min at 4 °C and wash twice with 150 μL perm buffer.
9. Resuspend pellets in 20 μL FoxP3-PerCp/Cy5.5 diluted 1:100 in perm buffer and incubate for 30 min in the dark at 4 °C.
10. Add 150 μL perm buffer and centrifuge at 1,600 rpm (460 × *g*) for 5 min at 4 °C.
11. Resuspend the pellets in 150 μL FACS buffer, and analyze a defined volume (e.g., 100 μL) on a flow cytometer to calculate absolute cell numbers per BAL.
12. Eosinophils are distinguished from other leukocyte populations by their $CD45^{med}Ly\text{-}6G^{low}$ Siglec-F^{high} phenotype.

3.6 Basophil Activation Test

Cell-bound IgE has a long half-life (weeks to months) compared to the short half-life of free IgE (~6 h). Therefore, the BAT assay which detects cell-bound IgE provides additional information to assays which detect free IgE in serum.

Similar to mast cells, basophils express the high-affinity receptor (FcεRI) against the constant region of IgE on their surface and release mediators from granules upon cross-linking of FcεRI. The BAT measures IgE-dependent basophil activation in whole-blood samples by flow cytometry.

Basophils are identified as $IgE^{+}CD19^{-}CD4^{-}$ PBMCs. In humans, CD69 or CD203c are used as activation markers. In mice, CD200R is used which is expressed at low levels in resting basophils. The cross-linking of FcεRI-bound IgE (induced by an allergen or an anti-IgE antibodies) induces up-regulation of CD200R. The increase in CD200R expression after allergen challenge therefore reflects the basophil activation in response to an allergen (*see* Fig. 2).

1. For restimulation of the basophils prepare 50 μL RPMI with 40 μg/mL recombinant Bet v 1.0101, as negative control 50 μL RPMI, and as positive control 50 μL RPMI with 4 μg/mL anti-mouse IgE per well (*see* **Note 11**).
2. Add 50 μL fresh heparinized blood (0.15 mg/mL heparin) to each well and resuspend by pipetting up and down. Incubate the plate for 2 h at 37 °C, 95 % relative humidity, and 7 % CO_2.
3. After incubation centrifuge the samples for 5 min at 1,200 rpm (259 × *g*) at 4 °C.
4. Tap out the plate, and resuspend the cells in 30 μL of ice-cold staining mix consisting of anti-mouse IgE-FITC, anti-mouse CD19-PE/

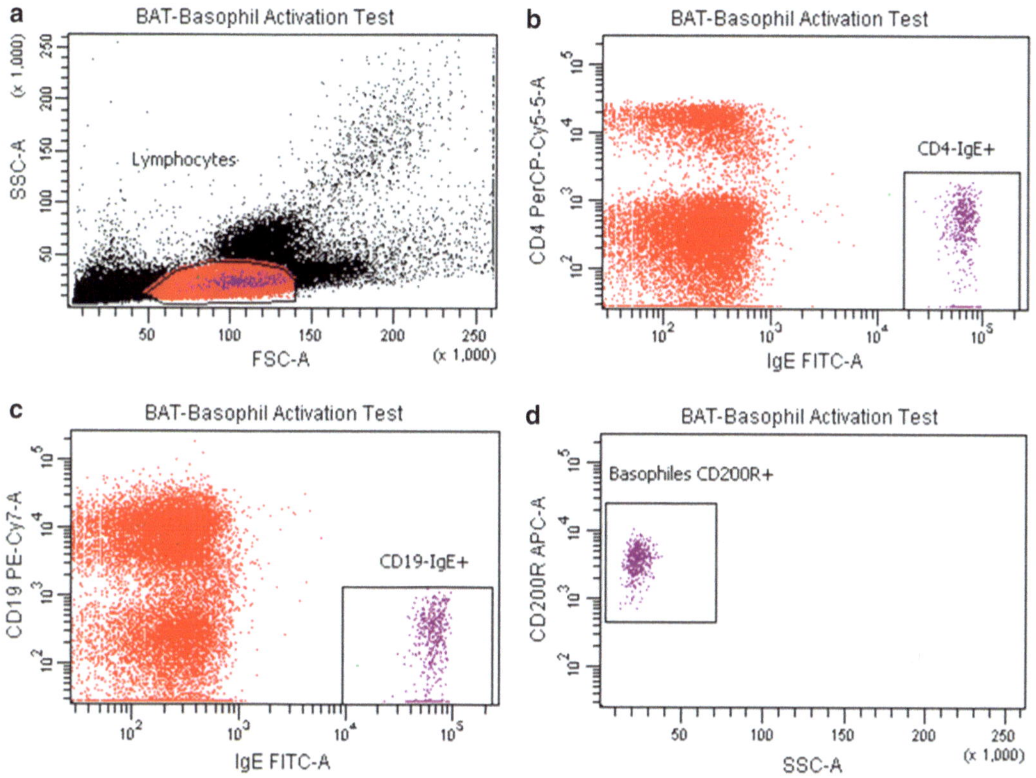

Fig. 2 Example plots of a BAT: (**a**) Gating of lymphocytes. (**b**) Exclusion of CD4+ T cells. (**c**) Exclusion of CD19+ B cells. (**d**) Analysis of CD200R on basophils

Cy7, anti-mouse CD200R-APC, and anti-mouse CD4-PerCp/Cy5.5 (all antibodies diluted 1:200 in FACS buffer).

5. Incubate samples for 10 min on ice in the dark.
6. Add 100 μL ice-cold FACS buffer for washing.
7. Centrifuge the samples for 5 min at 1,200 rpm (259 × *g*) at room temperature.
8. Tap out the plate.
9. Next resuspend the cells in 100 μL 1 × BD FACS lysing solution to get rid of erythrocytes.
10. Incubate for 5 min at room temperature.
11. Centrifuge for 5 min at 1,500 rpm (404 × *g*) at RT.
12. Tap out the plate, and resuspend the pellet in 100 μL FACS buffer.
13. Centrifuge for 5 min at 1,500 rpm (404 × *g*) at RT (*see* **Note 12**).
14. Resuspend the sample in 100 μL FACS buffer, and transfer the sample into a FACS tube.
15. Analyze on a flow cytometer.

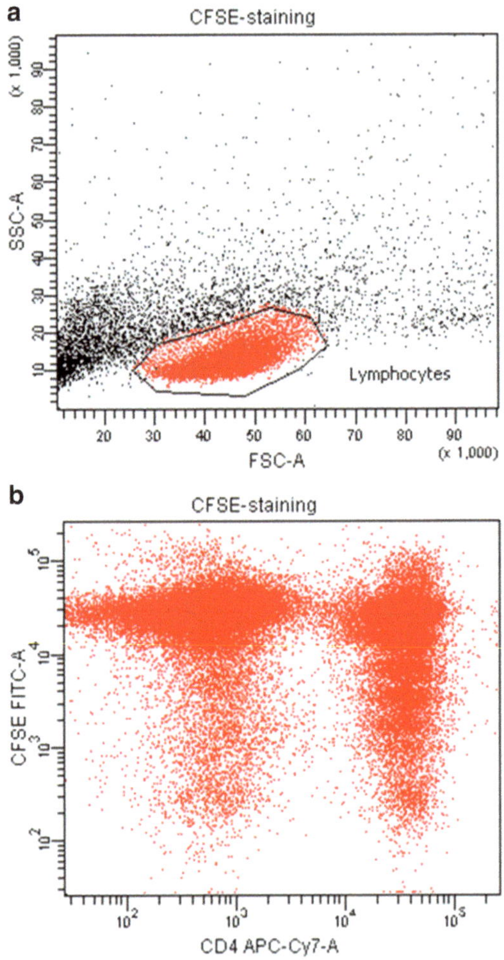

Fig. 3 Example plots of a CFSE proliferation assay: (**a**) Gating of live lymphocytes. (**b**) Proliferation of CD4 T cells

16. Gate cells based on forward and side scatter (FSC/SSC) properties of lymphocytes (*see* Fig. 2a).
17. Gate on CD19⁻IgE⁺ cells in PE/Cy7 vs. FITC plot (*see* Fig. 2b).
18. Gate on CD4⁻IgE⁺ cells in PerCp/Cy5.5 vs. FITC plot (*see* Fig. 2c).
19. Plot CD200R APC vs. SSC of the gated basophils (*see* Fig. 2d).
20. Open the statistic window, and display the mean fluorescence intensity (MFI).

3.7 CFSE Proliferation Assay and Intracellular FoxP3 Staining

At the end of the experiment, splenocytes are prepared. Next the cells are labeled with the fluorescent dye CFSE. Those cells that proliferate in response to antigen show a reduction in CFSE intensity (*see* Fig. 3). Cell division results in sequential halving of fluorescence signal. Therefore, it can be used to detect cell proliferation which is determined by flow cytometric analysis. Furthermore, transcription factor FoxP3 can be stained intracellularly, which is expressed by regulatory T cells which have a suppressive function and maintain homeostasis of the immune system.

1. Remove the spleens, and mince them in 500 μL culture medium using the back end of a plunger of a 2 mL syringe.
2. Let aggregates sediment for 5 min in a 1.5 mL Eppendorf tube.
3. Take 400 μL of the monodisperse supernatants and add to 7 mL ACK lysis buffer.
4. After 7-min incubation at room temperature, add 7 mL medium and wash the cells two times by centrifugation at 300 × *g*.
5. Resuspend the cells in 2 mL culture medium, and count the cells.
6. Take 2×10^7 splenocytes and wash twice with 15 mL PBS and resuspend in 1 mL pre-warmed PBS (*see* **Note 13**).
7. For 1 mL cells add 1 μL CFSE (5 mM) (*see* **Note 14**).
8. Incubate for 15 min at 37 °C in a water bath.
9. During incubation time mix the samples from time to time to make sure that the CFSE is evenly distributed.
10. After incubation fill up to 15 mL with PBS/5 % FCS and centrifuge at 1,200 rpm (259 × *g*) for 10 min.
11. Remove supernatant, and resuspend the pellet in 1 mL PBS/5 % FCS and incubate for 30 min at 37 °C on a shaker at 180–200 rpm.
12. Wash cells with 15 mL PBS/5 % FCS and centrifuge at 1,200 rpm (259 × *g*) for 10 min (*see* **Note 15**).
13. Resuspend cells in 2 mL culture medium, transfer 1 mL cells per well to 24-well flat-bottom tissue culture plates, and restimulate with 1 mL of 40 μg/mL protein solution or without protein as control (*see* **Note 16**).
14. Analyze cells 4–7 days later by flow cytometer.
15. Remove supernatant, and harvest cells in 2 mL FACS buffer in a 15 mL tube.
16. Add 12 mL PBS to the cells.
17. Centrifuge samples for 7 min at 1,200 rpm (259 × *g*).
18. Aspirate the supernatant, resuspend the cells in 100 μL FACS buffer, and transfer to 96-well V-bottom plate for staining.
19. Centrifuge the plate for 5 min at 1,200 rpm (259 × *g*).
20. Add 20 μL surface staining mix consisting of anti-mouse CD4-APC-eFluor 780 and anti-mouse CD25-PE/Cy7 all diluted 1:100 in FACS buffer.

21. Incubate for 15 min at 4 °C in the dark.
22. Wash each well with 150 μL FACS buffer and centrifuge at 1,200 rpm (259 × *g*) for 5 min.
23. Add 150 μL fix/perm per well and incubate for 30 min at 4 °C in the dark.
24. Centrifuge the plate for 10 min at 1,600 rpm (460 × *g*) followed by two washing steps with 150 μL perm buffer.
25. Add 20 μL anti-mouse FoxP3-PerCp/Cy5.5 diluted 1:100 in perm buffer per well and incubate for 30 min at 4 °C in the dark.
26. Wash the wells with 150 μL perm buffer per well and centrifuge at 1,600 rpm (460 × *g*) for 5 min.
27. Resuspend the pellets in 150 μL FACS buffer, transfer into FACS tubes, and analyze on a flow cytometer.

4 Notes

1. LAMP-I can be used as an alternative to LIMP-II, because it shows a clear localization in the lysosomes which we could not demonstrate with a Bet-LIMP-II construct.
2. Before use, gently mix the contents to ensure homogeneity by inverting the tube ten times. Mixing too vigorously may cause a loss of sensitivity.
3. After proper injection blisters will be formed and persist for some hours. If they are missing, a subcutaneous injection has been given.
4. Depending on the allergen used, a third sensitization applied 1 week after the second might be necessary to achieve high-serum IgE titers.
5. Alternatively, isoflurane-anesthetized mice can be challenged intranasally by pipetting 5 μg recombinant allergen in 40 μL PBS onto both nostrils.
6. In order to test if the mouse is in deep anesthesia, you can perform a paw pinch test. If no reflex is present, you can proceed.
7. Make sure that there are no air bubbles in the esophageal tube. Place the end of the tube into the mouth of the mouse. Open the three-way valve, and flush the tube with a small amount of liquid by pressing the plunger of the syringe on the pressure transducer to make sure that no air gets into the tube. Pull out the tongue of the mouse with a pair of curved forceps, and pull the mouse via the tongue towards the tubing so that the tube will slip into the esophagus.

8. Sometimes the mice try to breath by themselves despite the anesthesia. To avoid this situation you can inject again 100 μL ketamine (20 mg/mL in 0.9 % NaCl) without xylazine.
9. BAL fluid is easily accessible; however, for analyzing tissue-resident cell types, lungs can also be digested and analyzed by FACS.
10. After the centrifugation the supernatants can be carefully taken off and stored at −20 °C for later cytokine analysis of the BAL fluid.
11. With a BAT you can also show the presence of blocking IgG. For this purpose add 50 μL fresh heparinized blood and 50 μL heparinized RPMI in a V-bottom plate. Then centrifuge the plate for 5 min at 1,200 rpm (259 × *g*), and take off carefully 50 μL serum. Again add 50 μL heparinized RPMI medium and resuspend. Repeat this step further two times. Afterwards you can proceed with the protocol at **step 1**.
12. If necessary the lysis step can be repeated, but it has to be considered that each washing step also results in a loss of cells.
13. It is important to get rid of the FCS in the medium as proteins would inhibit CFSE labelling.
14. The optimum concentration of CFSE needs to be determined. If the concentration is too high the cells will not divide because of toxicity.
15. Do not expose the labelled cells to light for extended periods of time. Otherwise you will lose staining intensity.
16. Depending on the immunogenicity of the allergen concentrations can vary. The optimal concentration has to be determined for each allergen by titration.

Acknowledgments

This work was supported by the Austrian Science Fund, grant # W1213.

References

1. Weiss R, Durnberger J, Mostbock S et al (1999) Improvement of the immune response against plasmid DNA encoding OspC of Borrelia by an ER-targeting leader sequence. Vaccine 18:815–824
2. Haddad D, Liljeqvist S, Stahl S et al (1998) Differential induction of immunoglobulin G subclasses by immunization with DNA vectors containing or lacking a signal sequence. Immunol Lett 61:201–204
3. Bauer R, Scheiblhofer S, Kern K et al (2006) Generation of hypoallergenic DNA vaccines by forced ubiquitination: preventive and therapeutic effects in a mouse model of allergy. J Allergy Clin Immunol 118:269–276
4. Rodriguez F, An LL, Harkins S et al (1998) DNA immunization with minigenes: low frequency of memory cytotoxic T lymphocytes and inefficient antiviral protection are rectified by ubiquitination. J Virol 72:5174–5181

5. Ogata S, Fukuda M (1994) Lysosomal targeting of Limp II membrane glycoprotein requires a novel Leu-Ile motif at a particular position in its cytoplasmic tail. J Biol Chem 269:5210–5217
6. Sandoval IV, Arredondo JJ, Alcalde J et al (1994) The residues Leu(Ile)475-Ile(Leu, Val, Ala)476, contained in the extended carboxyl cytoplasmic tail, are critical for targeting of the resident lysosomal membrane protein LIMP II to lysosomes. J Biol Chem 269:6622–6631
7. Rodriguez F, Harkins S, Redwine JM et al (2001) CD4(+) T cells induced by a DNA vaccine: immunological consequences of epitope-specific lysosomal targeting. J Virol 75:10421–10430
8. Weiss R, Scheiblhofer S, Thalhamer J (2006) DNA vaccines for allergy treatment. In: Saltzman WM, Shen H, Brandsma JL (eds) Methods in molecular medicine. DNA vaccines methods and protocols, 2nd edn. Humana Press, New Jersey, USA
9. Weinberger, E. E., Isakovic, A., Scheiblhofer, S., et al. (2013) The influence of antigen targeting to sub-cellular compartments on the anti-allergic potential of a DNA vaccine. *Vaccine*
10. Boshart M, Weber F, Jahn G et al (1985) A very strong enhancer is located upstream of an immediate early gene of human cytomegalovirus. Cell 41:521–530
11. Carswell S, Alwine JC (1989) Efficiency of utilization of the simian virus 40 late polyadenylation site: effects of upstream sequences. Mol Cell Biol 9:4248–4258
12. Chou B, Hiromatsu K, Okano S et al (2012) Antiangiogenic tumor therapy by DNA vaccine inducing aquaporin-1-specific CTL based on ubiquitin-proteasome system in mice. J Immunol 189:1618–1626
13. Marques ET Jr, Chikhlikar P, de Arruda LB et al (2003) HIV-1 p55Gag encoded in the lysosome-associated membrane protein-1 as a DNA plasmid vaccine chimera is highly expressed, traffics to the major histocompatibility class II compartment, and elicits enhanced immune responses. J Biol Chem 278: 37926–37936

Chapter 16

Immunotherapy for Alzheimer's Disease: DNA- and Protein-Based Epitope Vaccines

Hayk Davtyan, Irina Petrushina, and Anahit Ghochikyan

Abstract

Active immunotherapy for Alzheimer's disease (AD) is aimed to induce antibodies specific to amyloid-beta (Aβ) that are capable to reduce the level of Aβ in the CNS of Alzheimer's disease patients. First clinical trial AN-1792 that was based on vaccination with full-length $A\beta_{42}$ showed that safe and effective AD vaccine should induce high titers of anti-Aβ antibodies without activation of harmful autoreactive T cells. Replacement of self-T cell epitope with foreign epitope, keeping self-B cell epitope intact, may allow to induce high titers of anti-Aβ antibodies while avoiding the activation of T cells specific to Aβ.

Here we describe the protocols for evaluation of AD DNA- or multiple antigenic peptide (MAP)-based epitope vaccines composed of $A\beta_{1-11}$ B cell epitope fused to synthetic T cell epitope PADRE ($A\beta_{1-11}$-PADRE). All protocols could be used for testing any epitope vaccine constructed in your lab and composed of other T cell epitopes using the appropriate peptides in tests for evaluation of humoral and cellular immune responses.

Key words Alzheimer's disease, Epitope vaccine, Antibody, ELISA, ELISPOT, Brain pathology, Immunohistochemistry

1 Introduction

Alzheimer's disease (AD) is the most common form of dementia in the elderly and is characterized clinically by an insidious onset and progressive cognitive decline that impacts memory, language, judgment, and orientation to time and space. The neuropathological features of the disease include deposition of amyloid-β (Aβ) in senile plaques, neurofibrillary tangles (NFT), and neuronal loss in affected brain regions [1, 2]. These pathological changes result in a profound loss of synapses over the course of the disease, thereby contributing to a progressive reduction in the functional capacity of the patient. Accumulated preclinical studies in mice transgenic for the human amyloid precursor protein (APP) have demonstrated that antibody-mediated

Monica Rinaldi et al. (eds.), *DNA Vaccines: Methods and Protocols*, Methods in Molecular Biology, vol. 1143,
DOI 10.1007/978-1-4939-0410-5_16, © Springer Science+Business Media New York 2014

reduction of Aβ deposits may prevent the cognitive impairment/improve cognitive function of these mice [3–8]. Based on these data first in-human trial (AN-1792) was conducted by Elan and Wyeth-Ayerst using as an antigen fibrillar $A\beta_{42}$ peptide formulated in QS21 adjuvant. However AN-1792 immunotherapy vaccine trial was halted when a subset of immunized individuals (6 %) developed adverse events in the central nervous system in the form of meningoencephalitis [9]. Although the exact cause of that is still unknown, postmortem examination of the brains from vaccinated AD patients with neuroinflammation led to the assumptions centered on autoreactive T cell responses to self-epitopes within the $A\beta_{42}$ peptide; the QS-21 adjuvant; and/or reformulation of the vaccine (by the inclusion of polysorbate 80) prior to the phase IIa trial [10–13]. Importantly, the same case reports indicated that anti-Aβ antibodies were not responsible for the observed adverse effects after active vaccination and that the AN-1792 vaccine was beneficial for AD patients and significantly reduced amyloid-β-plaque load [14].

Another pitfall connected with the vaccination with full-length $A\beta_{42}$ peptide was the induction of low/moderate anti-Aβ antibody titers in a small subset of immunized patients (19.7 %) likely due to induction of immune tolerance to self-antigen [15, 16].

In summary, the results of the first vaccination of elderly AD patients with the AN-1792 vaccine raised concerns about the safety and the efficacy of the active immunization strategy with $A\beta_{42}$ self-peptide, suggesting that alternative immunotherapeutic strategies should be pursued. One solution to avoid the activation of autoreactive T cells and to overcome tolerance of immune system to self-antigen is using an epitope vaccine composed of the self-B cell antigenic determinant of $A\beta_{42}$ peptide and nonself-foreign T cell antigenic determinant. Numerous studies on mapping of B and T cell epitopes in $A\beta_{42}$ peptides revealed that dominant B cell epitope is localized at the N-terminal region of the peptide (major antigenic determinant is DAEFRH), whereas T cell epitopes that are recognized by several mouse MHC and HLA molecules analyzed up to date are localized at the C-terminal region of the peptide [17–29]. This gives us an opportunity to develop epitope vaccine based on self-B cell epitope fused with foreign T cell epitope.

As an example for generation of AD epitope vaccine, in this chapter we describe a vaccine that was generated in our lab and was successfully tested in different mouse strains and rabbits [30–34]. Our AD epitope vaccine is composed of $A\beta_{1-11}$ B cell epitope fused with promiscuous synthetic pan HLA-DR epitope, PADRE that is active at least in mice with $H\text{-}2^b$ and $H\text{-}2^d$ haplotypes and is recognized by 14 of 15 human DR molecules [35]. A vaccine could be designed as a linear peptide, multiple antigenic peptide, or DNA. Alternatively, smaller region from

$A\beta_{42}$ peptide containing B cell epitope (e.g., $A\beta_{1-6}$, $A\beta_{1-7}$, $A\beta_{1-8}$) as well as any other known foreign T cell epitope/s could be used for development of epitope vaccine for AD. Inclusion of several copies of B cell epitope may increase an efficacy of a vaccine [36].

Each type of vaccine has advantages and disadvantages. Peptide/protein-based vaccines require formulation in strong adjuvant for induction of potent immune responses. Only few adjuvants are approved for human use.

DNA-based vaccination provides a unique method of immunization when the antigen is synthesized and processed within the subject's own cells inducing persistent immune responses.

Easy manipulations with DNA vaccine allow rational inactivation or removal of sequences encoding potentially toxic protein domains as well as inclusion of molecular adjuvants, such as cytokines, which can enhance immune responses and direct T helper cell responses toward the desired pathway. Among many significant advantages of DNA immunization are high stability and less complicated and cost-effective manufacturing of DNA vaccines.

However, injection of DNA vaccine with conventional syringe is associated with poor efficiency of intracellular DNA uptake and antigen expression that leads to the induction of low levels of immune responses. To improve transfection efficiency of DNA vaccines investigators currently use various DNA delivery systems such as gene gun and electroporation devices [36–39].

Here we present the process of immunization of wild-type C57Bl/6 and Tg2576 mice with a DNA-based AD epitope vaccine using gene gun and a peptide-based AD epitope vaccine injected subcutaneously (s.c.) and testing of the immunogenicity and therapeutic efficacy of the vaccines. Our DNA-based vaccine is a minigene encoding mouse Ig k chain signal sequence (METDTLLLWVLLLWVPGSTG) fused to three copies of $A\beta_{11}$ B cell epitope (DAEFRHDSGYE) and universal T helper (Th) epitope PADRE (AKFVAAWTLKAAA) cloned into the mammalian expression vector pVAX1 (p3$A\beta_{11}$-PADRE). Peptide-based vaccine is composed of two copies of $A\beta_{11}$ fused with PADRE (aK-Cha-VAAWTLKAAa, where a is D alanine, and Cha is L-cyclohexylalanine) and synthesized as multiple antigenic peptide (2$A\beta_{11}$-PADRE-MAP).

If you design the new vaccine, we recommend testing it first in wild-type mice. If vaccine is capable of inducing sufficient titers of anti-$A\beta$ antibodies in wild-type mice, you may test its therapeutic potency in APP transgenic mice. Note that often vaccines induce high titers of antibodies in wild-type mice, but fail to induce the same titers in Tg mice. Be aware to use wild-type mice of the same haplotype (at least partially) as APP Tg mice to be sure that T cell epitope will be active in both strains of mice.

2 Materials

2.1 Preparation of DNA-Based Vaccine and Gene Gun Inoculation of DNA in Mice

1. Purified plasmid DNA encoding three copies of $A\beta_{11}$ B cell epitope fused to PADRE Th epitope (p3$A\beta_{11}$-PADRE) (3 mg/ml in water).
2. Gold microcarriers (1.0 μm diameter; Bio-Rad).
3. 0.05 M spermidine.
4. 1 M calcium chloride ($CaCl_2$).
5. Ethanol (100 %; fresh).
6. Sonicating water bath.
7. 15 ml polypropylene tubes.
8. Compressed nitrogen.
9. Tubing Prep Station (Bio-Rad).
10. Teflon tubing (Tefzel tubing, 1/8 diameter; Bio-Rad).
11. Silicone adapter tubing.
12. 10 ml syringes.
13. Razor blade.
14. 50 ml tubes.
15. Desiccant packs.
16. Parafilm.
17. Helios gene gun.
18. Compressed helium.
19. Electric clippers.
20. C57BL/6 mice, 6–8 weeks old ($n=8$ per group) (The Jackson Laboratory).
21. Anesthetic (optional).

2.2 Preparation of Protein-Based Vaccine and Injection of Mice

1. Peptide; 2$A\beta_{1-11}$-PADRE-MAP4 (Invitrogen).
2. Adjuvant; Quil-A (Brenntag, Denmark).
3. Phosphate-buffered saline (PBS) pH 7.4, 10×.
4. Dimethyl sulfoxide (DMSO).
5. C57BL/6 mice, 6–8 weeks old ($n=8$ per group) (The Jackson Laboratory).
6. Syringe; capacity 1 ml (Kendall monoject).
7. Needle; 25 G×5/8″ (Kendall monoject magellan).
8. Vortex; Fisher Vortex Genie 2 (Fisher Scientific).
9. Rotator; Nutator (Beckman Dickenson).

2.3 Blood Collection

1. Eppendorf tubes; 1.7 ml (Eppendorf).
2. Pasteur pipette; borosilicate glass, disposable, 9″ (VWR).
3. Centrifuge; Spectrafuge 24D (Labnet International, Inc).

2.4 Detection of Anti-Aβ Antibodies by ELISA

1. Peptide; $A\beta_{42}$ (American Peptide).
2. 6E10; monoclonal antibody (Covance).
3. Horseradish peroxidase (HRP)-conjugated anti-mouse IgG; secondary antibody (Jackson Immunoresearch Laboratories).
4. 3,3′,5,5′ tetramethylbenzidine (TMB); substrate (Pierce).
5. Stop solution; 2 N sulfuric acid (H_2SO_4).
6. 96-Well microplate; Immunolux HB, flat bottom (DYNEX Technologies).
7. SealPLate Film (MidSci).
8. Coating buffer; Carbonate–Bicarbonate Buffer; pH 9.6 (Sigma-Aldrich).
9. Washing buffer; Tris-buffered saline containing 0.5 % Tween-20 (TBST); pH 7.4.
10. Blocking buffer; 3 % dry, nonfat milk in TBST.
11. Dilution buffer; 0.3 % dry, nonfat milk in TBST.
12. Rotator; Nutator (Beckman Dickenson).
13. Incubator-shaker; Lab-line (Orbit).
14. Plate Washer; Auto Strip Washer ELx50 (BioTek).
15. Absorbance microplate reader (BioTek).

2.5 Detection of Cellular Immune Responses

1. PBS pH 7.4, 10×.
2. R10 media; RPMI-1640 supplemented with 10 % heat-inactivated fetal bovine serum (HI FBS), 2 mM glutamine, 1 mM sodium pyruvate, MEM nonessential amino acids (MEM NEAA), 100 U/ml penicillin, and 100 mg/ml streptomycin.
3. Lyses buffer; red blood cell (RBC) lysing buffer.
4. Alcohol.
5. Trypan blue.
6. Tubes, 15 and 50 ml.
7. Sterile scissors (Roboz).
8. Cell lifter (Costar).
9. Culture plates; sterile (MidSci).
10. Cell counter; Countless Automated cell counter (Invitrogen).
11. Mouse cytokine ELISPOT set; containing capture antibody (purified), detection antibody, streptavidin–HRP, ELISPOT plates (BD Bioscience).

12. PBS containing 0.5 % Tween-20 (PBST), pH 7.4.
13. Antigens; PADRE, $A\beta_{40}$, P30 (Genscript).
14. Pokeweed mitogen (Sigma-Aldrich).
15. AEC substrate solution (Sigma-Aldrich).
16. 5 mM 7-aminoactinomycin D (CFSE) stock (Invitrogen).
17. CTL-ImmunoSpot S5 Macro Analyzer (Cellular Technology Ltd).
18. 14 ml Polystyrene Round-Bottom Tube; BD Falcon (BD Bioscience).

2.6 Brain Collection

1. Nembutal (Akorn, IL) (or any other anesthetic approved by Institutional Animal Care and Use Committee (IACUC) in your institute).
2. PBS pH 7.4, 10×.
3. Sodium azide solution, 0.02 % (w/v, in PBS).
4. 26 G Needle and syringe.
5. Surgical scissors.
6. Pointed forceps.
7. Hemostat.
8. 4 % Paraformaldehyde in PBS (w/v) (see recipe).
9. Perfusion pump.

2.7 Preparation of Brains for Slicing

1. PBS pH 7.4, 10×.
2. Sodium azide solution, 0.02 % (w/v, in PBS).
3. Brain matrix (available from several suppliers sized for either rat or mouse brains).
4. Single straight-edge blade.
5. Mounting block.
6. Cyanoacrylate glue.
7. 35-mm Tissue culture dishes.
8. 50 ml tubes.

2.8 Slicing of the Tissue and Preparation of Slides

1. PBS pH 7.4, 10×.
2. Sodium azide solution, 0.02 % (w/v, in PBS).
3. Vibratome.
4. Vibratome blades.
5. Paintbrush.
6. 96-Well plates.
7. Superfrost Plus Positively Charged Microscope Slides (Shandon).

2.9 Immunohistochemistry

1. Tris: 0.1 M Tris, 0.85 % saline, pH 7.4–7.6.
2. Tris-A: 0.1 M Tris, 0.85 % saline, pH 7.4–7.6, 0.1 % Triton X-100 (1 g Triton X-100 to 1 l Tris solution).
3. Tris-B: 0.1 M Tris, 0.85 % saline, pH 7.4–7.6, 0.1 % Triton X-100, 2.0 % (20 g bovine serum albumin to 1 l Tris-A). It is best to filter, aliquot, and keep at 4 °C.
4. H_2O_2 solution: 6 parts of Tris buffer, 1 part of methanol, 1 part of H_2O_2 (30 % w/w).
5. ABC (Vectastain Elite PK-6100, from Vector Laboratories).
6. 10 ml Tris-B, 2 drops of A, mix, add 2 drops of B, mix, and let sit for 30 min–1 h.
7. DAB Peroxidase Substrate kit, SK-4100 (Vector Laboratories).
8. PBS, pH 7.4, 10×.
9. Caliber TBS with Tween (TBST), pH 7.4, 20× solution (Hoefer, Inc.).
10. Any type of glass or plastic slide incubation containers.
11. CoverWell incubation chambers for 200 μl volume (Grace Bio-Labs).
12. Rocking platform.
13. Ethanol.
14. Xylene or histoclear I and II.
15. DePex (Curr, BDH Laboratories) or different resinous mounting medium.
16. Microscope Cover Glass (Fisherbrand), 24x50-1.
17. Timer.

2.10 Histostaining

1. Thioflavin S (Sigma-Aldrich).
2. Ethanol (50 %).
3. Tris: 0.1 M Tris, 0.85 % saline, pH 7.4–7.6.
4. Vectashield coverslips (Vector Labs).
5. Microscope Cover Glass (Fisherbrand), 24x50-1.
6. Fluorescent microscope.

2.11 Prussian Blue Staining

1. 5 % Aqueous solution of hydrochloric acid: 25 ml concentrated hydrochloric acid in 475 ml distilled water.
2. 5 % Aqueous solution of potassium ferrocyanide: 25 g Potassium ferrocyanide, Trihydrate (Sigma) dissolved in 500 ml distilled water.
3. Working solution: Mix equal parts of 5 % hydrochloric acid and 5 % potassium ferrocyanide solution just before use.
4. Nuclear fast red solution.

5. Xylene.
6. Ethanol.

2.12 Biochemical Analysis

1. 50 mM Tris–HCl buffer containing 2 % SDS (pH 8.0).
2. 1.0 M Tris base/0.5 M NaH_2PO_4.
3. Cocktail of protease inhibitors (MP Biomedicals, OH).
4. 70 % Formic acid.
5. β-Amyloid ELISA kits (BioSource, CA).
6. Sonicator.
7. Mortar and pestle.
8. Liquid nitrogen.
9. Ultracentrifuge (with 100,000 ×*g* rotor).
10. ELISA kit (Biosource).
11. BCA assay (Pierce).
12. PVDF membrane (GE Healthcare, NJ).

3 Methods

3.1 Preparation of DNA-Based Vaccine and Injection of Mice by Gene Gun

3.1.1 Preparation of DNA-Coated Gold Beads

1. Purify plasmid DNA (*see* **Note 1**).
2. Adjust concentration of DNA to 3 mg/ml in deionized water.
3. Weigh 45 mg of 1 μm gold beads into 1.5 ml microcentrifuge tubes. This amount of gold will be enough for preparation of about 75–80 cartridges.
4. Add 100 μl of 0.05 M spermidine to the gold beads, and immediately sonicate the tube for 3–5 s.
5. Add 90 μl DNA (at concentration 3 mg/ml) to the gold/spermidine mixture and vortex briefly (*see* **Note 2**).
6. Add 100 μl of 1 M $CaCl_2$ to the tube dropwise while gently vortexing at a medium setting.
7. Incubate the mixture for about 5 min at room temperature that will allow the DNA to precipitate onto the gold beads.
8. Microcentrifuge the tube for 20 s at 4,500 ×*g* at room temperature, and then discard supernatant.
9. Wash pellet by adding 0.5 ml of fresh 100 % ethanol (from new opened bottle), resuspend pellet by vortexing, and microcentrifuge tube for 20 s at 4,500 ×*g* at room temperature.
10. Repeat washing step two more times.
11. Place the 5.3 ml 100 % ethanol in a 15 ml tube.
12. Use 300–500 μl of ethanol from 15 ml tube for resuspending DNA/gold pellet in the 1.5 ml tube.

13. Vortex briefly and then place tube in a sonicating water bath for 5 s.
14. Transfer the DNA/gold suspension to the 15 ml tube. Repeat as necessary to remove all of the gold from the 1.5 ml tube.
15. Use DNA/gold/ethanol solution immediately for preparation of cartridge or it can be stored for several months at −20 °C in the 15 ml tube sealed with Parafilm.

3.1.2 Preparation of Cartridges (Beads) Containing DNA-Coated Gold Beads

1. Attach the tubing prep station to the nitrogen gas regulator that was attached to the nitrogen tank.
2. Purge empty Teflon tubing with nitrogen gas for 15–20 min for removing humidity in the tubing.
3. Turn on the nitrogen gas to the pressure of 1–2 psi.
4. Allow gas to flow through the empty Teflon tube at a rate of 0.5 l/min.
5. Prepare DNA/gold/ethanol bead by vortexing and then sonicating for 5–10 s.
6. Cut a piece of dried tubing (just a bit longer than the length of the tube tuner).
7. Connect the Teflon tubing to the 10 ml syringe via the silicone adapter tubing.
8. Vortex the DNA/gold suspension again, and then quickly use the syringe to draw 2.7–3 ml of the suspension into Teflon tubing.
9. Be sure to have 2–3 in. gap at either end that contains no solution.
10. Do not remove the syringe.
11. Slide the loaded Teflon tube into the turner.
12. Allow the DNA/gold suspension to settle for 4 min 15 s (*see* **Note 3**).
13. Draw off the ethanol very slowly using syringe.
14. When all of the ethanol has been drawn off detach the syringe and begin rotating tube in the tube turner.
15. Rotate tube for 30 s. During this time gold should be smeared evenly around the inside of the tubing.
16. Allow the gas to flow inside the rotating tube at a rate of 0.35–0.4 l/min.
17. Allow the tube to dry completely for about 5 min.
18. Once the tube is dry, stop rotating the tube and turn off the nitrogen gas.
19. Remove the tube from the turner, and determine the section of the tube where the gold is spread evenly.

20. Using razor blade cut the tube into 0.5 in. sections.
21. Place the tubes (beads) into labeled 50 ml tubes with desiccant pack.
22. Seal the cap with Parafilm and store at 4 °C for 1 year.

3.1.3 Injection of Mice by Gene Gun

1. Attach gene gun to helium tank according to the manufacturer's instruction.
2. Anesthetize mice with Nembutal (*see* **Notes 4** and **5**).
3. Shave the abdominal area of mouse with electric clippers.
4. Place empty cartridge holder in gun and turn on helium.
5. Set the helium pressure gauge to the pressure of 400 psi.
6. Purge gun with helium by quickly pressing trigger 2–3 times.
7. Remove empty cartridge holder, and place DNA-coated gold bead-loaded cartridge holder into gun.
8. Hold the mouse: restrain by the scruff of the neck with one hand, exposing and stretching the shaved target skin.
9. Position the gun directly on the mouse at 90° angle relative to the shaved abdominal skin, and press the trigger.
10. Each mouse should get total three shots in the same place (to get total 9 μg plasmid).
11. After immunizations, there should be a brown spot on the skin at the center of the target site representing the DNA-coated gold beads.
12. Return the mouse to its cage (*see* **Note 6**).

3.2 Preparation of Protein-Based Vaccine and Injection of Mice

3.2.1 Preparation of Antigen

1. Add 20 μl DMSO to 1 mg peptide and vortex for 20 s. Then add 880 μl 1× PBS to peptide solution and vortex for 1 min (final concentration 1.1 mg/ml) (*see* **Note 7**).
2. Prepare stock solution for Quil-A in water at a concentration of 2 mg/ml.
3. Mix 450 μl of peptide (1.1 mg/ml) with 100 μl of Quil-A (2 mg/ml) and add 450 μl 1× PBS to get final 1 ml volume of solution. Final concentrations of peptide and Quil-A will be 0.5 and 0.2 mg/ml, respectively.
4. Rotate peptide/Quil-A solution for 1 h at 4 °C (*see* **Note 8**).

3.2.2 Injection of Mice

1. Inject mouse with 100 μl peptide (0.5 mg/ml) formulated in Quil-A subcutaneously in flank by conventional syringe using 25 G×5/8″ needle (*see* **Note 9**).
2. Immunize mice at least three times with 2-week intervals for immunogenicity study (*see* **Note 10**).

3.3 Blood Collection

1. Take blood on 12–14th day after immunization for testing the concentration of induced antibodies in sera of vaccinated mice. Blood could be collected from retro-orbital sinus using Pasteur pipette or from the tail vein (as described in your IACUC protocol). Collect 100–200 μl of blood from each mouse.
2. Incubate blood for 1 h at room temperature and then keep at 4 °C overnight. Next day centrifuge the samples for 10 min at 1,400 × *g*, at room temperature (RT), in a bench-top centrifuge. Carefully collect upper phase (sera should be yellow, redness indicates hemolysis) into the clean tube and centrifuge for 10 min at 4,500 × *g*. Transfer supernatant into the clean tube; do not touch red pellet. Aliquot sera, and keep samples at −80 °C.

3.4 Detection of Anti-Aβ Antibodies by ELISA

1. Prepare solution of $A\beta_{42}$ peptide in coating buffer (carbonate–bicarbonate buffer; pH 9.6). Add 1 ml of coating buffer into 1 mg of peptide (concentration will be 1 mg/ml). Rotate for 1 h at room temperature and keep at 4 °C overnight.
2. Next day add 100 μl of soluble $A\beta_{42}$ peptide (1 mg/ml) into 10 ml of coating buffer (concentration will be 10 μg/ml). Rotate solution at room temperature for 1 h.
3. Add 100 μl of $A\beta_{42}$ peptide solution into each well of 96-well microplate (1 μg peptide in each well) (*see* **Note 11**). Seal plate with film, and incubate plate at 4 °C overnight.
4. Next day discard the peptide solution from the wells and wash three times with TBST.
5. Block the plate with blocking buffer (3 % dry, nonfat milk in TBST, 300 μl/well) at 4 °C overnight.
6. Next day discard the blocking buffer and wash the plate three times with TBST.
7. Make fivefold serial dilutions of sera and add into wells in duplicates. Initial dilution could be 1:100.
8. Calculate the concentration of antibody using standard curve. As a standard antibody use mouse monoclonal antibody specific to N terminus of Aβ (for example 6E10). For 1 mg/ml 6E10 dilute 10,000 times (100 ng/ml final concentration). Use this concentration as the highest one, and do twofold serial dilutions (100, 50, 25, 12.5, 6.25, 3.125, 1.5625, and 0 ng/ml). Add each dilution into the plate in duplicates (*see* **Note 12**).
9. Incubate the plate with sera at 4 °C overnight.
10. Next day discard the samples and standard and wash the plate three times with TBST.

11. Prepare secondary antibodies: Dilute secondary antibody (anti-mouse IgG-HRP) into 10 ml dilution buffer to a final concentration of 0.3–0.4 μg/ml and rotate at room temperature for 30 min. Add 100 μl of solution with secondary antibodies into each well, and incubate plate at 37 °C for 1 h.
12. Discard the solution from the wells and wash three times with washing buffer.
13. Add 100 μl/well of substrate (TMB) and incubate for 3–5 min at room temperature.
14. Stop reaction by adding 100 μl of stop solution (2 N sulfuric acid).
15. Read optical density (OD) at 450 nm.
16. For determination of concentration of antibody use the dilution of sample that gave OD in the linear portion of standard curve.

3.5 Detection of Cellular Immune Responses

3.5.1 Collection of Spleens and Single-Cell Suspension Preparation

1. On the 7th day after the last immunization, euthanize the mice as described in your IACUC protocol and remove spleen in aseptic conditions using sterile scissors and forceps.
2. Place the spleen into 100-mm cell culture plate containing 10 ml sterile 1× PBS.
3. Tear the spleen with cell lifter to tiny pieces. Use plunger end of the syringe to mash the spleen. Collect cell suspension from the plate in 15 ml tube.
4. Spin down cell suspension at 350 × *g* for 7 min.
5. Discard supernatant, and vortex pellet for 1 min.
6. Add 3 ml of RBC lysing buffer into cell and incubate for 5 min at room temperature.
7. Add 12 ml 1× PBS and centrifuge at 350 × *g* for 7 min.
8. Discard supernatant, and wash pellet two more times with 15 ml 1× PBS.
9. After the last washing step discard supernatant and resuspend cell pellet in 4 ml R10 media.
10. Count cells stained with trypan blue by automated cell counter or using hemocytometer.
11. Resuspend lymphocytes to a concentration of 4×10^6 cells/ml.

3.5.2 Detection of Cytokine-Producing Cells by ELISPOT

1. Coat 96-well ELISPOT plates with purified anti-mouse IFNγ or IL-4 monoclonal antibodies as recommended by the manufacturer.
2. Incubate overnight at 4 °C.
3. Next day wash plates with R10 once and block with 200 μl R10 media for 2 h at room temperature.

4. Add splenocytes from individual mice into 20 wells (100 μl volume, 4×10^5 cells/well).
5. Add the following peptides at the final concentration of 10 μg/ml into four wells/each: PADRE, $A\beta_{40}$, and irrelevant peptide. Into four wells add pokeweed mitogen at a final concentration of 2.5 μg/ml for checking the viability of cells. Into four wells add R10 media only for detection of background level (*see* **Note 13**).
6. Incubate plates for 24 h at 37 °C with 5 % CO_2 conditions.
7. Next day wash plates four times with deionized water and then three times with PBST.
8. Incubate wells with biotinylated monoclonal anti-mouse IFNγ or IL-4 antibodies (detection antibodies) for 2 h at room temperature as recommended by the manufacturer.
9. Wash plates five times with PBST.
10. Add streptavidin–HRP as recommended by the manufacturer and incubate for 1 h at room temperature.
11. Wash plates five times with PBST and then three times with PBS.
12. Add 100 μl AEC substrate solution to develop the reaction. After 10–15 min add 100 μl deionized water into wells to stop the reaction.
13. Wash wells with deionized water once and air-dry overnight.
14. Count colored spots (cells producing cytokines) using a CTL-ImmunoSpot S5 Macro Analyzer or any dissecting microscope.
15. Subtract the number of spots in control wells (R10 medium only, non-stimulated conditions) from the number of spots in experimental wells (stimulated conditions).

3.5.3 Analysis of $CD4^+$ T Cell Proliferation by CFSE Assay

1. Resuspend freshly isolated cell culture in 0.5 ml PBS containing 5 % BSA (2×10^7 cells/ml) (*see* **Note 14**).
2. Dilute 5 mM CFSE stock solution 1/500 in 5 % BSA in PBS (10 μM CFSE solution).
3. While vortexing, immediately add 0.5 ml of 10 μM CFSE solution into 0.5 ml prepared cell suspension (from **step 1**).
4. Incubate for 5 min at room temperature in the dark.
5. Wash three times with 10 ml of PBS containing 5 % BSA.
6. After last wash, discard supernatant.
7. Resuspend cells in 2 ml of R10 (you will have 5×10^6 cells/well final concentration).
8. Add splenocytes from individual mouse into 20 wells (100 μl volume, 5×10^5 cells/well).

9. Add the following peptides into four wells/each: PADRE, $A\beta_{40}$, and irrelevant peptide, each to final a concentration of 10 μg/ml, and pokeweed mitogen (for checking the viability of cells, final concentration 2.5 μg/ml). Incubate four wells of culture with R10 media only (background level) (*see* **Note 15**).
10. Incubate plates for 72 h at 37 °C with 5 % CO_2 conditions.
11. After 72 h of incubation, collect samples into polystyrene round-bottom tube and spin down at 350×*g* for 7 min.
12. Wash cell suspension two times with 1× PBS.
13. Discard supernatant, and vortex pellet with cell culture for 1 min.
14. Add 2 μl anti-mouse CD4-VioBlue (0.5 mg/ml) into each tube and vortex for few seconds (*see* **Note 16**).
15. Incubate samples at 4 °C for 20 min.
16. Add 2 ml 1× PBS, and centrifuge at 350×*g* for 7 min.
17. Wash cell suspension two times with 1× PBS.
18. Discard supernatant, and add 300–500 μl 1× PBS. Samples are ready for analyzing by a flow cytometer.
19. Analyze proliferation of viable cells by any flow cytometer according to the manufacturer's instruction. Macs Quant from Miltenyi Biotec was used in our experiments.
20. Calculate stimulation index as a percent of proliferating $CD4^+$ T cells in restimulated culture minus that in non-stimulated culture.

3.6 Detection of Neuropathological Changes in Brains of Transgenic Animals (See Note 17)

3.6.1 Preparation of 4 % Paraformaldehyde (1 l)

1. Formaldehyde is toxic. Ware gloves and safety glasses, and make solutions inside a fume hood.
2. Heat 800 mL of 1× PBS in a glass beaker on a stir plate to ~60 °C.
3. Add 40 g of paraformaldehyde powder to the heated PBS solution.
4. Slowly raise the pH by adding 1 N NaOH dropwise from a pipette until the paraformaldehyde is dissolved and the solution clears.
5. Cool the solution, filter, and adjust the volume to 1 l with PBS.
6. Check the pH and adjust to 7.0.

3.6.2 Brain Collection

1. Heavily anesthetize mice with Nembutal (150 mg/kg intraperitoneally).
2. Put the mouse on its back on a foam block, over a sink.
3. Fix the paws with pins. Make a cut, and carefully open the chest.
4. Perform perfusion through the left ventricle, and puncture the left ventricle with pointed forceps.

5. Place the perfusion cannula into the ventricle and hold in place with hemostat.
6. Cut the right atrium.
7. Perfuse mice transcardially through the ventricular catheter with ice-cold PBS until the perfusate runs clear.
8. Surgically remove mouse's skullcap.
9. Quickly excise brain and split into two halves by a single mid-sagittal cut.
10. Snap freeze the left hemisphere in dry ice and keep at –80 °C for biochemical analyses. Fix the right hemisphere.
11. Place the right hemisphere into a bottle containing 10 ml PBS–4 % paraformaldehyde.
12. Incubate for 24 h at 4 °C.
13. Wash the hemibrains with 1× PBS.
14. Add to brains 19 ml of 0.02 % sodium azide solution.
15. Store at 4 °C.

3.6.3 Preparation of Brains for Slicing

1. Label the mounting blocks, 96-well plates (2 plates per brain), and 50 ml tubes.
2. Add 250 μl of 0.02 % PBS–sodium azide solution per well.
3. Move each brain to a Petri dish containing a few drops of 1× PBS, and, using a blade, cut out the cerebellum and the front part of the brain (olfactory bulb area) (you can keep them if needed in PBS–azide buffer for the long-term storage at 4 °C) (*see* **Note 18**).
4. Apply glue to the block. To be able to obtain the coronal sections of the brain, carefully attach the remaining central part of the brain containing hippocampus to the block vertically, so that the side that was adjacent to cerebellum is at the bottom.
5. Wait for 20–30 s, applying gentle pressure to the top of the brain. After making sure that the whole piece is glued to the block evenly and firmly, put the block back in pre-labeled 50 ml tube containing PBS–azide buffer.
6. Transfer to 4 °C refrigerator.

3.6.4 Slicing of the Tissue and Preparation of Slides

1. Slice the brain using the vibratome with 40–50 μM thickness setting (use ice-cold PBS buffer solution).
2. Put each slice using a small art brush in the separate well of 96-well plate with Tris–HCl buffer/0.02 % sodium azide.
3. Make sure that the plates are firmly covered to prevent evaporation of buffer during the long-term storage and keep at 4 °C.

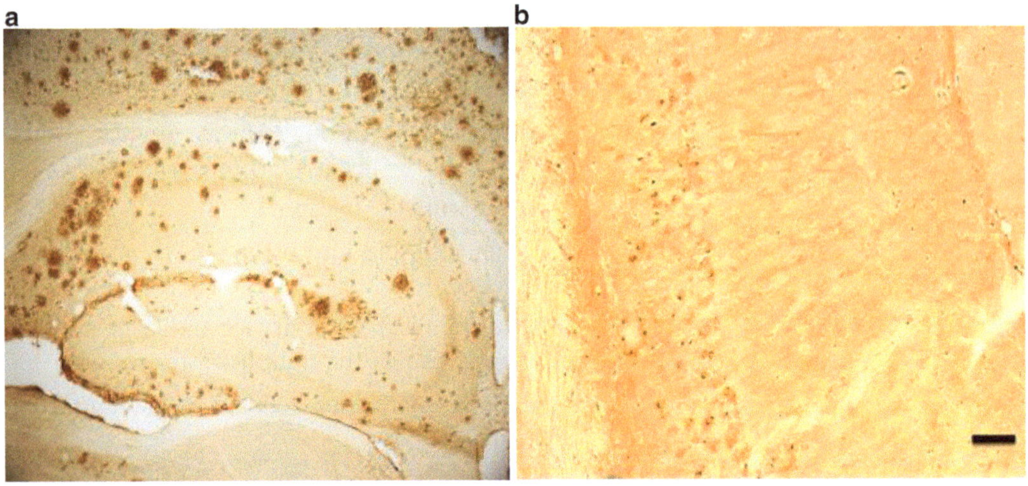

Fig. 1 (**a**) Staining of 16-month-old APP/Tg2576 mouse brain section with 6E10 anti-Aβ antibody. (**b**) Staining of 19-month-old 3xTg-AD mouse brain section with HT-7 anti-tau antibody

4. Attach tissue sections on the pre-rinsed SuperFrost glass slides if required for the IHC protocol, and let them dry for minimum of 2–3 h or overnight.
5. One extra slide should be prepared for the nonspecific staining control (secondary antibody only).

3.6.5 Immunohistochemistry: Staining of β-Amyloid and Tau (See ***Note 19****)*

An example image for staining of β-amyloid with 6E10 and Tau with HT-7 is showed in Fig. 1.

1. Epitope retrieval: For staining of extracellular (insoluble) β-amyloid protein, put the slides with the sections attached into 90 % formic acid for 4 min; for staining of intracellular (soluble) β-amyloid, use 10 % formic acid for 4 min (*see* **Note 20**).
2. Wash the slides in Tris buffer for 5 min.
3. Incubate slides in Tris/H_2O_2/methanol solution for blocking endogen peroxidase.
4. Wash sections in Tris for 5 min (twice).
5. Wash sections in Tris-A for 15 min.
6. Wash sections in Tris-B for 30 min.
7. Add primary antibody diluted in Tris-B, according to the manufacturer's instructions or as per optimization data.
8. Incubate for the predetermined optimal time period at 4 °C or RT (*see* **Note 21**).
9. Aspirate primary antibody off.
10. Wash solutions in Tris-A for 5 min twice.
11. Wash sections in Tris-B for 15 min.

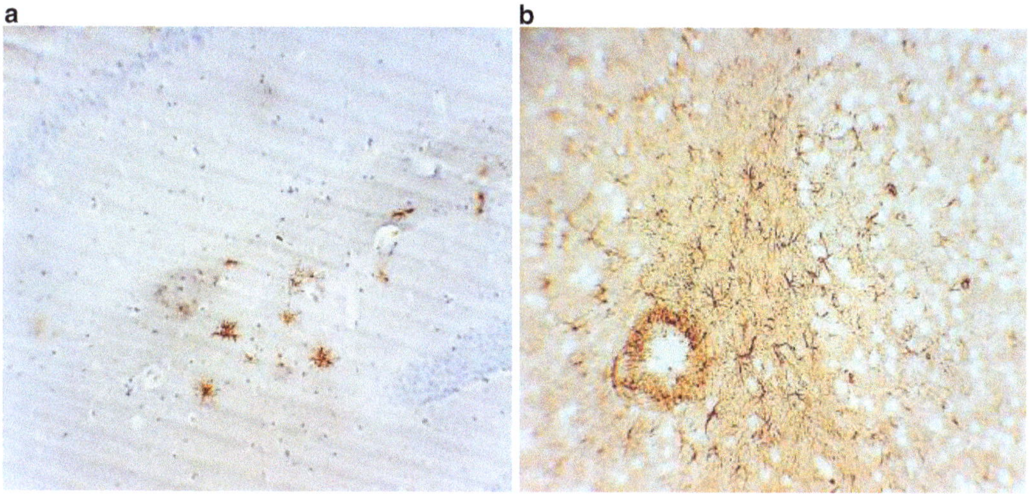

Fig. 2 (**a**) Staining of 16-month-old APP/Tg2576 mouse brain section with anti-MHC II antibody. (**b**) Staining of 16-month-old APP/Tg2576 mouse brain section with anti-GFAP antibody

12. Incubate sections for 1 h with secondary antibody diluted appropriately in Tris-B containing 4 % normal serum of the same species used for a production of the antibody.
13. Wash sections in Tris-A for 5 min (twice).
14. Wash sections in Tris-B for 15 min.
15. Incubate in freshly made avidin–biotin–HRP complex (ABC) for 1 h.
16. Wash sections in Tris for 5 min (two times).
17. Wash in DI water for 5 min.
18. Prepare DAB substrate solution and add enough to cover the slides.
19. Keep for 1–4 min or more, until the reaction is developed.
20. Wash sections in DI water for 5 min three times using the rocking platform.
21. Dehydrate slides through 50, 75, 95, and 100 % alcohol, 5 min each.
22. Transfer to xylene for 10 min.
23. Coverslip with resinous mounting medium.

3.6.6 Immunohistochemistry: MHC II/GFAP Staining (See ***Note 22****)*

An example image for staining of microglia with anti-MHC II and astrocytes with anti-GFAP is showed in Fig. 2.

1. For MHC II/GFAP staining use the same protocol as for staining of β-amyloid but without epitope retrieval.

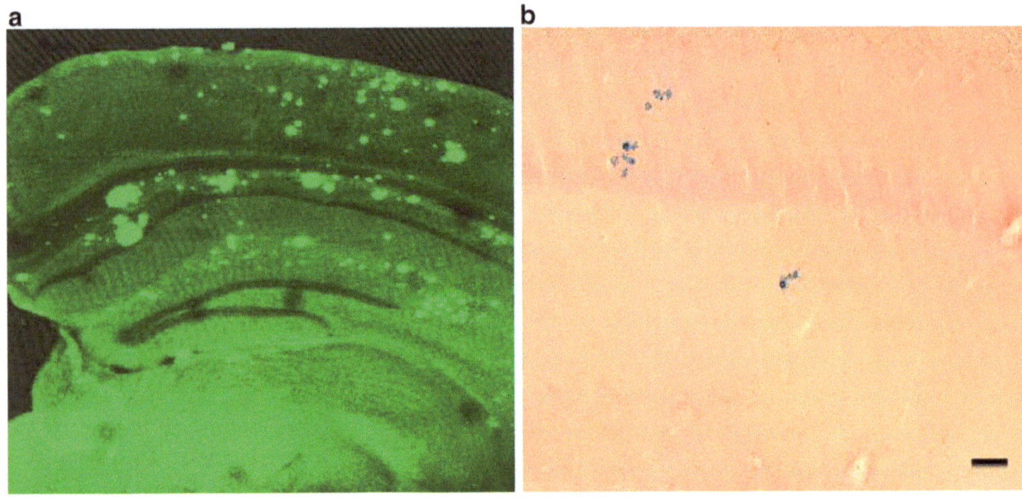

Fig. 3 (**a**) Staining of 16-month-old APP/Tg2576 mouse brain section with thioflavin S. (**b**) Detection of microhemorrhages by Prussian blue staining

3.6.7 Histostaining: Thioflavin S Staining Protocol (See ***Note 23****)*

An example image for staining of dense-core plaques with ThS is showed in Fig. 3a.

1. Prepare 0.5 % thioflavin S in 50 % ethanol (add 50 % ethanol to 0.5 g ThS).
2. Wash free-floating brain sections with Tris buffer.
3. Stain with a solution of 0.5 % thioflavin S in 50 % ethanol for 10 min.
4. Wash with 50 % ethanol twice.
5. Mount sections on silinated slide, and allow drying.
6. Cover using coverslips Vectashield (Vector Labs).
7. Store stained slides in cool, dark place.
8. Thioflavin S-bound amyloid plaques can be viewed under fluorescence microscope as green fluorescent plaques.

3.6.8 Detection of Microhemorrhages by Prussian Blue Staining Protocol for Iron (See ***Note 24****)*

An example image for detection of microhemorrhages by Prussian blue staining is showed in Fig. 3b.

1. Mount sections on silinated slides, and allow drying.
2. Immerse sections in HCl/potassium ferrocyanide solution for 20 min.
3. Wash in DI water, three times.
4. Counterstain with nuclear fast red for up to 5 min.
5. Rinse twice in DI water.
6. Dehydrate through 50, 75, 95, and 100 % alcohol.
7. Clear in xylene two times, 5 min each.
8. Coverslip with resinous mounting medium.

3.7 Biochemical Analysis

3.7.1 Preparation of Brain Homogenates

1. Weight the frozen hemibrain.
2. Put into the mortar with liquid nitrogen, and mash with pestle to make a powder. Transfer the powder into the tube with clean spatula.
3. Add Tris–HCl buffer (1 ml to 150 mg brain).
4. Centrifuge at 100,000 × *g*, for 1 h, at 4 °C.
5. Transfer supernatants into the clean tubes and store at −70 °C for further analysis of soluble β-amyloid.
6. Add 70 % formic acid to the pellets for extraction of SDS-insoluble β-amyloid.
7. Sonicate samples.
8. Centrifuge samples at 100,000 × *g*, for 1 h, at 4 °C, and store supernatants for analysis of insoluble β-amyloid.
9. Before performing ELISA, neutralize formic acid by adding 1.0 M Tris base/0.5 M NaH_2PO_4.
10. Measure concentrations of insoluble and soluble $A\beta_{40}$ and $A\beta_{42}$ using β-amyloid ELISA kits, according to the manufacturer's recommendations.

3.7.2 Immunoprecipitation/Western Blot (IP/WB) (See Notes 25 and 26)

1. Measure concentration of total protein using BCA assay.
2. Transfer each sample containing 200 μg total protein to the clean Eppendorf tube.
3. Adjust the volume to 500 μl with PBS.
4. To pre-clear the sample, add 30–40 μl protein G-sepharose beads and incubate at 4 °C for 2 h with rotation.
5. Centrifuge at 2,500 × *g* for 2 min.
6. Transfer the supernatant into the clean tube.
7. Add 1 μg anti-Aβ antibody (for example 6E10) and 40 μl protein G-sepharose.
8. Incubate overnight at 4 °C with rotation.
9. Next day, wash the beads three times with PBS (perform centrifugation on a benchtop: microcentrifuge at 2,500 × *g* for 2 min, carefully discard supernatant by pipette, add fresh PBS, mix gently, and centrifuge again).
10. After the last centrifugation remove carefully liquid and add 20 μl sample loading buffer for WB.
11. Incubate for 10 min at 70 °C or 2 min at 95–100 °C.
12. Load samples into the 4–12 % gradient Bis–Tris polyacrylamide gel and subject to electrophoresis in MOPS buffer (Invitrogen) (*see* **Note 27**).
13. Transfer electrophoretically the proteins to PVDF membrane (GE Healthcare, NJ).

14. Boil the membrane in PBS for 2 min and block with 5 % fat-free dry milk overnight at 4 °C.
15. Add biotinylated anti-Aβ monoclonal antibody (for example 6E10), diluted in 5 % fat-free dry milk. Incubate for 2 h at RT with rotation.
16. Wash the membrane three times (at least 5 min each wash) with TBST.
17. Add streptavidin–HRP diluted in 5 % fat-free dry milk. Incubate for 1 h at RT with rotation.
18. Wash the membrane three times with TBST and once with TBS.
19. Visualize bands using chemiluminescence substrate.
20. Scan western blots and convert into digital files.
21. Analyze Aβ oligomer bands by densitometry using NIH Image J software.

4 Notes

1. Purification of our DNA plasmid-encoding AD vaccine was performed by Aldevron. Alternatively plasmid could be purified using Qiagen EndoFree Plasmid Maxi kit.
2. 90 μl of 3 mg/ml plasmid should be sufficient for preparation of 75–80 cartridges with gold beads containing 3 μg of plasmid in each cartridge.
3. Make sure that nitrogen gas is turned off completely prior to sliding the tube into the tube turner, and syringe should be attached once the tube is loaded into the tube turner.
4. Anesthetization is optional, and it could be different based on different institutional animal-handling procedures.
5. All protocols using live animals must first be reviewed and approved by an IACUC and must follow officially approved procedures for the care and use of laboratory animals.
6. Mice should be observed for recovery from anesthesia.
7. Peptide solution could be prepared, aliquoted, and kept frozen at −20 °C.
8. 1 ml of antigen will be enough for immunization of eight mice.
9. Typically, use eight mice per group if you are testing only immunogenicity of a vaccine. Use 12–16 mice per group if you check the therapeutic efficacy of a vaccine in APP/Tg mice.
10. If you do protective immunization of APP/Tg mice, do three immunizations with 2-week intervals and then continue to do monthly immunizations.
11. 10 ml will be enough for coating one 96-well plate. For more plates multiply everything by the number of plates.

12. Standard should be used in each plate.
13. Depending on the number of antigen used for in vitro restimulation add four wells of cell culture for each peptide.
14. Concentration of cells could be from 10^6 to 2×10^7 cells/ml.
15. Depending on the number of antigens used for in vitro restimulation, add four wells with cell culture for each peptide.
16. Anti-mouse CD4 antibodies conjugated with any dye (PE, PerCP, APC), but not with FITC.
17. Tg2576 mice develop Aβ plaques at the age of 12 months. For preventive vaccination, start to immunize mice when 5–6 months old and continue to immunize monthly until mice are 14–15 months old.
18. It is helpful to use a matrix to make the starting cuts. Matrixes are metal or plastic blocks with cavities in the shape of a brain, and slots to guide a blade, used to make a starting cut approximately in a consistent plane, to reduce the need for adjustment in the cutting stage.
19. Perform tau staining only if your Tg mice develop tau pathology (for example 3xTg-AD mice).
20. No epitope retrieval was necessary for anti-Tau immunostaining.
21. It is convenient to use CoverWell incubation chambers for 200 μl volume (Grace Bio-Labs) to incubate slides with antibodies overnight.
22. Chronic inflammation is one of the main signs of Alzheimer's disease. Microglial cells surround amyloid plaques and dying neurons and may be helpful at the beginning, but when the inflammation becomes chronic with further development of disease, it can cause harm to the healthy cells. Antibody-mediated clearance of amyloid plaques also leads to decreasing of the inflammation. Therefore, detection of activated microglia and astrocytes in brains of vaccinated mice is an important measure of the vaccine efficacy and safety. The level of activated microglia and astrocytes could be detected by staining brain sections with antibodies specific to MHC class II molecules and GFAP, respectively.
23. Thioflavin S is a dye that is used to distinguish dense-core plaques of β-amyloid from diffused plaques. Dense-core plaques are fibrillar deposits of Aβ consisted of beta-sheet structure. Diffuse plaques are amorphous deposits. Thioflavin S binds to dense-core plaques but not to diffuse plaques.
24. Prussian blue reaction involves the treatment of sections in acid solutions of ferrocyanides. Any ferric ion (+3) present in the tissue will combine with the ferrocyanide and result in the formation of a blue pigment called "Prussian blue" or ferric ferrocyanide. This sensitive histochemical test will demonstrate

even a single granule of iron in blood cells. Hemorrhages in a brain will look like bright blue profiles on a pink background.

25. Analyze changes in oligomeric Aβ species by western blot.
26. Use regular WB apparatus and protocol that you use in your lab.
27. Load one well of the gel with pre-stained protein marker to control the duration of electrophoresis. Stop the electrophoresis when proteins included into the marker are divided but beware not to lose the smallest protein in a marker (6 kDa). Alternatively, you may use the electrophoresis apparatus and running buffer, which is routinely in use in your lab.

Acknowledgments

This work was supported by NIH grants AG20241, NS50895, NS057395, and NS065518 and from the Alzheimer's Association IIRG12239626.

References

1. Price DL, Sisodia SS (1994) Cellular and molecular biology of Alzheimer's disease and animal models. Annu Rev Med 45:435–446
2. Selkoe DJ (1991) The molecular pathology of Alzheimer's disease. Neuron 6:487–498
3. Hardy JA, Higgins GA (1992) Alzheimer's disease: the amyloid cascade hypothesis. Science 256:184–185
4. Schenk DJ, Barbour R, Dunn W et al (1999) Immunization with amyloid-beta attenuates Alzheimer-disease-like pathology in the PDAPP mouse. Nature 400:173–177
5. Morgan D, Diamond DM, Gottschall PE et al (2000) A beta peptide vaccination prevents memory loss in an animal model of Alzheimer's disease. Nature 408:982–985
6. Bard F, Cannon C, Barbour R (2000) Peripherally administered antibodies against amyloid beta-peptide enter the central nervous system and reduce pathology in a mouse model of Alzheimer disease. Nat Med 6:916–919
7. DeMattos RB, Bales KR, Cummins DJ (2001) Peripheral anti-A beta antibody alters CNS and plasma A beta clearance and decreases brain A beta burden in a mouse model of Alzheimer's disease. Proc Natl Acad Sci U S A 98:8850–8855
8. Dodart JC, Bales KR, Gannon KS et al (2002) Immunization reverses memory deficits without reducing brain Abeta burden in Alzheimer's disease model. Nat Neurosci 5:452–457
9. Nicoll JA, Wilkinson D, Holmes C et al (2003) Neuropathology of human Alzheimer disease after immunization with amyloid-beta peptide: a case report. Nat Med 9:448–452
10. Ferrer I, Rovira MB, Guerra MLS et al (2004) Neuropathology and pathogenesis of encephalitis following amyloid-beta immunization in Alzheimer's disease. Brain Pathol 14:11–20
11. Masliah E, Hansen L, Adame A et al (2005) Abeta vaccination effects on plaque pathology in the absence of encephalitis in Alzheimer disease. Neurology 64:129–131
12. Nicoll JA, Barton E, Boche D et al (2006) Abeta species removal after abeta42 immunization. J Neuropathol Exp Neurol 65:1040–1048
13. Patton RL, Kalback WM, Esh CL (2006) Amyloid-beta peptide remnants in AN-1792-immunized Alzheimer's disease patients: a biochemical analysis. Am J Pathol 169:1048–1063
14. Holmes C, Boche D, Wilkinson D et al (2008) Long-term effects of Abeta42 immunization in Alzheimer's disease: follow-up of a randomised, placebo-controlled phase I trial. Lancet 372: 216–223
15. Gilman S, Koller M, Black RS et al (2005) Clinical effects of Abeta immunization (AN1792) in patients with AD in an interrupted trial. Neurology 64:1553–1562
16. Bayer AJ, Bullock R, Jones RW et al (2005) Evaluation of the safety and immunogenicity of synthetic Abeta42 (AN1792) in patients with AD. Neurology 64:94–101
17. Cribbs DH, Ghochikyan A, Tran M et al (2003) Adjuvant-dependent modulation of

Thl and Th2 responses to immunization with beta-amyloid. Int Immunol 15:505–514

18. Bard F, Barbour R, Cannon C et al (2003) Epitope and isotype specificities of antibodies to beta -amyloid peptide for protection against Alzheimer's disease-like neuropathology. Proc Natl Acad Sci U S A 100:2023–2028
19. Lee M, Bard F, Johnson-Wood K (2005) Abeta42 immunization in Alzheimer's disease generates Abeta N-terminal antibodies. Ann Neurol 58:430–435
20. McLaurin J, Cecal R, Kierstead ME et al (2002) Therapeutically effective antibodies against amyloid-beta peptide target amyloid-beta residues 4–10 and inhibit cytotoxicity and fibrillogenesis. Nat Med 8:1263–1269
21. Town T, Tan J, Sansone N et al (2001) Characterization of murine immunoglobulin G antibodies against human amyloid-b 1–42. Neurosci Lett 307:101–104
22. Dickey CA, Morgan DG, Kudchodkar S et al (2001) Duration and specificity of humoral immune responses in mice vaccinated with the Alzheimer's disease-associated beta-amyloid 1–42 peptide. DNA Cell Biol 20(11):723–729
23. Lemere CA, Maron R, Selkoe DJ et al (2001) Nasal vaccination with beta-amyloid peptide for the treatment of Alzheimer's disease. DNA Cell Biol 20:705–711
24. Lemere CA, Maron R, Spooner ET et al (2000) Nasal Aß treatment induces anti-Aß antibody production and decreases cerebral amyloid burden in PD-APP mice. Ann N Y Acad Sci 920:328–331
25. Monsonego A, Zota V, Karni A (2003) Increased T cell reactivity to amyloid beta protein in older humans and patients with Alzheimer disease. J Clin Invest 112(3):415–422
26. Sigurdsson EM, Knudsen E, Asuni A et al (2004) An attenuated immune response is sufficient to enhance cognition in an Alzheimer's disease mouse model immunized with amyloid-beta derivatives. J Neurosci 24:6277–6282
27. Singh H, Raghava GP (2001) ProPred: prediction of HLA-DR binding sites. Bioinformatics 17:1236–1237
28. Singh H, Raghava GP (2003) ProPredl: prediction of promiscuous MHC Class-I binding sites. Bioinformatics 19:1009–1014
29. Das P, Chapoval S, Howard V et al (2003) Immune responses against Abetal-42 in HLA class II transgenic mice: implications for Abetal-42 immune-mediated therapies. Neurobiol Aging 24:969–996
30. Petrushina I, Ghochikyan A, Mktrichyan M et al (2007) Alzheimer's disease peptide epitope vaccine reduces insoluble but not soluble/oligomeric A{beta} species in amyloid precursor protein transgenic mice. J Neurosci 27:12721–12731
31. Movsesyan N, Ghochikyan A, Mkrtichyan M et al (2008) Reducing AD-like pathology in 3xTg-AD mouse model by DNA epitope vaccine- a novel immunotherapeutic strategy. PLos One 3:e21
32. Ghochikyan A, Davtyan H, Petrushina I et al (2013) Refinement of a DNA based Alzheimer's disease epitope vaccine in rabbits. Hum Vaccin Immunother 9:1002–1010
33. Davtyan H, Ghochikyan A, Petrushina I et al (2013) Immunogenicity, efficacy, safety, and mechanism of action of epitope vaccine (Lu AF20513) for Alzheimer's disease: prelude to a clinical trial. J Neurosci 33:4923–4934
34. Evans CF, Davtyan H, Petrushina I et al (2013) Epitope-based DNA vaccine for Alzheimer's disease: translational study in macaques. Alzheimer's Dement. [Epub ahead of print]
35. Alexander J, del Guercio MF, Maewal A (2000) Linear PADRE T helper epitope and carbohydrate B cell epitope conjugates induce specific high titer IgG antibody responses. J Immunol 164:1625–1633
36. Ghochikyan A (2009) Rationale for peptide and DNA based epitope vaccines for Alzheimer's disease immunotherapy. CNS Neurol Disord Drug Targets 8:128–143
37. Babiuk LA, Pontarollo R, Babiuk S et al (2003) Induction of immune responses by DNA vaccines in large animals. Vaccine 21:649–658
38. Gregersen JP (2001) DNA vaccines. Naturwissenschaften 88:504–513
39. Cribbs DH, Agadjanyan MG (2005) Immunotherapy for Alzheimer's disease: potential problems and possible solutions. Curr Immunol Rev 1:139–155

Chapter 17

Tetravalent DNA Vaccine Product as a Vaccine Candidate Against Dengue

Kevin R. Porter, Nimfa Teneza-Mora, and Kanakatte Raviprakash

Abstract

Dengue is the most important arbovirus worldwide and is the virus that causes dengue fever and the more severe dengue hemorrhagic fever. There are four serotypes of dengue with each possessing the ability to cause disease. Developing a preventive vaccine is the most efficient and effective way to prevent these diseases, and because immunity to one serotype does not protect against the other serotypes, a vaccine must provide tetravalent protection. We used DNA immunization as a platform to develop a tetravalent vaccine. In this chapter, we describe the laboratory, regulatory, and clinical methodology for evaluating a candidate tetravalent vaccine in a Phase 1 clinical trial.

Key words Dengue, Tetravalent, DNA vaccine, Clinical trial, Vaxfectin®, Adjuvant

1 Introduction

For over 50 years, scientists have attempted to develop a protective vaccine against dengue, currently the most important arbovirus worldwide. Traditional approaches have included attenuating wild-type virus by serial passage in tissue culture and inactivating live virus with formalin and purifying the proteins for use [1]. With advances in molecular biology, more recent approaches involved vectored vaccines [1] and live chimeric dengue vaccines consisting of the envelope proteins of dengue and nonstructural proteins derived from the 17D live attenuated yellow fever vaccine. Because there are four biologically distinct serotypes of dengue and immunity to one serotype does not provide lifelong protection against the other serotypes, any effective vaccine must provide tetravalent protection. While this means that the typical vaccine formulation must consist of four serotype-specific components, there have been two-component virus vectored formulations that provided tetravalent protection in animals [2].

Monica Rinaldi et al. (eds.), *DNA Vaccines: Methods and Protocols*, Methods in Molecular Biology, vol. 1143, DOI 10.1007/978-1-4939-0410-5_17, © Springer Science+Business Media New York 2014

Nucleic acid immunization has been used as a method of producing both cellular- and antibody-mediated immunity. The ability of naked closed circular plasmid DNA to elicit an immune response dates back to 1992 when Wolff et al. demonstrated that direct gene transfer into the muscle of mice [3] resulted in an immune response. Shortly after that, using genes that code for influenza proteins, Ulmer et al. showed that the immune response produced by injected plasmid DNA had the ability to protect ferrets against lethal challenge with live influenza virus [4]. Since then, DNA vaccines have been developed to target a variety of infectious organisms including HIV, tuberculosis, hepatitis C, and West Nile virus. We used DNA immunization as a platform to develop a tetravalent vaccine against dengue.

Beckett el al. conducted a pilot study using a monovalent dengue DNA vaccine against dengue serotype 1. The vaccine produced strong T cell responses in 10 of 12 volunteers in the high-dose group and modest anti-dengue-neutralizing antibody levels in 5 of 12 subjects in that group [5]. Subjects receiving a 1 mg dose failed to develop neutralizing antibodies. To satisfy the tetravalent requirement of an effective dengue vaccine, a four-component DNA vaccine was formulated consisting of plasmids that express the pre-membrane and envelope proteins of dengue virus serotypes 1–4. To improve anti-dengue-neutralizing antibody responses, the tetravalent vaccine was formulated in a proprietary adjuvant, Vaxfectin®, and evaluated for safety and immunogenicity in a Phase 1 clinical trial. The subject of this chapter is to report on the methodology and regulatory considerations for conducting a Phase 1 clinical trial of a tetravalent dengue DNA vaccine.

2 Materials

2.1 Laboratory Testing of Clinical Samples

2.1.1 Plaque Reduction Neutralization Assays

1. Low passage Vero tissue culture cells (ATCC, Gaithersburg, Maryland).
2. 2× EMEM media containing fetal bovine serum (FBS), L-GLUTAMINE, sodium bicarbonate, nonessential amino acids, penicillin/streptomycin.
3. Six-well tissue culture plates.
4. Low-melting-point agarose.
5. Neutral red vital stain.

2.1.2 Enzyme-Linked Immunosorbant Assay

1. Immulon-2 Flat bottom Microtiter plates (Dynex).
2. Inactivated whole dengue virions as the antigen.
3. Horseradish peroxidase-labeled conjugates to mouse, monkey, or human.

4. Polyoxyethylene-Sorbitan Monolaureate (TWEEN 20) (Sigma).
5. 10× Phosphate Buffered Saline (PBS) pH 7.4 (Quality Biologicals, Inc.).
6. Non-Fat Dry Milk (DIFCO, Becton Dickinson).
7. ABTS substrate system (Kirkegaard & Perry).

2.1.3 INF-γ Enzyme-Linked Immuno-Spot Assay

1. RPMI 1640 1× media modified with L-glutamine, FBS, heat inactivated, penicillin, streptomycin, L-glutamine, and nonessential amino acids.
2. PBS (pH 7.4 with 9.0 g/l NaCl, Quality Biological Inc.).
3. Enzyme-linked immuno-spot (ELISpot) plates (Millipore Plate either MAIPS4510 or MAIPSWU10).
4. Human IFN-γ ELISPOT Kit (Mabtec Lab.).
5. Guava cell counter (Guava Technologies).
6. Guava ViaCount Reagent, Cat. 4000-0040 (Guava Technologies).
7. 96-well round bottom tissue culture plates, polystyrene (Corning Incorporated Costa).
8. Pansorbin^{R} cells, Lyophilized (Calbiochem).
9. Dengue pre-membrane (prM) and envelope (E) protein peptide arrays of all four dengue serotypes (BEI Resources). Peptides from one serotype-specific protein were combined into one pool, stocked at >200 μg/ml in DMSO, stored in −20 °C, and thawed before use.
10. Tween-20, P1379-500 ml (Sigma).
11. 3-Amino-9-ethylcarbazole (AEC) substrate.
12. Automated ELISPOT Reader (AID Diagnostics).
13. 99 % ethanol.

2.2 Formulated DNA Vaccine

2.2.1 Tetravalent Dengue DNA Vaccine

1. Vialed tetravalent DNA vaccine product manufactured according to Good Manufacturing Practices (cGMP) specifications containing DNA plasmids that express the pre-membrane protein (prM) and envelope protein (E) of each of the four dengue virus serotypes (*see* **Note 1**).
2. Sterile saline solution manufactured under cGMP guidelines.

2.2.2 Vaxfectin® Adjuvant Prepared According to cGMP Specifications

1. Vaxfectin® was supplied as a cGMP-manufactured lipid film in sterile borosilicate vials (*see* **Note 2**).
2. Sterile cGMP-produced saline solution for reconstituting the Vaxfectin®.

3 Methods

3.1 Preparation of the Vaccine for Injection

1. 0.7 ml of saline was dispensed into the tetravalent dengue pDNA vaccine (TVDV) vial (0.7 ml, 2 mg/ml) and mixed by inverting the vial 3–4 times.
2. For formulation with Vaxfectin®, 1 ml of saline was dispensed into the vial containing dry lipid film. The vial is vortexed for 5 min, and 0.7 ml of resuspended Vaxfectin® was gently layered over TVDV. The mixture was rocked 3–4 times by inverting the vial.
3. Formulated vaccine preparations were used within 4 h of preparation.

3.2 Vaccine Administration

1. For the injections, either 1 mg (1 ml) or 2 mg (2×1 ml) of formulated TVDV was used, depending on the group to be immunized.
2. 1 ml was drawn from the formulated vaccine vial into a 1 ml syringe using a 22–25 gauge needle. All injections were given in the deltoid muscle. For the 2 mg total dose, both deltoid muscles were injected with 1 mg each.

3.3 Tetravalent DNA Vaccine Clinical Trial

3.3.1 Regulatory Requirements for the Tetravalent DNA Dengue Vaccine Product

1. An Investigational New Drug (IND) Application was submitted to the US Food and Drug Administration's Office of Vaccines Research and Review (*see* **Note 3**).
2. The clinical trial protocol was submitted to and approved by the Institutional Review Board for the protection of human subjects.

3.3.2 Inclusion Criteria

Subjects had to meet all of the following criteria to be included in the study:

1. Male or female, age 18–50 (inclusive) years old at the time of enrollment.
2. Had negative anti-dengue, -Japanese encephalitis, -West Nile, and -yellow fever enzyme-linked immunosorbant assay (ELISA) serological tests.
3. Was informed of the nature of the study and has to provide written informed consent.
4. If the subject was of child-producing potential, he/she agrees to practice adequate birth control or abstain from sex.
5. Had access to the clinical trial center, to be willing to attend all of the required follow-up visits for at least 270 days, and to be willing to refrain from participation in another investigational clinical trial involving an investigational product/device of any kind during the participation in this clinical trial.

6. Was in good general health and have no significant current or past disease as determined by the principal investigator based on the medical history, physical examination, and clinical laboratory evaluations (*see* **Note 4**).

3.3.3 Exclusion Criteria

Subjects meeting any of the following criteria were excluded from the study:

1. History of Flavivirus infection or history of Flavivirus vaccine (experimental or licensed product) including Japanese encephalitis, yellow fever, and dengue.
2. Had a known or a suspected hypersensitivity or adverse reaction to vaccines including anaphylaxis and related symptoms such as hives, respiratory difficulty, angioedema, and/or abdominal pain.
3. Had received a live-attenuated vaccine within 42 days prior to the initial injection on day 0 or a subunit or a killed vaccine within 30 days of the initial injection on day 0.
4. Had a positive screen for hepatitis B surface antigen (HBsAg), hepatitis C antibody, or HIV antibody.
5. Was pregnant or breastfeeding or had a positive urine β-human chorionic gonadotrophin (β-hCG) test at screening or prior to the initial injection on day 0.
6. Had donated or received blood, blood products, or plasma within 30 days prior to day 0 (or plan on donating blood or plasma within 60 days of their last injection of TVDV).
7. Had any acute illness, including an oral body temperature >100.4 °F, within 7 days before the initial injection on day 0 (if in the low-dose group, the subject could be moved into the high-dose group at the discretion of the principal investigator).
8. Had a past or a current history of malignant disease except for adequately treated basal cell or squamous cell skin cancer.
9. Had any occupational, social, or medical concerns that would impact subject safety, interfere with protocol adherence, or affect a subject's ability to give informed consent such as a psychiatric disease or a medical disease or condition that might compromise the cardiovascular, hematological, renal, hepatic, pulmonary, endocrine, central nervous, immune, or gastrointestinal systems (unless deemed not clinically significant by the principal investigator).
10. Psychiatric disease includes past or present psychoses, past or present bipolar disorder requiring therapy, and past or present suicidal ideation requiring therapy.
11. Exclusions included but were not limited to conditions pertaining to or evidence of immunodeficiency; allergies requiring

treatment with antigen injections within 14 days of the initial trial injection; autoimmune disease; severe migraine headaches occurring greater than twice a month and/or requiring prophylactic therapy; asthma that is unstable (e.g., use of oral, inhaled, or intravenous/injectable corticosteroids, emergency care, urgent care, hospitalization, or intubation during the past 1 year); hypertension, clinically significant cardiac arrhythmias, diabetes mellitus (type I or type II), including cases controlled with diet alone; thyroid disease, including a history of thyroidectomy and diagnoses requiring medication; bleeding disorder diagnosed by a doctor (e.g., factor VIII or IX deficiency, coagulopathy, or platelet disorder requiring special precautions); and seizure disorder.

12. A subject would not be excluded if the subject used inhaled corticosteroids and inhaled nonsteroidal medications for allergic rhinitis or had a remote history (over 3 years ago) of seizure and has not required medication for over 3 years if (a) the seizures were febrile seizures that occurred when the subject was under the age of two, (b) the seizures were secondary to alcohol withdrawal, or (c) the subject had a single seizure.
13. Had used immunomodulatory therapy for ≥14 days within the 6 months prior to the initial injection on day 0 or oral equivalent >0.5 mg/kg/day or high dose inhaled equivalent (≥2,000 beclomethasone equivalent units/day).
14. Had used medications or nutritional supplements known to or which potentially could affect organ functions within 30 days prior to the initial injection on day 0 unless approved by the principal investigator (e.g., antituberculosis prophylaxis, creatine workout supplements).
15. Had participated in an investigational drug, vaccine, or device study within a period of 30 days prior to day 0; for investigational products with a terminal elimination half-life greater than 10 days, this was extended to 60 days or 5 times the half-life, whichever was longer.
16. Was currently participating, or planned on participating, in an investigational drug, vaccine, or device study through day 360.
17. Clinical or laboratory evidence of significant anemia (hemoglobin <10 g/dL or hematocrit <35 %).
18. History of splenectomy.
19. Planned travel to dengue endemic areas during the study period.
20. Lack of informed consent.

3.3.4 Approach to Study Subject Retention Difficulties

1. Subjects who were enrolled, received at least one injection, and were withdrawn from the clinical trial for any reason were not replaced. Subjects who withdrew without receiving any injections were replaced with an alternate.

2. A subject was allowed to end his or her participation in the study at any time. If a subject withdrew, the investigator made a reasonable effort to determine the reason for the withdrawal from the study. Telephone calls, registered letters, and e-mail correspondence were considered reasonable effort. For subjects leaving the study, a targeted examination was performed if medically indicated and if permitted by the subject.

3.3.5 Clinical Trial Design

1. The study was designed as an open-label clinical trial whose primary objective was to evaluate the safety and tolerability of TVDV (with and without the adjuvant, Vaxfectin®).
2. The secondary objective was to evaluate TVDV's immunogenicity when delivered as a plasmid DNA vaccine alone and as a vaccine adjuvanted with Vaxfectin®.
3. The study endpoints were as follows: the incidence of adverse events (AEs) among the subjects during the first 7 days after receiving a vaccine dose; the incidence of serious adverse events (SAEs) throughout the study period, which may be related to the vaccination; the immunogenicity at specified time periods after receipt of the vaccine; and a comparison of the immune responses between the high- and low-dose groups.
4. A total of 119 volunteers were screened, and 40 subjects that met the inclusion criteria were enrolled and randomly assigned to one of the three groups: Group 1 consisted of 10 volunteers who received TVDV alone at a dose of 1.0 mg; group 2 consisted of 10 subjects who received TVDV with Vaxfectin® at a low dose (1.0 mg); group 3 consisted of 20 subjects who received TVDV with Vaxfectin® at a higher dose of 2.0 mg (two 1 mg injections).

3.3.6 Clinical Trial Execution

1. The study protocol was approved by the Naval Medical Research Center Institutional Review Board (NMRC/IRB) and conducted at the Walter Reed Army Institute of Research Clinical Trial Center (*see* **Note 5**).
2. Table 1 details all study-related activities including clinic visits, phlebotomy, and vaccine injections.
3. Study subjects were continuously monitored and evaluated using clinical and laboratory data to look for the occurrence of potential AEs. If identified, an AE was assigned a severity score based on FDA guidance [6], and its relationship with receipt of the vaccine was determined by the principal investigator. AEs were recorded in the volunteer's study file and were captured in the electronic case report forms.

3.3.7 Laboratory Evaluation of Clinical Samples

1. Humoral immunity was measured by ELISA (*see* **Note 6**) to determine the presence and level of all anti-dengue antibody and by plaque reduction neutralization assay (*see* **Note 7**) to

Table 1
Study activity schedule

Study day	Screen-ing day	−7	0	1	2	7	14	30	31	32	37	44	60	90	91	92	97	104	120	150	180	270	360
Visit number	1	2	3		4	5		6		7	8		9	10		11	12		13	14	15	16	
Written informed consent	x																						
Screening evaluation[a]	x	x[b]																					
Vaccine administration			x					x						x									
Pregnancy test	x		x					x						x								x	
Medical history	x		x					x						x									
Inclusion and exclusion criteria assessment	x		x		x	x		x		x	x		x	x		x	x		x	x	x	x	
Medication review	x		x		x	x		x		x	x		x	x		x	x		x	x	x	x	
Clinical evaluation[c]			x		x	x		x		x	x		x	x		x	x		x	x	x	x[d]	
Telephone follow-up				x			x		x			x			x			x					x
Safety																							
Memory aid completed			x	x	x	x		x	x	x	x			x	x	x	x						
ANA and dsDNA	x																					x	
Clinical chemistries[e]	x		x		x	x		x		x	x		x	x		x	x		x	x	x	x	
Hematology[f]	x		x		x	x		x		x	x		x	x		x	x		x	x	x	x	
Immunogenicity																							
ELISA		x						x					x	x					x	x	x	x	
MN50		x						x					x	x					x	x	x	x	
CMI assays[g]		x						x					x	x					x	x	x	x	

[a]Screening evaluation includes the following: (1) Complete medical history; (2) complete physical examination, including vital signs (blood pressure, pulse rate, respiratory rate, and body temperature), height, and weight; (3) clinical chemistries: blood urea nitrogen (BUN), creatinine, random glucose level, liver-associated enzymes, total bilirubin, and alkaline phosphatase; (4) Flavirirus serologies for dengue, yellow fever, Japanese encephalitis, and West Nile viruses (by ELISA); (5) serologies for HIV-1, hepatitis BsAg, and hepatitis C; (6) antinuclear antibody and double-stranded DNA tests; (7) urinalysis; and (8) pregnancy test

[b]During this visit, only assessments that need to be repeated based on the clinical investigator's initial evaluation are performed

[c]Symptom-oriented physical examination and vital signs including blood pressure, pulse rate, respiratory rate, and body temperature readings (routine, pre-, and post-injection) and assessment for adverse events (AEs)

[d]Body weight is recorded on the last study visit or day 270

[e]BUN, creatinine, liver-associated enzymes, total bilirubin, alkaline phosphatase

[f]Complete blood count

[g]Cell-mediated immunity assessments (ELISpot measurements and intracellular cytokine staining)

measure the ability of anti-dengue antibodies to neutralize dengue viruses type 1–4 in vitro [7].

2. Cellular immunity against dengue envelope proteins was measured by INF-γ enzyme-linked immuno-spot (ELISpot) assay as described by Beckett et al. [5] (*see* **Note 8**).

3.4 Data Analysis

Safety data was analyzed primarily by descriptive statistics. The frequency and severity of clinical and laboratory AEs were tabulated for each dose group. The incidence, intensity, and relationship of each solicited symptom (or laboratory abnormality) to the vaccine over the 7-day follow-up period were determined for each study group. Immunogenicity results were analyzed using descriptive statistics at each time point/study day. A comparison of the immunologic responses (humoral and cell mediated) between the high- and low-dose groups was performed using nonparametric methods. The time after vaccination to the peak geometric titer and the duration of antibody responses were determined.

4 Notes

1. DNA sequences encoding the pre-membrane protein (prM) and envelope protein (E) of each of the four dengue virus serotypes were cloned separately into the expression vector VR1012, described previously [8–10]. Vaccine preparations were validated by complete nucleotide sequence determination and expression of appropriate antigens in transfected HEK-293 cells. The DNA vaccine product was prepared from individual 2 mg/ml bulk preparations in 0.9 % sodium chloride and 20 mM sodium phosphate pH 7.2. The final tetravalent dengue (serotypes 1, 2, 3, and 4) pDNA vaccine (TVDV) consisted of each of the expression plasmids mixed together at a 1:1:1:1 mass ratio in a concentration of 2 mg/ml, 0.5 mg of each pDNA in PBS. TVDV was filled into sterile, depyrogenated 2 ml type I borosilicate glass vials to a volume of 0.7 ml per vial. The vials are closed with sterile butyl rubber stoppers in an ISO 5 (Class 100) filling area and then sealed with aluminum crimp seals.
2. Vaxfectin® was prepared according to cGMP specifications at Vical, Inc. as a 1:1 (molar) mixture of GAP-DMORIE [(±)-*N*-(3-aminopropyl)-*N*,*N*-dimethyl-2,3 bis (*cis-9-tetradecenyloxy*)-1-propanaminium bromide] and DPyPE (1,2-diphytanoyl-sn-glycero-3-phosphoethanolamine, Avanti Polar Lipids, Inc.). Vaxfectin® was supplied as a lipid film in sterile borosilicate vials. Each vial contained 0.97 mg GAP-DMORIE and 1.21 mg DPyPE.

3. The IND contained information in the following general areas: animal pharmacology and toxicology, product manufacturing information, clinical protocol, and investigator information [11]. The IND contained data from preclinical studies that demonstrated the safety and immunogenicity of the product when administered in animal studies.
4. A clinically significant medical history or process includes one or more of the following: a condition that is chronic or recurring, an immunosuppressive procedure or medication, a condition for which repeated injections or blood draws may pose additional risk to the subject, a condition that requires active medical intervention or monitoring, and a condition or a process in which signs or symptoms could be confused with reactions to vaccine.
5. The study protocol was reviewed in detail and approved by the Naval Medical Research Center Institutional Review Board (NMRC/IRB). The study was conducted in compliance with all US Federal Regulations governing the protection of human subjects.
6. Obtain the previously determined optimal antigen concentration for plate coating. Quick thaw the dengue virions and negative control antigen. Prepare dilutions of the positive virions and negative control antigens in 1× PBS. Place 100 μl of diluted antigen in each well using a 12-channel pipettor. Positive and negative antigens are placed in alternating rows with the positive starting from row A. Incubate at 4 °C overnight. Make 1:100 dilutions of sera and controls to be tested by making a masterplate utilizing the polypropylene microtubes. Add 500 μl of serum dilution/blocking buffer (90 ml deionized water; 10 ml 10× PBS; 100 μl Tween 20; 5 g nonfat dry milk) into each tube, and then add 5 μl of sera into each sample tube and negative control. Wash plates six times using the microplate auto washer. Place 200 μl of serum dilution/blocking buffer solution in each well. Cover the plates with lids or aluminum foil or plastic lid and incubate at 37 °C for 60 min. After washing plates, add 100 μl of pre-diluted samples or controls into the appropriate wells. Cover and incubate at 37 °C for 60 min. Determine previously optimized dilution of the HRP-labeled conjugate. Prepare the appropriate dilution using serum dilution buffer. Wash plates again, and then add 100 μl of the diluted conjugate to each well. Cover and incubate at 37 °C for 60 min. While the plates are incubating, prepare equal amounts of the ABTS peroxidase substrate and peroxidase solution B and put them separately into a conical centrifuge tube. Cover and place in the incubator. Wash the plates again. Mix the ABTS peroxidase substrate and peroxidase solution. Add 100 μl of the substrate mixture into each well.

Place the plates inside the 37 °C incubator and read at exactly 30 or 60 min as previously determined.

7. Heat inactivate all sera to be assayed in a 56 °C water bath for 30 min. Prepare serial twofold dilutions of test sera and positive and negative control sera. Prepare dilutions for the virus plaque dose in diluent. Using a sterile repeating pipettor add an equal volume (i.e., 250 μl) of the diluted virus stock to each of the serum dilution tubes, including controls for a final virus concentration of approximately 50 PFU/0.2 ml. Add virus to the diluent-only control first, then negative controls, then to the test sera, and finally to the positive controls; avoid contaminating the pipet with sera. Mix the tubes well by gentle vortexing or shaking. Incubate the tubes in a 37 °C water bath for 30 min. Label 6-well plates containing confluent Vero cell monolayers for inoculation (two 6-well plates will be required for each serum dilution series). Using a vacuum aspirator or pipet, remove all but about 0.2 ml of the culture supernatant fluid from each well, leaving enough residual fluid so that the cell monolayers do not become dry. Using a sterile tip add 200 μl of the virus–serum mixture and inoculate into each of the two wells of a properly prepared 6-well plate. Evenly distribute the inoculum by rocking the plate back and forth and from side to side. With a fresh tip, continue inoculation until all plates have been inoculated. Incubate the inoculated plates at 37 °C for 1 h in a 5 % CO_2 incubator to allow for virus absorption. Be sure that the plates are level so that the cell monolayers do not become dry. After the incubation period, remove the inoculum and add 2.5 ml of agarose-containing overlay medium (200 ml 2× EMEM; 20 ml FBS; 4 ml nonessential amino acids; 4 ml penicillin/streptomycin; 2 ml L-glutamine; 4 ml HEPES) to each well using a pipet or a Cornwall syringe or manostat. Set the plates at room temperature for 15–20 min to allow the agarose to solidify. Incubate the plates in a 37 °C, 5 % CO_2, incubator for 5–7 days to allow virus plaques to develop. After the previously determined incubation time prepare the second overlay containing the vital stain neutral red. Add 2 ml of the second overlay (300 ml sterile saline; 13.5 ml neutral red) to each well. Set the plates at room temperature for 15–20 min to allow the agarose to solidify. Incubate the plates, same environment as above, for a minimum of 18 h or up to 48 h to allow the cells to maximally take up stain and show visible plaques. Determine the PRNT by linear regression (probit) analysis using either probit paper or a computer software program (e.g., the SPSS software program for MS DOS/Windows is useful).

8. Coat the ELISpot plates with anti-INF-γ antibody. Make a 35 % alcohol solution by diluting the 99 % ethanol in sterile

deionized water. Add 25 μl per well of the 35 % alcohol to an ELISpot plate, and immediately decant the plate by hand-flipping the plate above a waste container. Wash the ELISpot plate immediately with sterile PBS six times by adding 200 μl of PBS and decanting the plate. After the last wash, decant the plate and immediately add the coating antibody, 10 μg/ml anti-IFN-γ, in PBS (100 μl/well). Incubate plate overnight in 4 °C. The plate must be wrapped with a wet paper towel to avoid evaporation. Wash plate 5× with plain RPMI. After the last wash, add 200 μl/well of the complete medium containing 10 % FBS to block the plate for at least 1 h at room temperature. Decant the blocking medium, and the plate is ready for use. Thaw vials of frozen PBMC by putting the cyrovials upright into a 37 °C water bath, and gently shake the vials. The water line must be below the cap of the vial. As soon as 80 % of the frozen content is thawed, transfer the content into a 15-ml conical tube pre-filled with 10 ml of plain RPMI. Be very fast with this step, because the freezing medium contains dimethyl sulfoxide (DMSO), which is toxic to cells. Wash cells twice in plain RPMI, and combine same samples to one tube. Resuspend each sample in 10 ml of complete medium and count on the Guava counter. Resuspend cells to appropriate concentration (2×10^6), and add 100 μl of cell suspension to the ELISpot plate. Following appropriate incubation, wash the ELISpot plate six times with PBS-T (Tween-20, 0.05 %) to remove cells. After the last wash, decant the plate and add the secondary antibody. Add 100 μl/well of biotinylated anti-IFN-γ antibody at 1:1,000 dilution. Cover the plate to avoid direct light, and incubate the plate for 2 h at room temperature. Wash the plate six times with PBS-T and add peroxidase-conjugated streptavidin at 1:200 dilution in PBS. Cover the plate and incubate for 1.5 h at room temperature. Wash six times with PBS-T. Add AEC substrate. Prepare AEC substrate right before use. Develop plates for 15–30 min; wash thoroughly with H_2O to stop reaction and dry plates upside down. Count the plates on the AID ELISpot Reader.

Acknowledgements

We thank Dr. Peifang Sun, Ms. Monika Simmons, and Mr. Dan Ewing for their input on the laboratory test procedures. The views expressed in this chapter are those of the author and do not necessarily reflect the official policy or position of the Department of the Navy, Department of Defense, nor the US Government. We are employees of the US Government. This work was prepared as part of my official duties. Title 17 U.S.C. article 105 provides that "Copyright protection under this title is not available for any work of the United States Government." Title 17 U.S.C. article 101 defines a US Government work as a work prepared by a military service member or employee of the US Government as part of that person's official duties. For the clinical trial described in this chapter, the NMRC Institutional Review Board, along with the US Army Human Subjects Research Review Board, reviewed and approved the study protocol in compliance with all applicable federal regulations governing the protection of human subjects.

References

1. Raviprakash K et al (2009) Advances in dengue vaccine development. Hum Vaccin 5:1–9
2. Raviprakash K et al (2006) A chimeric tetravalent dengue DNA vaccine elicits neutralizing antibody to all four virus serotypes in rhesus macaques. Virology 353:166–173
3. Wolff JA et al (1992) Long-term persistence of plasmid DNA and foreign gene expression in mouse muscle. Hum Mol Genet 1: 363–369
4. Ulmer JB et al (1993) Heterologous protection against influenza by injection of DNA encoding a viral protein. Science 259: 1745–1749
5. Beckett CG et al (2011) Evaluation of a prototype dengue-1 DNA vaccine in a Phase 1 clinical trial. Vaccine 29:960–968
6. (2007) Guidance for industry: toxicity grading scale for healthy adults and adolescent volunteers enrolled in preventive vaccine clinical trials. http://www.fda.gov/downloads/BiologicsBloodVaccines/GuidanceComplianceRegulatoryInformation/Guidances/Vaccines/ucm091977.pdf.
7. Putnak JR et al (2008) Comparative evaluation of three assays for measurement of dengue virus neutralizing antibodies. Am J Trop Med Hyg 79:115–122
8. Raviprakash K et al (2000) Dengue virus type 1 DNA vaccine induces protective immune responses in rhesus macaques. J Gen Virol 81: 1659–1667
9. Lu Y et al (2003) Dengue 2 PreM-E/LAMP chimera targeted to the MHC class II compartment elicits long-lasting neutralizing antibodies. Vaccine 21:2178–2189
10. Blair PJ et al (2006) Evaluation of immunity and protective efficacy of a dengue-3 premembrane and envelope DNA vaccine in Aotus nancymaae monkeys. Vaccine 24:1427–1432
11. (1995) Guidance for industry: content and format of investigational new drug applications (INDs) for phase 1 studies of drugs, including well-characterized, therapeutic, biotechnology-derived products. http://www.fda.gov/downloads/Drugs/GuidanceComplianceRegulatoryInformation/Guidances/UCM074980.pdf.

Chapter 18

DNA Vaccination as a Treatment for Chronic Kidney Disease

Yuan Min Wang and Stephen I. Alexander

Abstract

Chronic kidney disease is one of the major health problems worldwide. DNA vaccination delivers plasmid DNA encoding the target gene to induce both humoral and cellular immune responses. Here, we describe the methods of CD40 DNA vaccine enhanced by dendritic cell (DC) targeting on the development of Heymann nephritis (HN), a rat model of human membranous nephropathy.

Key words DNA vaccination, Chronic kidney disease, Dendritic cell, Active Heymann nephritis (HN), Adriamycin nephropathy (AN)

1 Introduction

Every year thousands of chronic kidney disease (CKD) patients develop to the end-stage kidney disease (ESKD) and depend on dialysis and kidney transplantation. Currently, the successful therapeutic strategies for delay CKD to ESKD are limited. New treatment approaches are necessary to benefit CKD patients and reduce the medical costs. While DNA vaccines have reached clinical use, in general they have been limited by poor immune responses, and this is a particular problem while generating responses to self-antigen. In addition, the potential of DNA vaccination as a therapeutic approach for CKD has not been fully assessed. Our previous studies have shown that DNA vaccination targeting T cell receptor (TCR) subsets can reduce proteinuria in Heymann nephritis (HN) [1]. Another study of DNA vaccination targeting the chemokine CCL2 (monocyte chemoattractant protein 1) has suggested that DNA vaccination can protect against adriamycin nephropathy (AN, a toxin-induced model of proteinuric renal disease) [2, 3]. These studies have demonstrated that therapeutic strategies of DNA vaccination are protective and can induce specific cellular and antibody responses against the targeting antigens. We have recently

Monica Rinaldi et al. (eds.), *DNA Vaccines: Methods and Protocols*, Methods in Molecular Biology, vol. 1143,
DOI 10.1007/978-1-4939-0410-5_18,

used a plasmid containing the gene encoding a single-chain Fv antibody specific for the dendritic cell-restricted antigen-uptake receptor DEC205 developed by Steinman laboratory [4]. By cloning a gene of interest, co-stimulatory molecule CD40, into this plasmid by electroporation, we have successfully demonstrated that this particular vaccine (DEC205-CD40) can prevent the development of HN [5, 6].

2 Materials

Prepare all solutions using ultrapure water (prepared by purifying deionized water to attain a sensitivity of 18 MΩ cm at 25 °C) and analytical grade reagents. Prepare and store all reagents at room temperature unless indicated otherwise.

2.1 Experimental Animals

Inbred male Lewis rats (aged 6 weeks and weighing 180–200 g); inbred male Wistar rats, and outbred male Sprague-Dawley (SD) rats (aged 8 weeks and weighing 200–220 g) were maintained under standard sterile conditions in animal house.

2.2 Preparation of Renal Tubular Antigen (RTA/ Fx1A)

1. Anesthetic solution: 0.75 ml Ketamine (100 mg/ml) and 0.25 ml xylazine (20 mg/ml) and 0.75 ml H_2O. Total 2 ml of anesthetic solution; every 0.2 ml per 100 g rat for intraperitoneal (i.p.) injection.
2. 150 mesh steel sieve (bore size 104–106 μm).
3. Ultracentrifugation: Bachman ultracentrifuge: rotor type: TFT70; rotor serial no.: 14790; rpm: 33 K.

2.3 DNA Vaccination and Induction of Active HN

1. Two plasmid DNA vaccination solution: scDEC-CD40 (300 μg in 150 μl sterile H_2O) and scControl-CD40 (300 μg/150 μl sterile H_2O).
2. HN induction of Fx1A emulsion solution containing 15 mg of Fx1A, 1 mg mycobacterium tuberculosis HRa37 (TB, Difco, Detroit, MI), 100 μl of Incomplete Freund's adjuvant (IFA) (Sigma-Aldrich, St. Louis, MO), and 100 μl of PBS.
3. Complete Freund's adjuvant (CFA) (Sigma-Aldrich, St. Louis, MO).
4. 0.75 % bupivacaine (1 μl/g body wt; Sigma, St. Louis, MO).
5. Square wave electroporation at the injection site using BTX830 two-needle array electrodes (BTX Harvard Apparatus, San Diego, CA).

2.4 Enzyme-Linked Immunosorbent Assay

1. Enzyme-linked immunosorbent assay (ELISA) "Ensemble" kit for the detection of rat primary antibody (Alpha Diagnostic International, Texas).

2. C96 Maxisorp MicroWell™ (NUNC, Thermo Fisher Scientific, MA) plates.
3. Recombinant mouse CD40 protein (R&D system, Minneapolis, MN).
4. Multiskan EX Microplate photometer (Thermo Fisher Scientific Inc., MA).

2.5 Renal Function

1. Colorimetric assay kit (Bio-Rad, Hercules, CA).
2. Automated chemistry analyzer VITROS (Ortho Clinical Diagnostics, Johnson & Johnson).

2.6 Renal Histology and Immunohistochemistry (IHC) Staining

1. 10 % neutral-buffered formalin (Sigma, Australia).
2. Periodic acid-Schiff's (PAS) reagent and hematoxylin (Sigma, Australia).
3. Optimal Cutting Temperature (OCT) Compound (SakuraFinetek Inc, Torrance, USA).
4. ScanScope digital slide scanner (Aperio Technologies, Inc. Vista, CA).
5. Image J software (NIH, Bethesda, MD).
6. DeltaVision Core Microscope (Applied Precision, Inc, Washington).

2.7 Antibodies for IHC and Flow Cytometry

1. Mouse anti-rat CD4 (OX35) (Serotec, Oxford, UK).
2. Mouse anti-rat CD8a (OX8) (eBioscience, CA).
3. Mouse anti-rat CD68 (ED1) (Serotec, Oxford, UK).
4. Mouse anti-rat CD45RA (OX33) (BD Pharmingen™, San Diego, CA).
5. Biotinylated goat anti-mouse immunoglobulin (Zymed Laboratories, San Francisco, CA).
6. Anti-mouse/rat CD40 antibody (clone: HM40-3) (eBioscience, CA).
7. PE anti-mouse CD40 antibody (FGK45.5) was kindly provided by Dr Irina Caminschi from Burnet Institute, Melbourne, Victoria, Australia. Propidium iodide (Merck Millipore Biosciences, Darmstadt, Germany).
8. FITC-goat anti-rat IgG (Zymed Laboratories, San Francisco, CA).
9. APC-mouse anti-rat CD4 and PE-mouse anti-rat CD8 (BD Pharmingen™, San Diego, CA). APC-mouse anti-rat CD45RA (OX33) (eBioscience, CA).
10. FACSCanto™ or FACSCalibur flow cytometer (Becton Dickinson, Mountain View, CA).
11. Flowjo software (Tree Star, Inc., Ashland, Oregon).

3 Methods

3.1 Preparation of Renal Tubular Antigen (RTA/ Fx1A)

1. Ten adult male SD rats were used for Fx1A preparation. Rats were anesthetized with anesthetic solution by i.p. injection at 0.2 ml per 100 g rat.
2. Kidney perfusion: Open rat abdomen, and isolate both kidneys and abdominal aorta. Puncture a hole in both kidney veins, and then inject saline or phosphate-buffered saline (PBS) (pre-warmed at 25 °C, pH 7.3 for PBS) through the abdominal aorta until both kidneys become white.
4. The kidney cortices were dissected and pushed through a 150 mesh steel sieve (bore size 104–106 μm) in PBS.
5. Filtrates were transferred to a 50 ml tube and centrifuged twice with PBS at 400 × *g* for 10 min at 4 °C.
6. The supernatants, which contain the tubular fraction, were washed three times with distilled water (cold) and ultracentrifuged at 33,000 × *g* for 45 min at 4 °C (Beckman ultracentrifuge: rotor type: TFT70; rotor serial no.: 14790; rpm: 33 K).
7. Each time, the supernatants were discarded and the pellet was resuspended in ultrapure distilled water (cold) and then ultracentrifuged at 79,000 × *g* for 45 min at 4 °C (pellet from brown to clear).
8. The sediment was lyophilized and then stored at −80 °C. Ten SD rats generate about 1,000 mg of lyophilized Fx1A.

3.2 DNA Vaccination and Induction of Active HN

1. Rats were divided into four groups: DEC-CD40-HN group: rats vaccinated with scDEC-CD40 followed by Fx1A immunization; con-CD40-HN group: rats vaccinated with scControl-CD40 followed by Fx1A immunization; HN group: non-vaccinated rats with Fx1A immunization; and CFA group: non-vaccinated rats with CFA immunization.
2. Rats were immunized with DNA by intramuscular (i.m.) injection in conjunction with electroporation in the anterior tibialis (TA) muscles of the right hind leg.
3. 1 Week before DNA immunization, rats were injected i.m. with 0.75 % bupivacaine (1 μl/g body wt), which was followed by two i.m. DNA injections (3 weeks apart with 300 μg/150 μl sterile H_2O/rat/injection) at the same location as bupivacaine injection.
4. Each DNA injection was followed immediately by square wave electroporation at the injection site using BTX830 two-needle array electrodes (BTX Harvard Apparatus, San Diego, CA). The distance between the electrodes was 10 mm, and the array was inserted longitudinally relative to the muscle fibers. In vivo electroporation parameters were 100 V/cm; 50-ms pulse length; and 6 pulses with reversal of polarity after 3 pulses.

5. 2 Weeks after the second DNA injection, HN was induced. To induce HN, Lewis rats were immunized subcutaneously (s.c.) in both hind footpads. Each footpad was injected with 100 μl of emulsion containing 15 mg of Fx1A, 1 mg mycobacterium tuberculosis HRa37, 100 μl of IFA, and 100 μl of PBS.
6. CFA control rats were immunized with emulsion without Fx1A.
7. Alternative protocol for more severe disease of HN (*see* **Note 1**).

3.3 Enzyme-Linked Immunosorbent Assay

1. Serum anti-CD40 antibody level was evaluated using ELISA "Ensemble" kit for the detection of rat primary antibody.
2. C96 Maxisorp MicroWell™ plates were used, and recombinant mouse CD40 protein was used for coating of the plates at the concentration of 1 μg/ml in 100 μl of coating buffer.
3. All the sera were diluted 1:10 with sample diluent as provided in the kit.
4. Absorbance was measured at 450 nm with Multiskan EX Microplate photometer.
5. Absorbance at 570 nm was used as background correction.

3.4 Renal Function Assay

1. Blood and 16-h urine samples were collected (*see* **Notes 2** and **3**) every 2 weeks after the induction of HN.
2. Urine protein concentration was measured using colorimetric assay (Bio-Rad, Hercules, CA).
3. Urine creatinine, serum albumin, and serum creatinine were analyzed using an automated chemistry analyzer VITROS (Ortho Clinical Diagnostics, Johnson & Johnson).

3.5 Renal Histology and Immunohistochemistry

1. Coronal sections of the kidney were fixed in 10 % neutral-buffered formalin and embedded in paraffin.
2. 4 μm section of paraffin block was stained with PAS reagent and counterstained with hematoxylin (Fig. 1a, b).
3. Immunohistochemical staining was performed to determine the infiltrations of CD4+ and CD8+ T cells and macrophages in the kidney (Fig. 1c, d).
4. For staining of CD4+ T cells, frozen sections (cut at 5 μm from kidney tissues embedded in OCT compound) were used. And for staining of CD8+ T cells and macrophages, paraffin sections were used.
5. Sections were incubated with primary antibody (16 h, 4 °C) followed by secondary antibody incubation (30 min, RT).
6. Immune cell infiltration was quantified by counting ten consecutive high-power fields per animal and expressed as cells per 200× field.

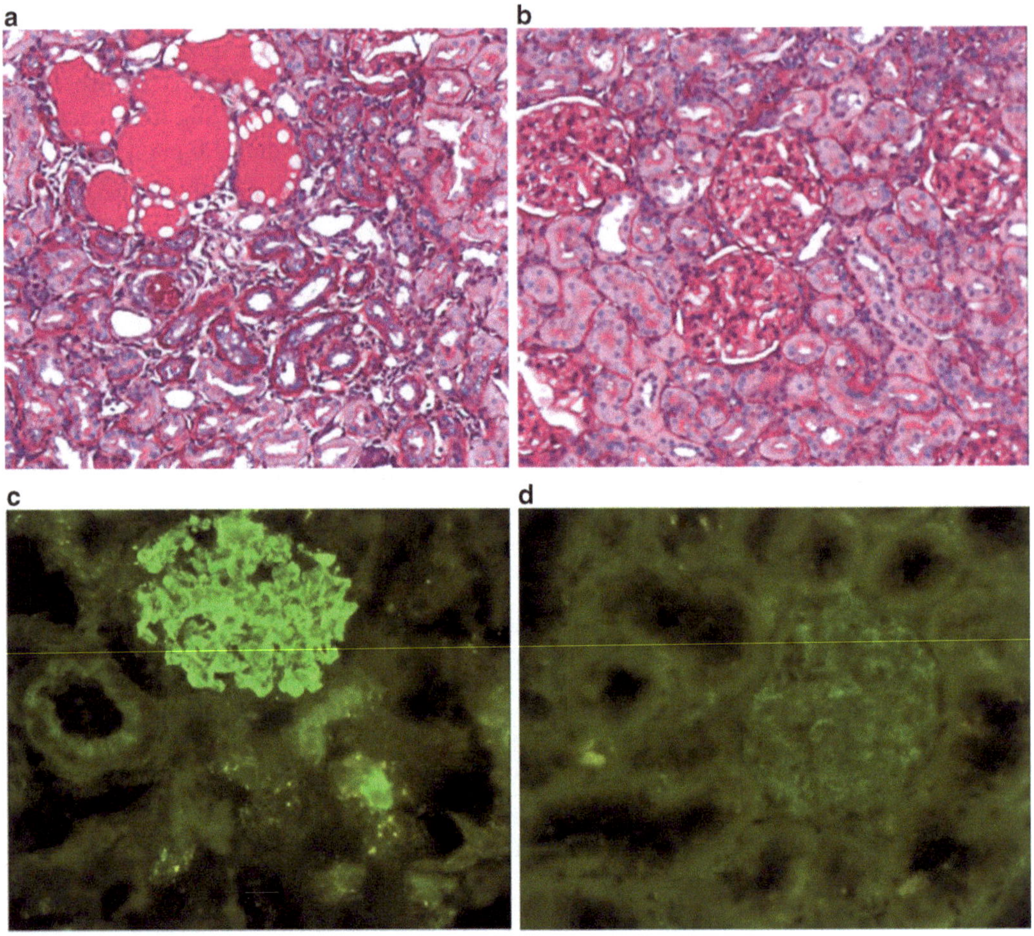

Fig. 1 CD40 DNA vaccination reduces renal structural injury and glomerular IgG deposition. Histology of renal representative sections of kidney under PAS staining showed renal injury on HN rat (**a**) and reduced renal damage in CD40 DNA-vaccinated rat (**b**). Immunohistochemistry staining showed subepithelial glomerular immune deposits in representative sections of kidney in HN rat (**c**) and less deposition in kidney for CD40 DNA-vaccinated rat (**d**) (magnification, 400×)

7. Number of OX33-positive B cells in the spleen were quantified by counting three random high-power fields per animal and expressed as cells per 400× field.
8. Slides were scanned using ScanScope digital slide scanner. Image analysis was performed using Image J software.
9. Immunofluorescence staining of IgG in kidney sections was performed to assess IgG deposition on glomeruli. Frozen sections were incubated with goat serum (15 min, RT) followed by FITC goat anti-rat IgG antibody (2 h, RT). Images (magnification 400×) were taken using DeltaVision core microscope. Fluorescence density was quantified using Image J software.

3.6 Flow Cytometry Analysis

1. Single-cell suspension was directly stained with APC- and PE-conjugated mouse anti-rat mAb to cell surface antigens.
2. All samples were analyzed on FACSCanto™ or FACSCalibur flow cytometer.
3. Flowjo software was used for analysis.

4 Notes

1. Alternative protocol (more severe disease): Rats were boosted once at 2 weeks after the first Fx1A injection with a half dose of Fx1A with CFA.
2. Accurate urine collection without contamination and animal injury is very important. For HN, rats were fasted and water deprived whilst in the metabolic cages for 16 h to collect urine at baseline and weeks 8, 10, and 12.
3. Commercial metabolic cages were useful for collecting urine, but specific attention has to be paid to their limitations—water leakage and evaporation of urine during overnight collection.

Acknowledgement

This work was supported by research grants from the National Health and Medical Research Council of Australia (NHMRC grant 571343 and 632517). We thank the animal house staff in the Westmead Hospital Animal House Facility for care of the animals. We thank Godwin Nchinda (The Rockefeller University) for kindly providing the DEC205 vectors. We thank Virginia James (Westmead Millennium Institute, Australia) for her technical assistances in histology.

References

1. Wu H, Walters G, Knight JF et al (2003) DNA vaccination against specific pathogenic TCRs reduces proteinuria in active Heymann nephritis by inducing specific autoantibodies. J Immunol 171:4824–4829
2. Wu H, Wang Y, Tay YC et al (2005) DNA vaccination with naked DNA encoding MCP-1 and RANTES protects against renal injury in adriamycin nephropathy. Kidney Int 67:2178–2186
3. Zheng G, Wang Y, Xiang SH et al (2006) DNA vaccination with CCL2 DNA modified by the addition of an adjuvant epitope protects against "nonimmune" toxic renal injury. J Am Soc Nephrol 17:465–474
4. Nchinda G, Kuroiwa J, Oks M et al (2008) The efficacy of DNA vaccination is enhanced in mice by targeting the encoded protein to dendritic cells. J Clin Invest 118:1427–1436
5. Wang Y, Wang YM, Zheng G et al (2013) DNA vaccine encoding CD40 targeted to dendritic cells in situ prevents the development of Heymann nephritis in rats. Kidney Int 83: 223–232
6. Wang YM, Zhou JJ, Wang Y et al (2013) Daedalic DNA vaccination against self antigens as a treatment for chronic kidney disease. Int J Clin Exp Pathol 6:326–333

Index

Monica Rinaldi et al. (eds.), *DNA Vaccines: Methods and Protocols*, Methods in Molecular Biology, vol. 1143, DOI 10.1007/978-1-4939-0410-5,

T

U

V

Y

MIX
Papier aus verantwortungsvollen Quellen
Paper from responsible sources
FSC® C105338

If you have any concerns about our products,
you can contact us on
ProductSafety@springernature.com

In case Publisher is established outside the EU,
the EU authorized representative is:
Springer Nature Customer Service Center GmbH
Europaplatz 3, 69115 Heidelberg, Germany

Printed by Libri Plureos GmbH
in Hamburg, Germany